Mathematics in Engineering Sciences

Mathematical Engineering, Manufacturing, and Management Sciences

Series Editor:
Mangey Ram, Professor
Department of Mathematics, Computer Science & Engineering,
Graphic Era (Deemed to be University), Dehradun, India

The aim of this new book series is to publish the research studies and articles that bring up the latest development and research applied to mathematics and its applications in the manufacturing and management sciences areas. Mathematical tool and techniques are the strength of engineering sciences. They form the common foundation of all novel disciplines as engineering evolves and develops. The series will include a comprehensive range of applied mathematics and its application in engineering areas such as optimization techniques, mathematical modeling and simulation, stochastic processes and systems engineering, safety-critical system performance, system safety, system security, high assurance software architecture and design, mathematical modeling in environmental safety sciences, finite element methods, differential equations, reliability engineering, etc.

Sustainable Procurement in Supply Chain Operations
Edited by Sachin Mangla, Sunil Luthra, Suresh Jakar,
Anil Kumar, and Nirpendra Rana

Mathematics in Engineering Sciences
Novel Theories, Technologies, and Applications
Edited by Mangey Ram

For more information about this series, please visit: https://www.crc-press.com/Mathematical-Engineering-Manufacturing-and-Management-Sciences/book-series/CRCMEMMS

Mathematics in
Engineering Sciences
Novel Theories, Technologies, and Applications

Edited by
Mangey Ram

CRC Press
Taylor & Francis Group
Boca Raton London New York

CRC Press is an imprint of the
Taylor & Francis Group, an **informa** business

CRC Press
Taylor & Francis Group
6000 Broken Sound Parkway NW, Suite 300
Boca Raton, FL 33487-2742

First issued in paperback 2021

© 2020 by Taylor & Francis Group, LLC
CRC Press is an imprint of Taylor & Francis Group, an Informa business

No claim to original U.S. Government works

ISBN-13: 978-0-367-77664-0 (pbk)
ISBN-13: 978-1-138-57767-1 (hbk)

Library of Congress Cataloging-in-Publication Data

Names: Ram, Mangey, editor.
Title: Mathematics in engineering sciences: novel theories, technologies, and
    applications / edited by Mangey Ram.
Description: Boca Raton : Taylor & Francis, a CRC title, part of the Taylor &
    Francis imprint, a member of the Taylor & Francis Group, the academic
    division of T&F Informa, plc, 2019. | Includes bibliographical references.
Identifiers: LCCN 2019019739 | ISBN 9781138577671 (hardback : alk. paper) |
    ISBN 9781351266321 (e-book)
Subjects: LCSH: Engineering mathematics.
Classification: LCC TA330 .M327 2019 |
    DDC 620.001/51—dc23 LC record available at https://lccn.loc.gov/2019019739

Visit the Taylor & Francis Web site at
http://www.taylorandfrancis.com

and the CRC Press Web site at
http://www.crcpress.com

Contents

Preface

Mathematics is one of the oldest sciences in the world, and is the base and strength of engineering sciences. Mathematics is used widely in a range of professions such as engineering. We can make anything possible in engineering sciences just by adding some mathematical techniques. This book *Mathematics in Engineering Sciences: Novel Theories, Technologies, and Applications* contains plenty of mathematical tools and techniques applied in several fields of engineering sciences.

In Chapter 1, the fine-tuning methodology was applied to the case study, followed by the application of particle swarm optimization (PSO) and genetic algorithm (GA) to find the alternatives of mold design, material, and production conditions that maximize eco-efficiency, i.e. maximize different eco-efficiency indicator, being used alone or together as cost functions.

Chapter 2 proposes two distance-based multiobjective decision-making (MODM) approaches called goal programming (GP) and distance-based function for ideal solutions. It also provides a methodology to find a process setting that leads to the most desirable values of quality characteristics, also called responses.

Chapter 3 establishes the applicability of engineering-based analysis in biomedical sector by conjugating many spheres of science into one. The finite element method (FEM) analysis enables industry to save cost and provide insight into the points where the product can fail.

Chapter 4 briefly discusses about the magnetic nanofluids along with its constitutions, the cause of production, and various applications. Further, the mathematical modeling and its solution is explained for the rotating/ stationary/stretchable rotating disk problem.

In Chapter 5, a brief exposition of the qualitative theory of mathematical models involving various types of differential equations is presented. Many approaches are used to develop the qualitative theory in terms of the concepts of differential inequalities, comparison principle, iterative techniques using lower and upper solutions, and stability and perturbation theory.

Chapter 6 investigates a numerical problem of steady, incompressible magneto hydro dynamics (MHD) stagnation point flow of an electrically conducting Cu–water nanofluid in a porous medium over a stretching/ shrinking plate with radiation and heat source/sink impacts and inspected.

Chapter 7 investigates the fundamental flows of a Ree–Eyring fluid between infinitely parallel plates. The effects of porous medium, magnetic field, thermal radiation, viscous dissipation, and Joule heating have been considered in all the mentioned cases.

Chapter 8 establishes an equivalence between the two well-known graph theory problems, the minimum traveling salesman tour (MTST), and the minimum spanning tree (MST).

Chapter 9 studies one- and two-dimensional spatial patterns in a predator–prey model with hunting cooperation under zero-flux boundary conditions and circular initial conditions.

Chapter 10 discusses a detailed study on solving multiobjective transportation problem (MOTP) under cost reliability using utility function approach.

In Chapter 11, threshold in terms of environmental pollutants is computed for each stakeholder. Stability is worked out. The sensitivity analysis is carried out to classify critical model parameters. The model is validated through numerical simulation.

In Chapter 12, robust controllers are developed for quadruple tank multiple input multiple output (MIMO) system in the presence of matched uncertainties. The sliding mode controllers are designed using two different approaches, i.e. (i) exponential reaching law and (ii) nonswitching reaching law.

In Chapter 13, a unique compensation technique is proposed for nonlinear networked control system in the discrete-time domain, which compensates the deterministic type of network-induced delay. The communication delay within networks between sensor, controller, and actuator channel is nullified by Thiran's approximation rule.

This book will be very beneficial to undergraduate and postgraduate students of engineering, engineers, and research scientists/academics involved in the engineering sciences.

Mangey Ram
Graphic Era (Deemed to be University), India

MATLAB® is a registered trademark of The MathWorks, Inc. For product information,

please contact:
The MathWorks, Inc.
3 Apple Hill Drive
Natick, MA 01760-2098 USA
Tel: 508-647-7000
Fax: 508-647-7001
E-mail: info@mathworks.com
Web: www.mathworks.com

Acknowledgments

The editor acknowledges CRC press for this opportunity and professional support. My special thanks to Cindy Renee Carelli, Executive Editor, CRC Press—Taylor & Francis Group, for the excellent support she provided to complete this book. Thanks to Erin Harris, editorial assistant to Cindy, for her follow-up and aid. Also, I would like to thank all the chapter authors and reviewers for their availability for this work.

Mangey Ram
Graphic Era (Deemed to be University), India

Editor

Dr. Mangey Ram received his Ph.D. degree major in mathematics and minor in computer science from G.B. Pant University of Agriculture and Technology, Pantnagar, India. He has been a faculty member for around ten years and has taught several core courses in pure and applied mathematics at undergraduate, postgraduate, and doctorate levels. He is currently a professor at Graphic Era (Deemed to be University), Dehradun, India. Before joining Graphic Era, he was a deputy manager (probationary officer) with Syndicate Bank for a short period. He is an editor-in-chief of *International Journal of Mathematical, Engineering and Management Sciences* and a guest editor and member of the editorial board of various journals. He is a regular reviewer for international publishers, including IEEE, Elsevier, Springer, Emerald, John Wiley, Taylor & Francis, and many other publishers. He has published 150 plus research publications in IEEE, Taylor & Francis, Springer, Elsevier, Emerald, World Scientific, and many other national and international journals of repute and also presented his works at national and international conferences. His fields of research are reliability theory and applied mathematics. Dr. Ram is a Senior Member of the IEEE, life member of Operational Research Society of India, Society for Reliability Engineering, Quality and Operations Management in India, Indian Society of Industrial and Applied Mathematics, member of International Association of Engineers in Hong Kong, and Emerald Literati Network in the United Kingdom. He has been a member of the organizing committee of a number of international and national conferences, seminars, and workshops. He has been conferred with the Young Scientist Award by the Uttarakhand State Council for Science and Technology, Dehradun, in 2009. He received the Best Faculty Award in 2011, the Research Excellence Award in 2015, and most recently, the Outstanding Researcher Award in 2018 for his significant contribution in academics and research at Graphic Era Deemed to be University, Dehradun, India.

Contributors

Masar Al-Rabeeah
Department of Mathematical and
    Geospatial Sciences
School of Sciences, RMIT University
Melbourne, VIC, Australia
and
Faculty of Sciences, Department of
    Mathematics
Basrah University
Basrah, Iraq

Ana Rita Alves
IDMEC, Instituto Superior Técnico
Universidade de Lisboa
Lisbon, Portugal

Hamed Baziyad
Department of Information
    Technology Engineering
Tarbiat Modares University
Tehran, Iran

Shahin Behboudi
Department of Industrial
    Engineering & Management
    Systems
Amirkabir University of Technology
Tehran, Iran

Santosh Chaudhary
Department of Mathematics
Malaviya National Institute of
    Technology Jaipur
Jaipur, Rajasthan, India

Zahia Drici
Department of Mathematics
Illinois Wesleyan University
Bloomington, Illinois

Ramu Dubey
Department of Mathematics
J. C. Bose University of Science &
    Technology, YMCA
Faridabad, Haryana, India

Sahil Garg
Engineering Department
Apeejay Institute of Management
    Technical Campus
Jalandhar, Punjab, India

Taha-Hossein Hejazi
Department of Industrial
    Engineering
Amirkabir University of
    Technology, Garmsar Campus
Tehran, Iran

Diogo Pina Jorge
Efficiency Rising (Erising)
Lisbon, Portugal

Vimal Kumar Joshi
Department of Mathematics
Dronacharya Group of
    Institutions
Greater Noida, Uttar Pradesh, India

KM Kanika
Department of Mathematics
Malaviya National Institute of
    Technology Jaipur
Jaipur, Rajasthan, India

Mohit Kumar
Auxein Medical Private Limited
Sonipat, Haryana, India

Santosh Kumar
Department of Mathematics and
   Statistics
University of Melbourne
Parkville, VIC, Australia
and
Department of Mathematical and
   Geospatial Sciences
School of Sciences, RMIT
   University
Melbourne, VIC, Australia

Gurupada Maity
Department of Applied
   Mathematics with Oceanology
   and Computer Programming
Vidyasagar University
Midnapore, West Bengal, India

Dharmadas Mardanya
Department of Applied
   Mathematics with Oceanology
   and Computer Programming
Vidyasagar University
Midnapore, West Bengal, India

Farzana A. McRae
Department of Mathematics
Catholic University of America
Washington, District of Columbia

Axaykumar Mehta
Electrical Department
Institute of Infrastructure Technology
   Research and Management
Ahmedabad, Gujarat, India

Elias Munapo
Department of Statistics and
   Operations Research
School of Economics and
   Decision Sciences, North
   West University
Mafikeng, South Africa

Mohit Pant
Mechanical Engineering
   Department
National Institute of Technology
Hamirpur, Himachal Pradesh India

Dhruv Patel
Instrumentation and Control
   Department
Sardar Vallabhbhai Patel Institute of
   Technology
Anand, Gujarat, India

Paulo Peças
IDMEC, Instituto Superior Técnico
Universidade de Lisboa
Lisbon, Portugal

Paras Ram
Department of Mathematics
National Institute of Technology
Kurukshetra, Haryana, India

Katta Ramesh
Department of Mathematics
Symbiosis Institute of Technology,
   Symbiosis International
   University
Pune, Maharashtra, India

Sankar Kumar Roy
Department of Applied
   Mathematics with
   Oceanology and Computer
   Programming
Vidyasagar University
Midnapore, West Bengal, India

Moksha H. Satia
Department of Mathematics
Gujarat University
Ahmedabad, Gujarat, India

Dipesh Shah
Instrumentation and Control
    Department
Sardar Vallabhbhai Patel Institute
    of Technology
Anand, Gujarat, India

Nita H. Shah
Department of Mathematics
Gujarat University
Ahmedabad, Gujarat, India

Caston Sigauke
Department of Statistics
University of Venda
Thohoyandou, South Africa

Teekam Singh
Department of Mathematics
Graphic Era (Deemed to be
    University)
Dehradun, Uttarakhand, India

Foram A. Thakkar
Department of Mathematics
Gujarat University
Ahmedabad, Gujarat, India

Jonnalagedda Vasundhara Devi
Department of Mathematics
GVP - Prof. V. Lakshmikantham
    Institute for Advanced Studies,
    Gayatri Vidya Parishad College
    of Engineering (Autonomous)
Madhurawada, Visakhapatnam
    India

1

Application of Optimization Techniques to Support Eco-Efficiency Assessment in Manufacturing Processes

Paulo Peças and Ana Rita Alves

IDMEC, Instituto Superior Técnico Universidade de Lisboa

Diogo Pina Jorge

Efficiency Rising (Erising), Lisboa

CONTENTS

1.1 Introduction

The last two decades have shown an increase in awareness of sustainability, which has been reflected in the establishment of stringent environment-related demands from governmental bodies and needs of resources efficiency increasing, fostering the design and manufacture of greener products to satisfy customers' needs and keep competitiveness.

1

The manufacturing industry, in general, and the decision makers of those companies, in particular, play an important role towards sustainability, since their activity and decisions influence the resource consumption efficiency and effectiveness and the type and intensity of emissions (Ingarao, 2016; Sproedt, Plehn, Schönsleben, & Herrmann, 2015; Duflou et al., 2012). This influence is (i) direct by the production system, i.e., the type of processes and its energy consumption, the type and level of wastes and disposals as well as the kind and intensity of emissions from manufacturing processes and materials used, and is also (ii) indirect by the product produced, namely the ones related with product design decisions, i.e., the materials used in the product, and the energy and resources consumed by products across their life cycle (Duflou et al., 2012).

Towards a sustainable development, three main pillars have to be taken into account: economic growth, environmental protection, and social equality. Two of those three are the focus of eco-efficiency, a management philosophy that encourages business to search for environmental improvements that yield parallel economic benefits (Lyubomirsky, Kurtz, & Lyubomirsky, 2009).

From there comes the necessity of developing general and proper metrics to enable designers, engineers, and managers to guide their decision processes, from the factory planning to operations, management, and control. The World Business Council for Sustainable Development (WBCSD) proposed an appropriate platform or evaluation method that exactly agrees with the original definition of eco-efficiency: create more value with less impact, Eq. (1.1) (Verfaillie & Bidwell, 2000).

$$\text{Eco-efficiency ratio} = \frac{\text{Production or service value}}{\text{Environmental influence}} \qquad (1.1)$$

Conceptually, eco-efficiency is well established and easily understood. Yet the number of potential indicators for the two sides of the eco-efficiency ratio is very high, e.g., monetary, volume, lifetime related to the numerator and material-, energy-, emission-related, and/or lifetime related for the denominator (Peças, Götze, Bravo, Richter, & Ribeiro, 2018; Lourenço et al., 2018). In addition, there are several techniques to measure the economic and environmental impact for each type of indicator (Passetti & Tenucci, 2016). Therefore, a great amount of combination can be used, meaning that there are several available combinations of eco-efficiency ratios, which may lead to different interpretations and a loss of accuracy on tracing eco-efficiency (Peças et al., 2018; Lourenço et al., 2018). In Table 1.1, a few examples are given, where it is also evident that the ratios are different depending on the focus of assessment. At company level, the eco-efficiency ratios must account for a macro-quantification of the company activities' value and environmental impacts for a specific period. At the product level, the ratios should account for the

TABLE 1.1

Examples of Eco-Efficiency Ratios

Company Level Related Eco-Efficiency Ratios
$\dfrac{\text{Gross value} - \text{added of year } X}{\text{Overall environmental impact of year } X}$
$\dfrac{\text{Gross value} - \text{added of year } X}{\text{Environmental impact of materials or energy of year } X}$
$\dfrac{\text{Earnings before interest, taxes, depreciation, and amortization of year } X}{\text{Production environmental impact of year } X}$
$\dfrac{\text{Total number of products produced in year } X}{\text{Production environmental impact in year } X}$

Product Level Related Eco-Efficiency Ratios
$\dfrac{\text{Parts produced per year}}{\text{Production-driven environmental impact per year}}$
$\dfrac{\text{Profit per unit produced per year}}{\text{Production-driven environmental impact per year}}$
$\dfrac{\text{Profit of tool or machine sale}}{\text{Environmental impact of tool or machine production}}$
$\dfrac{\text{Tool or equipment life cycle profit}}{\text{Total life cycle environmental impact}}$

Source: Adapted from Peças et al. (2018) and Lourenço et al. (2018).

value generated by the specific product and the respective specific environmental impact, using a fixed period or the product life cycle time spam.

Despite this abundance of possible eco-efficiency ratios, WBCSD (Lehni, 2000; WBCSD, 2000a,b) proposes guidelines to select a set of value- and environmental-related indicators that represent the behavior of the product, system, or company: they should change with the variation of design parameters of the object under analysis. Following WBCSD, understandable and coherent eco-efficiency ratios should be used, namely using indicators for the same time period and for the same functional unit that represent some meaning for the decision makers, e.g., added value in euros per megajoules of energy, units produced per each ton of $CO_{2eq.}$ emitted.

Following a continuous improvement strategy fostered by WBCSD (2000a,b) and other normative documents on eco-efficiency (ISO, 2012; Jasch, 2000), the aim is to increase the eco-efficiency by increasing the figures of the selected eco-efficiency ratios, i.e., increase value more (keeping or decreasing environmental influence) or decrease environmental influence (keeping or increasing value). But, as stated in Table 1.1, the coexistence of several eco-efficiency ratios to improve might cause a mismatch of design combinations regarding the different eco-efficiency ratios that come to blur and complicate decision making (Peças et al., 2018; Lourenço et al., 2018). In other words, one solution may be good for one ratio but unfavorable for the others. This defines eco-efficiency as a complex multiobjective problem (MOP) with multidimensional external performances (An, Cui, & Qi, 2010), where the designer has a Multi-Criteria Decision Making (MCDM) to handle (Miettinen, 2008). The MOP approaches are used to find an optimal solution for a large or infinite set of alternatives, being usually used to support decision making in problems associated with network design, transportation planning, scheduling, and with allocation problems (Banasik, Bloemhof-Ruwaard, Kanellopoulos, Claassen, & van der Vorst, 2016; Arbiza et al., 2008; Zohal & Soleimani, 2016).

Therefore, this book presents a novel approach for optimizing eco-efficiency of manufacturing processes and products. As an MOP, characterized by many eco-efficiency ratios whose solutions are predictably clashing, the main contribution of this work comprehends the application of optimization methods to support the decision making in early design phases. It also contributes for the eco-efficiency assessment by analyzing and comparing different metaheuristics, considering different design variables and different eco-efficiency indicators, while accounting for the trade-offs that come along with them. Furthermore, the need to fine-tune the metaheuristic parameters applied resulted in a user guide for future applications and for a better understanding of metaheuristic dynamics. The approach is applied to the injection molding process, aiming to identify the mold design characteristics and injection molding parameters that maximize eco-efficiency ratios. This application allows an understanding of how the approach can be applied, contributing to its application in other injection molding cases and also in other types of processes/products.

1.2 Problem Statement and Solution

The proposed approach aims to find the combinations of process or design parameters that optimize eco-efficiency performance, being the eco-efficiency measured in (one or more) ratios between value and environmental performance indicators. The case study selected to demonstrate the benefits of applying the proposed approach can be specifically described as identification of the combination(s) of mold design and injection molding process characteristics that optimize a set of eco-efficiency ratios found relevant by the user (decision maker). Eco-efficiency, as an MOP, is characterized by having a multiplicity of solutions, i.e., there is not a single best solution (the global optimum), but a set of particular feasible solutions called Pareto set or *nondominated* solutions, that are superior to others when considering all objectives. The multiplicity of solutions is explained by the fact that the objectives are conflicting ones (Quariguasi Frota Neto, Walther, Bloemhof, van Nunen, & Spengler, 2009).

Moreover, eco-efficiency quantification is, itself, a two-goal concept that concerns about environmental and economic benefits—the main issue of coupling economic growth with environmental influence is that designers and engineers come across a lot of trade-offs (Banasik et al., 2016; Quariguasi Frota Neto et al., 2009; Li, Kara, & Qureshia, 2015). This multifacetted and contradictory nature of eco-efficiency ratios causes difficulty in the decision-making process (Sproedt et al., 2015). As MCDM refers to a general class of Operations Research (OR) (Pohekar & Ramachandran, 2004), an appropriate procedure must be followed to study the problem (Table 1.2). OR is the discipline that deals with the application of advanced analytical methods to support decision making. It encompasses a wide range of problem-solving techniques and methods applied in the pursuit of improved decision making and efficiency (Hillier & Lieberman, 2015).

The first step of the *OR procedure* consists in defining the problem that needs to be solved and to gather the data that portrays it. The problem of interest is to maximize eco-efficiency ratios that depend on combinations/parameters of the product design and/or manufacturing process. The specific application case is a plastic part produced by injection molding, being possible to produce it with different mold designs and different types of injection machines. The relevant data was gathered in industrial environment and described in Section 1.3.

The second step of the *OR procedure* is the modeling of the system to be optimized. In this case, this was obtained through a Process-Based Model (PBM). The PBM comprises physical and empirical relations that bridge the design choices and the resources inventory from where the costs, environmental impact, and value are calculated. The PBM used includes the correlations required to estimate the injection molding cycle time, setup time, and plastic material required, depending on the injection mold design characteristics, namely the cavities' geometry, the number of cavities, and the type of plastic injection system. The PBM also selects the required injection molding machine power depending on mold size and cavity geometry, meaning that

TABLE 1.2

OR Procedure: Guidelines to Perform an OR Study

OR Procedure
1. Define the problem of interest and gather relevant data.
2. Formulate a mathematical model to represent the problem.
3. Develop a computer-based procedure for deriving solutions to the problem from the model.
4. Test the model and refine it as needed.
5. Prepare for the ongoing application of the model.
6. Implement.

Source: Pohekar and Ramachandran (2004).

the consumed energy can be further estimated. The resources consumed for a specific mold design are then transformed in cost by the use of cost factors based on industrial gathered information. The costs are then transformed in monetary indicators of value, e.g., profit, earnings before interest, taxes, depreciation and amortization (EBITDA), etc. using company-specific data. The environmental impact, namely from the mold and injected part materials, the energy consumed in the mold and plastic manufacturing, and the emissions related with the material and energy production are accounted by the use of end-point eco-indicator ReCiPe method derived from the Simapro software that uses the EcoInvent database (George & Bressler, 2017; Korol, Burchart-Korol, & Pichlak, 2016).

The different design alternatives are to be assessed by each eco-efficient ratio. However, trying all design possibilities may not be effective because there might be many design configurations and many eco-efficiency indicators (EEI), increasing error susceptibility. One of the purpose of this work is to avoid considering all design possibilities to minimize time consumption.

So, to accomplish the third step, a computer-based procedure was developed, consisting of the proposed approach described in Section 1.4. The problem characterization of the proposed approach is illustrated in Figure 1.1 resorting to OR.

Mathematically, the problem under consideration, a MOP, is defined as in Eq. (1.2) (Coello, Dhaenens, & Jourdan, 2010). At this point, the decision

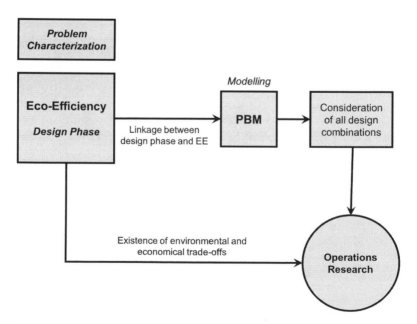

FIGURE 1.1
A general approach to the problem.

variables, the constraints, and the objective function should be defined (Coello, Lamont, & Van Veldhuizen, 2007).

$$(\text{MOP})\begin{cases}\text{Optimize } f(x) = \big(f_1(x), f_2(x), \ldots, f_k(x)\big) \\[2mm] \text{with } x \in D\end{cases} \qquad (1.2)$$

where x is the vector of decision/design variables, D represents the set of feasible solutions, and $f(x)$ is a vector of objective/cost functions (Talbi, 2009b). In this case, the vector $f(x)$ is the PBM whose output is the eco-efficiency value, and its elements are the eco-efficiency indicators (EEI).

After formulating the problem in Eq. (1.2), a computer-based procedure was developed for deriving solutions to the problem from the model, according to the step 4 of *OR procedure* listed in Table 1.2. Metaheuristics were applied in MOPs, allowing the computation of a set of nondominated solutions in terms of the predefined criteria (Eq. 1.2). This set of solutions can be used by the decision maker to select the one that represents the best compromise among the different objectives (Arbiza et al., 2008). Metaheuristics represent a family of approximate optimization techniques that gained a lot of popularity in the past two decades. According to Glover (2009), they are among the most promising and successful techniques. This significant growth of interest in metaheuristic domain is due to the fact that metaheuristics provide "acceptable" solutions in a reasonable time for solving hard and complex problems in science and engineering (Glover, 2009).

The fifth and sixth steps are not included in this study because they are being applied under the funded project that supports this study (GLN, IST, & ISQ, 2017).

1.3 Case Study

The case study has its origin in industrial context, comprising several alternatives of plastic part injection molding. For the case study, a plastic part with three alternative thicknesses and two alternative materials was considered to be obtained by injection molding. In addition, there are alternatives regarding the injection process, namely two types of injection machines and different types of mold design (six different number of cavities per mold and two types of runners). These are alternatives considered to assess the proposed approach performance, and are discussed in the following paragraph. Nevertheless, the PBM potential allows a higher multitude of alternatives (Figure 1.2), namely changings in the part design, in the mold manufacturing processes, and also in factors related to production (production volume,

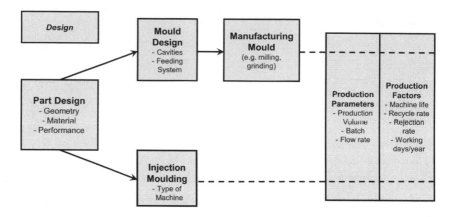

FIGURE 1.2
Set of design variables available in PBM. Any change in one of these variables causes a recalculation of the time and resources needed to inject the required volume of plastic parts.

batch, recycling rate, etc.). It was decided not to include all these alternatives to avoid losing accuracy in the phase of validation of the proposed approach. In addition, variables like production volume and batch size largely depend on the market; rejection rate and material life variation (i.e., decrease) will obviously decrease eco-efficiency, so their inclusion in the decision-making process was considered irrelevant for the study.

The use of different materials will affect the eco-efficiency performance as well as thickness of the part. The range of thickness considered does not affect the performance of the part. These questions of geometry and material type are the most common ones in eco-efficiency decision making. The type of machine where the mold is used for the injection molding process has impacts in the resource consumption. The choice of the machine has to be mainly based on the clamping force, meaning the force applied to the mold to keep it closed during the injection phase (Bravo, 2015). Regarding the machine operating systems, there are three groups: hydraulic, electrical, and hybrid. Hybrid machines, which are a combination of electrical and hydraulic, have advantages of both power systems. Given the machine availability in the company where the case study was developed, only the first two are considered in the present work. Regarding the mold design, the feeding system comprises runners that can be cold or hot. The first ones require lower investment and have the ability to accommodate a wide variety of polymers; however, even though the removed runners can be recycled, there is material waste in the runners' channels. Contrarily, hot runners have potential for faster cycle times and lower material waste (Bravo, 2015). The number of cavities is a mold design parameter that greatly influences several aspects in the injection molding process. The increase in the number of cavities decreases the number of cycles needed to produce the necessary production volume, thereby reducing the time of production and increasing the life

TABLE 1.3

Main Characteristics of the Part Used in the Case Study

Material	PP vs. PBT
Part Volume	4,126.79 mm^3
Part Life	10
Annual production volume	100,000,000
Recycle rate	0%

Image and volume for 1 mm thickness.

of the mold. However, greater the number of cavities, higher the clamping force, the total energy consumption, and the level of material wasted in the runners (Bravo, 2015).

The application of the proposed method is applied to a specific part produced by injection molding. The part is illustrated in Table 1.3 with its main characteristics.

1.3.1 Metaheuristics Selection

Metaheuristics are high-level strategies for exploring search spaces using different methods (Coello et al., 2010). All of them have a common purpose of avoiding the generation of poor quality solutions by introducing general mechanisms that extend problem-specific, single-run algorithms like greedy construction heuristics or iterative improvement local search. The differences among the available metaheuristics concern (i) the techniques employed to avoid getting stuck in suboptimal solutions and (ii) the type of trajectory followed in the space of either partial or full solutions (Dorigo & Stützle, 2003).

The way metaheuristics are included in the model is presented in Figure 1.3. The problem is an MOP with a multiplicity of solutions. So, the selected

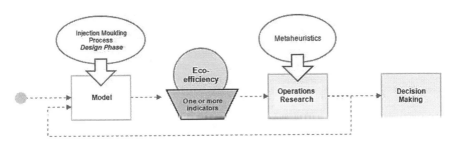

FIGURE 1.3
Overview of the problem, including the use of metaheuristics.

metaheuristic algorithm must take into account the need of manipulations of several candidate solutions in every iteration, meaning that P-metaheuristics, i.e., population-based search metaheuristics, should be used. The main advantage of a population method against the single-state metaheuristic is regards the final stage, where the first has a possibility to retrieve many (near) optimal solutions. Even for a single-objective problem, this is particularly useful when, for instance, inside the injection molding industry, the designer already has a feasible mold produced and does not need to spend resources to produce an optimal one. Inside P-metaheuristics, where population characteristics guide the search, are included the Evolutionary Algorithm (EA) and Swarm Intelligence (Talbi, 2009a).

1.3.1.1 Evolutionary Algorithms

EAs have shown to be a useful tool for approximating the nondominated set of multiobjective optimization problems (Talbi, 2009b; Laumanns, Laumanns, & Kitterer, 2002) and are the most studied population-based algorithms (Talbi, 2009a). EAs are based on the notion of competition. They represent a class of iterative optimization algorithms that simulate the evolution of species throughout generations (Figure 1.4; Talbi, 2009a).

Historically, there are three main algorithmic developments within the field of EA: evolution strategies, evolutionary programming, and genetic algorithms (GA). Common to these approaches are the population-based algorithms that use operators inspired by population genetics to explore the search space—the most typical genetic operators are reproduction, mutation, and recombination (Dorigo & Stützle, 2003).

Differences among the different EA concern the particular representations chosen for the individuals (representing the variables) and the way the genetic operators are implemented. For example, GAs typically use binary or discrete-valued variables to represent information in individuals, and they favor the use of recombination, while evolution strategies and evolutionary programming often use continuous variables and put more emphasis on the mutation operator (Dorigo & Stützle, 2003).

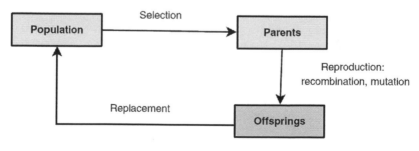

FIGURE 1.4
Generation in EAs (Talbi, 2009a).

Within these three approaches, the GA was chosen for its aptness of dealing with binary or discrete-valued variables, existent in this case study. GA is the most studied population-based algorithm. Being a nature-inspired algorithm, crossover, mutation, and selection operators are responsible for producing new solutions in every iteration. GA initially generates a population of feasible pseudorandom candidate solutions. In every step, called generation, the algorithm evaluates each individual in population, selects parents in the population, applies crossover to obtain new candidate solutions from selected parents, and applies mutation to favor diversification in population. Achieving the termination criteria, the best individual(s) in the last current population is/are returned as optimal solution(s) (Nicoară, 2012).

1.3.1.2 Swarm Intelligence

Algorithms inspired from the collective behavior of species such as ants, bees, wasps, termites, fishes, and birds are referred as swarm intelligence algorithms. Swarm intelligence originated from the social behavior of those species that compete for food.

Swarm intelligence's problem-solving techniques present several advantages over more traditional ones. These techniques are simple, robust, and provide a solution without centralized control or the provision of a global model. Among these algorithms, two successful swarm intelligence methods are the ant colony optimization (ACO) and particle swarm optimization (PSO) techniques (Talbi, 2009a; Saka, Doğan, & Aydogdu, 2013).

The main difference between ACO and PSO is that the first is a greedy algorithm and the latter is an iterative algorithm. Greedy algorithms start from an empty solution, and at each step, a decision variable of the problem is assigned until a complete solution is obtained. In an iterative algorithm, the start is with a complete solution (or population of solutions) that is transformed at each iteration using some search operators (Glover, 2009).

Greedy algorithms mostly (but not always) fail to find the globally optimal solution, because these usually do not operate exhaustively on all the data. Commitments to certain choices are taken too early, which prevent them from finding the best overall solution. Backtracking or dynamic programming is preferable to greedy algorithms (Malik, Sharma, & Saroha, 2013). After considering the referred pros and cons, the PSO type was selected to be used in this study.

PSO was developed by Kennedy and Eberhart in 1995, which mimics the metaphor of bird flocking. Due to the simplicity in implementation and efficiency in solving optimization problems, it has been used for solving MOP problems (Yang, 2012). In fact, PSO seems particularly suitable for MOP, mainly because of the high speed of convergence that the algorithm presents for single-objective optimization (Coello, Pulido, & Lechuga, 2004).

In PSO, particles "flow" through the searching space and change their positions by considering both cognitive and social components, thus leading the population to an emergent behavior. PSO particles are initially given a random position and null velocity. At each iteration of the algorithm, each particle moves with a velocity that is a weighted sum of three components: the old velocity, a velocity component that drives the particle toward the location in the search space where it previously found the best solution so far, and a velocity component that drives the particle toward the location in the search space where the neighbor particles found the best solution so far (Saka et al., 2013).

Having in mind these considerations, GA and PSO algorithms are the ones to integrate in the system under analysis. The algorithm's performance was assessed in both situations, using just one of them and integrating them. The results are described and discussed in the next section.

1.4 Solution Development—Modeling and Integration

The solution proposed to assess eco-efficiency in injection molding based on metaheuristics comprises three stages. The first stage comprehends the modeling of the system. The second stage is the metaheuristic integration with the model. These two stages make up the solution itself, i.e., the solution development, and correspond to the step 2 of the OR procedure presented in Table 1.2. The last stage, presented and discussed in Section 1.4.2, is the solution application where metaheuristics are fine-tuned and tested with different scenarios, while varying design variables and cost functions (eco-efficiency indicators) correspond to the step 3 of the OR procedure presented in Table 1.2.

1.4.1 Modeling—PBM

The PBM was developed to make the metaheuristic algorithm manipulation possible. It was based on an industrial context, with a real case study, allowing the validation of the proposed approach and models (Figure 1.5).

The PBM was built in the MATLAB® tool and was based on Bravo's work (Bravo, 2015), where the influence of part and mold design on the mold cost and price and on the injection molding time and resource consumption estimation was modeled. The PBM is a multidesign variable model based on simple technological relations, where for each type of mold and part material there is a specific injection molding cycle time, material consumed per cycle (part material and wasted per cycle), rejection rate, setup time, batch size, labor per machine per cycle, etc.

The model allows to estimate the equipment time required for an inputted annual production volume, the labor time required, and the plastic material required. The number of injection cycles (mold closing and opening operations) is obtained by the ratio between the effective production volume and

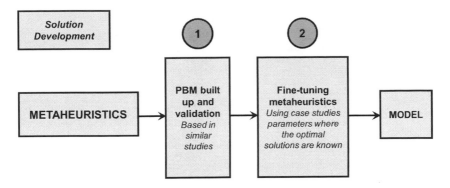

FIGURE 1.5
Solution development comprising two stages of solution development.

the number of cavities per mold. The effective production volume is the imposed production volume plus the estimated rejection percentage. The PBM was built for the following part and mold design alternatives:

- Part material: polypropylene (PP) and PBT;
- Number of cavities per mold: 4, 8, 16, 32, 48, 64;
- Feeding system: hot runners and cold runners;
- Type of machine: electric or hydraulic.

The developed PBM is a multidesign variable model that includes specific technological relations regarding mold maintenance operations used in the company's production realm. For PP and PBT, the company does a maintenance operation of 5h for each 11,836 injection cycles of the mold. This information regarding time and resources consumed is used by PBM to calculate the total production cost, together with mold cost and indirect production cost, for each mold and material alternative. The value indicators (e.g., EBITDA) are then calculated by PBM. Based on the resources consumed, material, energy, and consumables, PBM uses eco-indicators based on ReCiPe database to estimate the environmental impact (Goedkoop et al., 2009; Pre Sustainability, 2014). In Tables 1.4 and 1.5, the cost and environmental impact for a specific alternative are showed.

The results of PBM for one of the eco-efficiency indicators, the ratio between EBITDA and Environmental impact (EEI_1), are showed in Figure 1.6, where the solutions linked by a dashed line have the same number of cavities. When considering 64 cavities, there aren't available electric machines that can support the clamping force needed. The results are presented with EEI_1 acting as the cost value, by using the relation in Eq. (1.3):

$$Q_{sol} = \frac{\text{Solution}}{\text{Known global solution}} \quad (1.3)$$

TABLE 1.4

Production Costs for a Specific Alternative: Mold with 32 Cavities, Hot Runners, and Using Electric Machine in Injection Molding

Cost (€)/Year			
Material cost	618,634	Tooling cost	68,530
Energy cost	20,992	Building cost	1,068
Labor cost	12,233	Maintenance cost	13,300
Machine cost	30,773	TOTAL	765,530

The part material is PP. Annual production volume considered: 1 × 108 parts.

TABLE 1.5

Environmental Impact for a Specific Alternative: Mold with 32 Cavities, Hot Runners, and Using Electric Machine in Injection Molding

		Resources	Specific envioronmental impact (EI)	Environmental Impact
Mold production	Material necessary	1,934.75 kg	1.575 Pts/kg	3,044.03 Pts
	Recycling	393.20 kg	−0.150 Pts/kg	−58.88 Pts
	Consumables	1,720.65 kg	0.264 Pts/kg	454.09 Pts
	Energy	35,376.60 MJ	0.0158 Pts/MJ	557.16 Pts
Injection process	Material injected	441,881.44 kg	0.246 Pts/kg	108,586.38 Pts
	Recycling	4,441.72 kg	−0.227 Pts/kg	−1,008.27 Pts
	Energy	539,797.91 MJ	0.016 Pts/MJ	8,501.49 Pts
	Material wasted	4,441.70 kg	0.499 Pts/kg	2,217.07 Pts
Total				122,293.1 Pts

The part material is PP. Annual production volume considered: 1 × 108 parts. Environmental impact is obtained by the product of the resource and the specific eco-indicator.
Source: Goedkoop et al. (2009) and Pre Sustainability (2014).

In addition, in Figure 1.6, the solutions that have cost values equal to 98% of the global optimum are highlighted with a dashed pink line. Then it is possible to have a visual perception about the solutions that are closer to the maximum.

By analyzing the domain of solutions regarding EEI_1 cost value, the difference between the global optimum and the solution that has the lowest cost value is about 0.25. The solutions are concentrated on the top of the graphic—when evaluating their cost values, about 81% of them are above 0.85, and approximately 10% of them are above 0.98. Numerically, the results are very close to each other. This proximity between solutions that have high cost values has influence on metaheuristic performance as they assure to provide feasible solutions. These results are just for this indicator, meaning that concurrent results can be obtained for other eco-efficiency indicators or combinations of them, acting as cost values. Particularly when evaluating

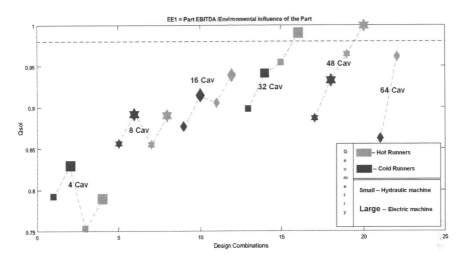

FIGURE 1.6

Cost values (regarding eco-efficiency indicator EEI_1) of some of the design combinations, compared using Q_{sol}. Annual production volume 1×109 parts. Material, PP.

alternatives with EEI_1, it came up with the same global optimum configuration: 48 cavities, hot runners, and electrical machine. However, as it can be seen, the mold composed of 32 cavities, hot runners, and electrical machine is also a good option. Hence, it is of designer's interest to know those good solutions that are very close to the optimum as well. This is also an advantage of P-metaheuristics, whose purpose is to provide not only global optimum, but also other good solutions. In this case, it was straightforward to take these two optimal solutions for the problem. Further on, as the design variables and cost functions were added, the problem got more complex, and the solutions and performance of metaheuristics were different.

1.4.2 Methodology—Fine-Tuning Metaheuristics

Metaheuristics have proved to be powerful tools in optimization tasks. All metaheuristic algorithms have algorithm-dependent parameters that have to be fine-tuned to maximize the efficiency and effectiveness of the performance. So, one of the challenging issues is deciding what values of parameters need to be used in an algorithm (Yang & Karamanoglu, 2013).

Efforts were made by researchers to find performance indicators that could provide relevant outcomes while changing and testing different set of parameters. A study was found where the *t*-test (hypothesis testing) was successfully used to assess and compare the effectiveness (finding the true global optimum) and efficiency (computational cost) of two searching algorithms, the GA and PSO (Hassan & Cohanim, 2005). While efficiency has an intuitive definition, effectiveness was the focus of attention. It can be defined as the ability of the algorithm to repeatedly find the known global solution or arrive at sufficiently close solutions when the algorithm is started from many random different points in the design space (Hassan & Cohanim, 2005).

The same methodology was used in this study, and the purpose of the *t*-test usage was to assess and compare the effectiveness of PSO and GA while changing their parameters. The results are the eco-efficiency indicators, normalized using Eq. (1.3).

In hypothesis testing, a null hypothesis, H_o, will be correctly accepted with a significance level $(1-\alpha)$ (Hassan & Cohanim, 2005). The null hypothesis, H_o, was defined as analyzing the effectiveness of the set of parameters that provide results whose mean of Q_{sol} of Eq. (1.3) is equal or above 99% of the global optimum, as in Eq. (1.4):

$$H_o : \mu_{Q_{sol}} \geq 0.99 \text{ vs. } H_a : \mu_{Q_{sol}} < 0.99 \tag{1.4}$$

The considered significance level, α, is 5%.

Along with the definition of the performance indicator, it is important to know which parameter can be changed for each algorithm. The GA and PSO algorithms are both population based, so the number of elements of the population and the number of iterations are parameters in common.

The specific parameters that characterize and define the dynamics of the algorithms (operators that transform solutions at each iteration), in this case for GA and PSO, and the considered values are presented in Table 1.7. For PSO, they control the velocity of each particle. For GA, they control the mutation and crossover intensities (Table 1.6).

The methodology developed towards fine-tuning of PSO and GA parameters, promotes the application of operators at each iteration leading to optimized solution(s). So parameters have to be first fine-tuned to guarantee that the algorithms are having good performance and are reliable. The fine-tuning process is illustrated in Figure 1.7, with the aim of supporting the use of metaheuristics in other applications as a user guide with code. A code of colors was used to distinguish the different phases of this fine-tuning process.

This methodology comprises different fine-tuning phases while varying the number of iterations (N_{It}) and the number of elements in the population (N_{Pop}). It is a comprehensive strategy where the sets of parameters that do not provide convergent solutions according to the hypothesis test of Eq. (1.3) are dropped off and do not pass to the next phase. It is an eliminative test in

TABLE 1.6

Specific Parameters of the Metaheuristics Applied

PSO	GA
$w = \{0.5, 0.7, 0.9, 1.1, 1.4\}$	$cf = \{0, 0.2, 0.5, 0.7, 1\}$
$c_1 = \{1.5, 2\}$	$sc = \{1, 2\}$
$c_2 = \{1.5, 2\}$	$sh = \{0.01, 1\}$

TABLE 1.7

Methodology for Fine-Tuning the Algorithms

START		EXPLORE		EXPLOIT	
N_{It} and N_{Pop} should be high enough to achieve convergence but low enough to not spend too much computational time. By trial and error.		Decrease N_{It} and increase N_{Pop}—more exploration, less exploitation. Consider the set of parameters that weren't rejected in the previous phase.		Increase N_{It} and decrease N_{Pop}—more exploitation, less exploration.	

Passage to the Next Phase: Regardless of the Set of Parameters ... Convergence of

≤80% of the solutions?		≤30% of the solutions?			
Yes	No	Yes	No	Yes	No
Go to EXPLORE phase. Reject the set of parameters whose solutions doesn't converge.	Increase N_{It} and N_{Pop}. Repeat.	Reject the set of parameters whose solutions doesn't converge. Decrease N_{It} while maintaining or increasing N_{Pop} and repeat; OR go to EXPLOIT phase.	Results are not meaningful as N_{It} is too small. No set of parameters is rejected and go to EXPLOIT phase.	Reject the set of parameters whose solutions doesn't converge. Increase N_{It} while maintaining or decreasing N_{Pop}.	Reject the set of parameters whose solutions doesn't converge. *Optional action:* Decrease N_{It} and N_{Pop}.

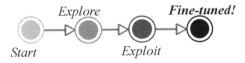

Explore Fine-tuned!

Start Exploit

FIGURE 1.7
Color code for distinguishing fine-tuning phases.

which the survival of the fittest sets of parameters considering N_{It} and N_{Pop} variations will dictate which ones are to be used to find the optimal solutions by changing the design variables and for different cost functions (and combinations of them). This strategy guarantees the choice of the most versatile and adjustable set(s) of parameters that can fit for different N_{It} and N_{Pop}.

1.4.3 User Guide for Fine-Tuning PSO and GA

The overall process is illustrated in Figure 1.8 and should be consulted along with the following explanation. The methodology uses a trial-and-error approach and is highly dependent on the complexity of the problem.

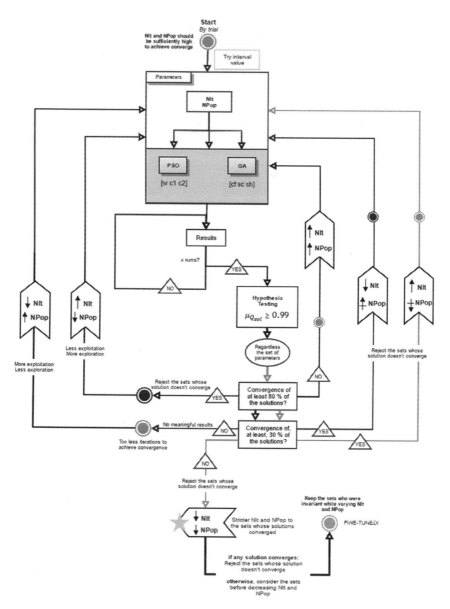

FIGURE 1.8
Methodology for fine-tuning metaheuristics.

The user will gain sensibility and empirical knowledge for the fine-tuning process while trying and experimenting different N_{It} and N_{Pop}.

The process starts (blue circle) with the definition of the hypothesis testing regarding PSO and GA effectiveness, like the one presented in Eq. (1.4). The user can define other conditions for hypothesis testing. The number of runs (x in Figure 1.8) varies in the different fine-tuning phases, being that the sample size depends on it:

$$\text{Sample size} = \text{Runs} \times N_{Pop} \tag{1.5}$$

The sample size (group of solutions) that is to be evaluated is created by collecting the results from every run and is meant to be equal for all experiments. If N_{Pop} is different, the number of runs x will vary proportionally to have the same sample size always (Eq. 1.5).

As a rule of thumb, based on the experience applying the method, the user should perform 3–5 iterations and use a population of 10–20 elements to be more efficient in achieving results. In addition, assuming that the hypothesis is accepted/rejected no matter the number of runs, the lowest amount of runs should be chosen to save computational time. Using the expected largest N_{Pop} and using the lowest amount of runs that resulted from the previous point, the sample size is fixed (according to Eq. 1.5). So, the number of runs x can be computed by defining N_{Pop} in every fine-tuning phases and using the fixed sample size. The fine-tuning methodology is summarized in Table 1.8 and Figure 1.8.

After fine-tuning, PSO and GA can be used to search quickly and more accurately, i.e., more effectively, for optimal solution(s).

1.5 Results and Discussion

The fine-tuning methodology was applied to the case study, followed by the application of PSO and GA to find alternatives of mold design, material, and production conditions that maximize eco-efficiency, i.e., maximize different eco-efficiency indicator, being used alone or together as cost functions.

1.5.1 Finding the Fine-Tuning Parameters

Towards applying the methodology by following the user guide (Section 1.4.3), two EEI were used as cost functions (Eqs. 1.6 and 1.7).

$$EEI_1 = \frac{\text{Part EBITDA}}{\text{Environmental influence of the part}} \tag{1.6}$$

TABLE 1.8

Representation of Rejected Solutions (Left) and the Best Solution (Right)

Set of Parameters Whose Solutions Are Rejected	Set of Parameters Whose Solutions Are the Most Invariant to N_{It} and N_{Pop} Changes
[1 0] [1 1] [0 1]	[0 0]

$$\text{EEI}_2 = \frac{\text{Parts produced/mold life cycle}}{\text{Mold LCA}} \tag{1.7}$$

EEI_1 assesses the relation between the product value; in this case, a part produced by injection molding process and its environmental impact. The value is translated by EBITDA and the environmental impact by the ReCiPe method using the SimaPro software (Goedkoop et al., 2009; Pre Sustainability, 2014). This indicator considers the industrial context regarding the annual demand of parts. The second indicator, EEI_2, assesses the relation between the number of parts that a mold can produce (without considering the demand) and the environmental impact of mold production and mold use phase. Again, the LCA (Life Cycle Assessment) was developed using the SimaPro software and applying the ReCiPe method.

From hypothesis testing, the results are presented as [EEI_1 EEI_2] in such a way that if H_o is accepted as being true at a 99% significance level, the result is 0, otherwise is 1 (Table 1.7). It is irrelevant to present the design variables at this stage.

By following the user guide of Section 4.3, the hypothesis testing is equally defined as in Eq. (1.4). Regarding the number of runs to be performed, the sample size n, Eq. (1.5), is meant to be equal for all experiments, to be always evaluating the same number of elements in hypothesis testing. By fixing n, if N_{Pop} is different, the number of runs x will vary proportionally, according to Eq. (1.5).

To establish a minimum number of runs, the results were compared with 3, 5, and 10 runs as shown in Table 1.9. By evaluating each row of Table 1.9 regarding the two EEI considered and by looking at the results while varying the number of runs, it can be seen highlighted with colors that the acceptance/rejection of the hypothesis is the same for 3, 5, or 10 runs. The conclusion taken from here is that three runs are enough to eliminate the randomness feature of the algorithm dynamics. However, as aforementioned, there is also a desire of having large sample sizes.

The sample size was fixed as being $n = 20 \times 3 = 60$ because the largest N_{Pop} is equal to 20, and three runs allowed the results to be invariant. Then, the

TABLE 1.9

Results vs. Number of Runs

			\[EEI$_1$ EEI$_2$\] HYP.1 $- H_o : \mu_{Q_{sol}} \geq 0.99$								
Runs			**3**			**5**			**10**		
w	c_1	c_2	HYP 1	Mean	Std. Dev.	HYP 1	Mean	Std. Dev.	HYP 1	Mean	Std. Dev.
0.9	1.5	1.5	[1 0]	[0.9847 0.9926]	[0.0225 0.0100]	[1 0]	[0.9835 0.9924]	[0.0302 0.0112]	[1 0]	[0.9860 0.9928]	[0.0283 0.0115]
1.4	1.5	1.5	[1 1]	[0.9617 0.9858]	[0.0644 0.0195]	[1 1]	[0.9554 0.9835]	[0.0682 0.0203]	[1 1]	[0.9626 0.9868]	[0.0596 0.0174]

number of runs, x in Figure 1.8, varies and is computed according to Eq. (1.8), by considering a sample size of 60, in Eq. (1.5).

$$60 = \text{Runs} \times N_{Pop} \qquad (1.8)$$

With the number of runs settled, the fine-tuning methodology is developed by following the instructions described in Figure 1.8. All the possible combinations of parameters presented in Table 1.6 were considered and passed through the methodology. Here the results are only shown from the exploitation phase of PSO, where the sets of parameters that survived the previous fine-tuning phases are presented. Results showed that only 25% (less than 30%) of the remaining sets of parameters were convergent (Table 1.10). According to the methodology, Figure 1.8, this means that the optional action ("orange star") can be done, or the algorithm can be considered as being fine-tuned. In this case, as only one set of parameters converged, the algorithm is considered as being fine-tuned. The same procedure was undertaken for GA, and the most invariant parameters setting to N_{It} and N_{Pop} variations are presented in Table 1.11.

After fine-tuning both algorithms, the model is completed, and the metaheuristic search for the best solution(s) takes place.

1.5.2 Finding the Best(s) Alternative(s)

At this stage, metaheuristics integrated with PBM allowing the interaction between design inputs and eco-efficiency outcomes. This is the third stage

TABLE 1.10

Results from Exploitation Phase for PSO Algorithm

⬤ Runs = 20				$\left[\text{EEI}_1 \ \text{EEI}_2\right]$	
N_{Pop}				3	
w	c_1	c_2	$H_o : \mu_{Q_{sol}} \geq 0.99$	Mean	Std. Dev.
0.5	1.5	1.5	[1 0]	[0.9739 0.9873]	[0.0235 0.0096]
0.7	1.5	1.5	[1 1]	[0.9484 0.9754]	[0.0521 0.0206]
		2	[1 0]	[0.9526 0.9861]	[0.0473 0.0111]
	2	1.5	[0 0]	[0.9671 0.9900]	[0.0418 0.0121]
≤30% of Convergence?				NO (1/4)	
⬤				$\{w, c_1, c_2\} = \{0.7, 2, 1.5\}$	

TABLE 1.11

Final Parameter Setting for GA Algorithm

⬤ $\{cf, sc, sh\} = \{0.7, 2, 0.01\}$

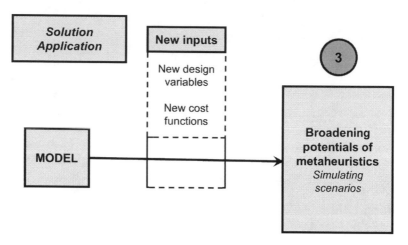

FIGURE 1.9
Solution application.

of eco-efficiency assessment, where the impact of different alternative design variables is calculated using different eco-efficiency indicators as cost functions. Several cost functions, i.e., eco-efficiency indicator of performance, and their combinations represent different scenarios of decision making. The simulation of these scenarios broadens the potential of metaheuristics application to the eco-efficiency realm (Figure 1.9).

Several design parameters and cost functions are presented in Table 1.12. In this table, the experiments conducted are also explained. In all experiments, the number of cavities, the type of machine, and the type of runners were considered as design variables, because these are the ones that influence the performance. The annual production volume was kept constant in each experiment. In experiment 1, the thickness and the part's material were not considered as design variables; in experiment 2, the thickness was used as a design variable; and in experiment 3, thickness was kept constant and part's material was considered as a design variable.

The nomenclature used and the domain of solutions of each design variable are in Table 1.13. These are based on the industrial case study described in Section 1.3. These design variables are the elements of vector x referred in Eq. (1.2).

Each experiment described in Table 1.12 must be run for a specific scenario. In experiments 2 and 3, several scenarios were considered as interesting to test, and in experiment 1, only one scenario was considered. In each scenario, a specific cost function is used (B and C) or a specific set of cost functions is used (A and D scenarios). It means that experiment 2 was run for B scenario as well as for C and D scenario. The "cost function scenarios" of Table 1.12 are the elements of vector $f(x)$ of Eq. (1.2). The scenarios A and D have more than one EEI_i to be maximized, i.e., the cost function is the maximization of

TABLE 1.1

Different Experiments for Exploring Metaheuristics Potentials

Experiment	1	2	3
Focus		Injection Process	Part's material
Design variables		Cavities Type of machine Runners	
		Part's thickness	Part's material
Cost functions	A*	B, C*, D	B, C, D*
Annual production volume	50,000,000	100,000,000	100,000,000

Cost functions for each scenario

A	B	C	D
$EE_1 = \dfrac{\text{Part's profit}}{\text{Part EI}}$	$EE_1 = \dfrac{\text{Part's profit}}{\text{Part EI}}$	$EE_2 = \dfrac{\text{Parts produced mold life cycle}}{\text{Mold LCA}}$	$EE_3 = \dfrac{\text{Part's profit}}{\text{Energy consumption in injection process}}$
$EE_2 = \dfrac{\text{Parts produced/mold life cycle}}{\text{Mold LCA}}$			$EE_4 = \dfrac{\text{Part's profit}}{\text{Material consumption in injection process}}$
			$EE_5 = \dfrac{\text{Mold's profit}}{\text{Mold LCA}}$

The ones signaled with * are the ones presented and discussed in this chapter.

TABLE 1.13

Nomenclature Used and Domain of Solution of Design Variables

Variables	Domain of Solution
Cavities	{4, 8, 16, 32, 48, 64}
Type of machine	{H (Hydraulic), E (electric)}
Runners	{H (Hot), C (Cold)}
Material	{PP, PBT}
Thickness	{1; 1.25; 1.75}

several eco-efficiency indicators, meaning that these can be considered as MOPs. The B and C scenarios are single optimization problems.

The user of the proposed methodology must design the experiment and the scenario that most matches the aim of the ongoing decision-making process. In this chapter, three experiments/scenarios combinations were shown for the sake of understanding the potential of the proposed methodology: experiment 1 for scenario A, experiment 2 for scenario C, and experiment 3 for scenario D.

In experiment 1, the material and thickness are fixed (PP and 1 mm), along with the annual production volume. The aim is to identify what is (are) the combination(s) of number of mold cavities, types of machines, and the type of runners that maximizes two eco-efficiency indicators, EEI_1 and EEI_2, simultaneously. The first assesses the eco-efficiency of the injected part and the second, the eco-efficiency of the injection mold (Table 1.14).

This experiment is very similar to the ones that were made during the fine-tuning of metaheuristics in Section 1.4.2. Nevertheless, a different annual production volume is used, 5×10^7 instead of 10^8 units per year. For the previous annual production volume, the best design alternative was the same for each of the two eco-efficiency indicator used (EEI_1 and EEI_2). In this experiment and scenario, EEI have different optimal solutions [32 E C] and [48 E H],

TABLE 1.14

PSO and GA Solutions and Respective Cost Values Regarding Experiment 1, Scenario A (for PP and 1 mm Thickness)

Scenario A	PSO	GA				
Solutions	Frequency	Frequency	$EE_1 = \dfrac{\text{Part's profit}}{\text{Part EI}}$		$EE_2 = \dfrac{\text{Parts produced/mold life cycle}}{\text{Mold LCA}}$	
[16 E H]	0	5/30	0.823		0.990	
[16 E C]	0	6/30	0.961		0.986	
[32 E H]	24/30	3/30	0.922		0.993	
[32 E C]	6/30	6/30	1		0.978	
[48 E H]	0	6/30	0.908		1	
[48 E C]	0	4/30	0.962		0.973	

respectively. This result is positive, meaning that the algorithms react and provide solutions. From Table 1.14, it is possible to conclude all the solution spotlight electric machines. Also, molds with 4, 8, and 64 cavities are out of the solutions. The high production volume and the lower energy consumption of electrical machines can be considered as main reasons for this result (the mold with 64 cavities cannot be used in electrical machines!!). PSO outputted two out of six solutions. It suggested to use 32 cavities and electric machine. Regarding the runners, they can be hot or cold; however, 80% of the population on the three runs performed preferred hot runners. GA outputted all the six solutions. All of them were approximately outputted in the same proportions. Curiously, by comparing the solutions that were outputted by both algorithms, the solution that was most preferred by the PSO was less preferred by GA, and the solution that was less preferred by PSO was one of the most preferred by GA. Among this group of solutions, the last two columns of Table 1.14 provide the cost values and are extremely important for decision making. The better option for EEI_1 is a mold with 32 cavities with cold runners and using electric machines [32 E C]. The better option for EEI_2 is a mold with 48 cavities with hot runners and using electric machines [48 E H], but the difference between all alternatives of the EEI is not significant, giving to the decision maker the indication that, for this EEI, these design parameters have no special influence. In fact, this EEI measures mold eco-efficiency and this is not highly affected by mold design parameters, since those ones mainly affect the eco-efficiency of the plastic part production process. Evaluating both EEI together, cold runners were more favorable for the some number of cavities, (the difference is more notorious when looking to EEI_1).

As regards the comparison of the performance of two algorithms, GA provided more solutions. On one side, this can be advantageous because the user has more feasible solutions, but on the other side, the algorithm provided solutions that were less good. For instance, the solution [16 E H] has a considerable low value regarding the EEI_1, so it can be disregarded. The algorithm selected that solution because of the high value regarding EEI_2; however, there are solutions in which EEI_2 is equally weighted and that doesn't have such a low cost for EEI_1 as [16 E H], for instance, solutions [32 E H] and [48 E H]. Although PSO provided less solutions by varying only the runners type, it provided solutions whose weights are differently distributed.

Despite several knowledge about the design alternatives, the user of this methodology can benefit from this experiment if one design alternative that has to be elected as the best one is a mold with 32 cavities, with cold runner, and used in an electric machine [32 E C]. There is an alternative way to represent the results to facilitate decision making, the use of Pareto frontier (Figure 1.10). The performance of each cost function allows for a more visual-based comparison.

Regarding experiment 2 with scenario C, there are six different solutions pointed out—four solutions by PSO and two solutions by GA (Table 1.15). PSO offered a more variable set of solutions, whereas GA only offered

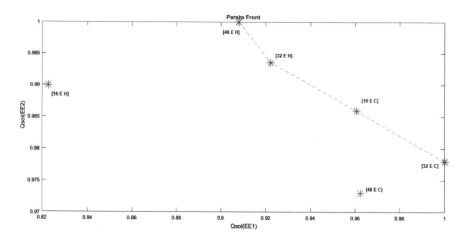

FIGURE 1.10
Pareto frontier of experiment 1, scenario A.

TABLE 1.15

PSO and GA Solutions and Respective Cost Values Regarding Experiment 2, Scenario C

Scenario C	PSO	GA	$EE_2 = $	Parts produced/mold life cycle
Solutions	**Frequency**	**Frequency**		**Mold LCA**
[32 E H 1]	18/30	0		0.994
[32 E H 1.25]	8/30	0		0.992
[48 E H 1]	2/30	0		1
[48 E H 1.25]	0	20/30		0.999
[48 E H 1.75]	0	10/30		0.996
[64 H H 1]	2/30	0		0.972

solutions that does not correspond to a global optimum. Evaluating the cost value of solutions, high weights are attributed from all solutions as they are greater than 95% of the optimal solution, so all solutions (design alternatives) are considered as being eco-efficient (assuming EEI_2 as the indicator of eco-efficiency). By looking at the design alternatives, 32 and 48 cavities are the ones with best performance and using electrical injection machines. Nevertheless, in this case, only molds with hot runners are among the best design alternatives, mainly because of the very high production volume. Since EEI_2 is used as cost function, mainly measuring the eco-efficiency of the mold, the influence of part thickness is not significant. Again, the methodology gives important information for the decision-making process.

Finally, the results of experiment 3 with scenario D are to be presented, where the part thickness (1 mm) and annual production volume (1×10^8 units per year) are kept constant. This scenario aims to find a design alternative that allows the optimization of three EEI_i: EEI_5 that measures the overall eco-efficiency of the product mold, the EEI_4 that measures the eco-efficiency using the plastic part profit and the environmental impact of the material consumption; and the EEI_3 that measures the plastic part profit and the environmental impact of the energy consumed. In this case, besides the optimization of the three EEI_i included in the scenario (as cost functions), a macro eco-efficiency cost function was also included, which is represented by the optimization of the accumulated result of the three EEI_s. Besides, performance accumulation causes loss of information regarding each specific cost function, and in some cases might be useful for the decision-making process to have a more aggregated result (it allows to find the "best of the best design alternative"). The solutions outputted by the algorithms are exposed in Table 1.16. In this case, there are three EEI as objective functions, so the chance of having conflicting solutions is more likely.

In this MOP case, Table 1.16, there is an extensive list of solutions, because there are three EEI plus the macro EEI as objective functions. Hence, there are more combinations of weights for the EEI to have different feasible solutions. However, after analyzing solutions and their respective cost values, some of them can be overlooked (the ones with lower normalized values in all EEI_i, like [4 E C PP] and [4 H H PP]).

There are some additional interesting outputs. There is a common feature between solutions—almost all of them point out to electric machines and all of them choose PP as the most suitable material. If only the mold eco-efficiency-related indicator, EEI_5, is considered, the best solutions are the ones with low number of cavities. In fact, the profit of smaller molds is higher than the one of the larger molds. When EEI_3 and EEI_4 are considered individually, the best design alternatives are the ones with higher number of cavities (but not the one with 64 cavities, because there is only one hydraulic machine option available, causing much higher energy consumption). For EEI_4, the best alternatives are the ones with cold runners, since this EEI gives preference to the reduction of environmental impact of energy consumption (less energy to the runners compensates for a higher cycle time). For EEI_3, the best alternatives are the ones with hot runners, since EEI gives preference to the reduction of material consumption, and this type of runners minimize it. When considering all EEI_i, using the macro EEI, the higher impact of the 10^8 plastic part in the energy consumption and in material use (EEI_3 and EEI_4 related performance) than the unique mold profit and life cycle impact (EEI_5) makes best design alternatives for maximization of the three EEI_i to be the ones with higher cavities per mold [48 E H PP] and [32 E H PP], found by the GA algorithm. Nevertheless, the other options with hot runners and electric machines, with 16 and 8 cavities per mold are not far from the score of those with higher number of cavities. These ones were found by the P3O algorithm.

In summary, the performances of the algorithms were good. They provided many feasible solutions that give designers useful information and different

TABLE 1.16

PSO and GA Solutions and Respective Cost Values Regarding Experiment 3, Scenario D, Including Simultaneously Three EEI_i

Cost Functions / Solutions	Mold's profit — EE5 = Mold LCA				Part's profit — EE3 = Energy consumption in injection process	Part's profit — EE4 = Material consumption in injection process			
	EEI5	EEI3	EEI4	(Σ EEI_i / nΣ EEI_i)	Solutions	EEI5	EEI3	EEI4	(Σ EEI_i / nΣ EEI_i)
[4 E H PP]	1	0.429	0.679	(0.703)	[16 E C PP]	0.271	0.852	0.984	(0.702)
GA	6/30	PSO	3/30	0.93	GA	0	PSO	6/30	0.93
[4 E C PP]	0.598	0.627	0.867	(0.697)	[32 E H PP]	0.327	0.919	0.974	(0.740)
GA	1/10	PSO	0	0.92	GA	2/30	PSO	0	0.98
[4 H H PP]	0.969	0.321	0.666	(0.652)	[32 E C PP]	0.212	0.773	1	(0.662)
GA	3/30	PSO	0	0.86	GA	6/30	PSO	1	0.88
[8 E H PP]	0.633	0.700	0.833	(0.722)	[48 E H PP]	0.295	1	0.971	(0.755)
GA	3/30	PSO	6/30	0.96	GA	3/30	PSO	0	1
[16 E H PP]	0.429	0.847	0.930	(0.734)	[48 E C PP]	0.192	0.719	0.987	(0.633)
GA	0	PSO	12/30	0.97	GA	6/30	PSO	3/30	0.84

In the table (Σ EEI_i) is the macrovalue for the cost function; the nΣ EEI_i is the normalized value regarding the possible solutions (possible design alternatives).

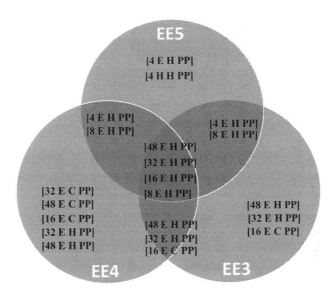

FIGURE 1.11
Experiment 3, scenario D: Optimal solutions according to EEI; normalized criteria (normalized ≥ 0.95).

alternatives to have an eco-efficient scenario. Among the solutions afore-mentioned, the fourth design alternatives with similar performances were complementarily outputted by both algorithms, which mean that both inter-ventions were determinant and not redundant. This emphasizes the fact that the user needs to collect the results of both algorithms instead of compar-ing the results of each one of them. For practical application, in Figure 1.11, a graphical representation of the solutions outlined is presented. Depending on the weights that the designer wants to prize, those solutions are allocated in the most appropriate areas of the figure regarding the highest EEI weighting. So, the performance of the algorithm achieved the global optima of EEI and provided many feasible solutions, which gives the designer useful informa-tion and different alternatives to have an eco-efficient decision.

1.6 Conclusions

This chapter proposes the assessment of eco-efficiency of manufacturing processes through metaheuristics, aiming to support design decisions.

A case study in the injection molding industry was followed, with the defi-nition of different design alternatives not only regarding the part and mold design, but also operational design variables. After defining and modeling the system to be optimized in a mathematical model (process base modeling was used), metaheuristics were integrated, where the designer can set the targets regarding design variables, constraints, and cost functions (entries that have influence for decision making).

Metaheuristics need to be fine-tuned in order to provide reliable solutions. This study proposes a methodology for fine-tuning the algorithms that were successfully applied to this case study, as the results obtained presented alternatives with high eco-efficiency ratios.

GA and PSO algorithms were used to assess which of them would be more adequate. The application of both algorithms was advantageous, as they provided coherent and complementary solutions with high eco-efficiency ratios, offering the user a good span of possibilities. In an eco-efficiency context with different dimensions of analysis and parallel indicators, this is an advantage to other methods that only present the best solution regardless of how close it is from the others. The solutions were also accompanied by their cost values, which help the designer to know the weights that are being attributed to each cost function.

The application of both algorithms was important as, for example, in some experiments, GA provided a smaller group of solutions but with a global optimum, whereas PSO provided a wider group of solutions comprising the near optimal ones. In fact, "GA vs. PSO" initial idealization switched to "GA & PSO" concretization, meaning that the algorithms are not competing against each other, but complement each other. Nonetheless, using both metaheuristics puts away the advantage of having lower computational times, as the tasks have to be always replicated for the other algorithm.

Acknowledgements

This work was supported by FCT, through IDMEC, under LAETA, project UID/EMS/50022/2019.

Authors gratefully acknowledge the funding of Project POCI-01-0247-FEDER-017637, cofinanced by Programa Operacional Competitividade e Internacionaliza-ção and Programa Operacional Regional de Lisboa, through Fundo Europeu de Desenvolvimento Regional (FEDER) and by National Funds through FCT—Funda-ção para a Ciência e Tecnologia.

References

An, X. H., Cui, Y. M., & Qi, E. S. (2010). Study on eco-efficiency evaluation of manufacturing system based on circular economy. *Proceedings -2010 IEEE 17th International Conference on Industrial Engineering and Engineering Management, IE and EM2010*, Xiamen, China, 591–593. doi:10.1109/ICIEEM.2010.5646545.

Arbiza, M. J., Bonfill, A., Guillén, G., Mele, F. D., Espuña, A., & Puigjaner, L. (2008). Metaheuristic multiobjective optimisation approach for the scheduling of multiproduct batch chemical plants. *Journal of Cleaner Production, 16*(2), 233–244. doi:10.1016/j.jclepro.2006.08.028.

Banasik, A., Bloemhof-Ruwaard, J. M., Kanellopoulos, A., Claassen, G. D. H., & van der Vorst, J. G. A. J. (2016). Multi-criteria decision making approaches for green supply chains: A review. *Flexible Services and Manufacturing Journal*, 1–31. doi:10.1007/s10696-016-9263-5.

Bravo, R. (2015). Analysis of influence of design factors in eco-efficiency of the injection moulding process. MSc Thesis, IST.

Coello, C. A. C., Dhaenens, C., & Jourdan, L. (2010). Multi-objective combinatorial optimization: Problematic and context. *Studies in Computational Intelligence, 272*, 1–21. doi:10.1007/978-3-642-11218-8_1.

Coello, C. A. C., Lamont, G. B., & Van Veldhuizen, D A. (2007). *Evolutionary Algorithms for Solving Multi-Objective Problems*. Springer. doi:10.1007/s11276-014-0817-8.

Coello, C. A. C., Pulido, G. T., & Lechuga, M. S. (2004). Handling multiple objectives with particle swarm optimization. *IEEE Transactions on Evolutionary Computation, 8*(3), 1942–1948. doi:10.1007/s11721-007-0002-0.

Dorigo, M., & Stützle, T. (2003). *Ant Colony Optimization*. Cambridge, MA: The MIT Press.

Duflou, J. R., Sutherland, J. W., Dornfeld, D., Herrmann, C., Jeswiet, J., Kara, S., … Kellens, K. (2012). Towards energy and resource efficient manufacturing: A processes and systems approach. *CIRP Annals - Manufacturing Technology, 61*(2), 587–609. doi:10.1016/j.cirp.2012.05.002.

George, M., & Bressler, D. C. (2017). Comparative evaluation of the environmental impact of chemical methods used to enhance natural fibres for composite applications and glass fibre based composites. *Journal of Cleaner Production, 149*, 491–501. doi:10.1016/j.jclepro.2017.02.091.

GLN, IST, & ISQ. (2017). Projeto Eco2Plast - Desenvolvimento de sistema de produção ECOeficiente para componentes PLASTicos. Retrieved from www.gln.pt/pt/ECO2Plast/.

Glover, F. (2009). Common concepts for metaheuristics. In E.-G. Talbi (Ed.), *Metaheuristics from Design to Implementation* (pp. 1–86). ISBN: 978-0-470-27858-1.

Goedkoop, M., Heijungs, R., Huijbregts, M., Schryver, A. De, Struijs, J., & Van Zelm, R. (2009). ReCiPe 2008. *Ministerie van Volkshuisvesting, Ruimtelijke ordening en Milieubeheer*. doi:10.029/2003JD004283.

Hassan, R., & Cohanim, B. (2005). A comparison of particle swarm optimization and the genetic algorithm. *1st AIAA Multidisciplinary Design Optimization Specialist Conference*, Austin, TX, 1–13. doi:10.2514/6.2005-1897.

Hillier, F., & Lieberman, G. (2015). *Introduction to Operations Research*. Mcgraw-Hill Publishing Company. ISBN: 9780071238281.

Ingarao, G. (2016). Manufacturing strategies for efficiency in energy and resources use: The role of metal shaping processes. *Journal of Cleaner Production*. doi:10.1016/j.jclepro.2016.10.182.

ISO. (2012). ISO 14045:2012 Preview environmental management - Eco-efficiency assessment of product systems - Principles, requirements and guidelines. Retrieved from www.iso.org/standard/43262.html.

Jasch, C. (2000). Environmental performance evaluation and indicators. *Journal of Cleaner Production, 8*(1), 79–88. doi:10.1016/S0959-6526(99)00235-8.

Korol, J., Burchart-Korol, D., & Pichlak, M. (2016). Expansion of environmental impact assessment for eco-efficiency evaluation of biocomposites for industrial application. *Journal of Cleaner Production, 113*, 144–152. doi:10.1016/j.jclepro.2015.11.101.

Laumanns, N., Laumanns, M., & Kitterer, H. (2002). Evolutionary multi-objective integer programming for the design of adaptive cruise control systems. *Proceedings of the Fifteenth International Conference on Industrial & Engineering Applications of Artificial Intelligence & Expert Systems (IEA/AIE-2002)*, Cair, Australia, 200–210.

Lehni, M. (2000). *Eco-Efficiency. Creating More Value with Less Impact* (pp. 1–32). Geneva, Switzerland: World Business Council for Sustainable Development.

Li, W., Kara, S., & Qureshia, F. (2015). Characterising energy and eco-efficiency of injection moulding processes. *International Journal of Sustainable Engineering, 8*(1), 55–65. doi:10.1080/19397038.2014.895067.

Lourenço, E. J., Moita, N., Esteves, S., Peças, P., Ribeiro, I., Henriques, E., ... Oliveira, L. (2018). Multi perspective eco-efficiency assessment to foster sustainability in plastic parts production: An integrated tool for industrial use. In M. Uthayakumar, S. Aravind Raj, T. J. Ko, S. Thirumalai Kumaran, & J. P. Davim (Eds.), *Handbook of Research on Green Engineering Techniques for Modern Manufacturing* (p. 380). Hershey, PA: IGI-Global. doi:10.4018/978-1-5225-5445-5.

Lyubomirsky, B. S., Kurtz, J., & Lyubomirsky, S. (2009). Eco-efficiency learning module.

Malik, A., Sharma, A., & Saroha, V. (2013). Greedy algorithm. *International Journal of Scientific and Research Publications, 3*(8), 1–4.

Miettinen, K. (2008). Introduction to multiobjective optimization: Noninteractive approaches. *Lecture Notes in Computer Science (Including Subseries Lecture Notes in Artificial Intelligence and Lecture Notes in Bioinformatics), 5252 LNCS*, 1–26. doi:10.1007/978-3-540-88908-3-1.

Nicoară, E. (2012). Population-based metaheuristics: A comparative analysis. *International Journal of Science and Engineering Invetigations, 1*(8), 84–88.

Passetti, E., & Tenucci, A. (2016). Eco-efficiency measurement and the influence of organisational factors: Evidence from large Italian companies. *Journal of Cleaner Production, 122*, 228–239. doi:10.1016/j.jclepro.2016.02.035.

Peças, P., Götze, U., Bravo, R., Richter, F., & Ribeiro, I. (2018). Methodology for selection and application of eco-efficiency indicators fostering decision-making and communication at product level – The case of moulds for injection moulding. In M. Ram & P. J. Davim (Eds.), *Advanced Applications in Manufacturing Engineering* (1st ed., p. 278). Sawston, UK: Woodhead Publishing, Elsevier. Retrieved from www.elsevier.com/books/advanced-applications-in-manufacturing-enginering/ram/978-0-08-102414-0.

Pohekar, S. D., & Ramachandran, M. (2004). Application of multi-criteria decision making to sustainable energy planning - A review. *Renewable and Sustainable Energy Reviews, 8*(4), 365–381. doi:10.1016/j.rser.2003.12.007.

Pre Sustainability. (2014). *SimaPro database manual. PRe'.* doi:10.1017/CBO9781107415324.004.

Quariguasi Frota Neto, J., Walther, G., Bloemhof, J., van Nunen, J. A. E. E., & Spengler, T. (2009). A methodology for assessing eco-efficiency in logistics networks. *European Journal of Operational Research, 193*(3), 670–682. doi:10.1016/j.ejor.2007.06.056.

Saka, M. P., Doğan, E., & Aydogdu, I. (2013). Analysis of swarm intelligence-based algorithms for constrained optimization. *Swarm Intelligence and Bio-Inspired Computation*, 25–48. doi:10.1016/B978-0-12-405163-8.00002-8.

Sproedt, A., Plehn, J., Schönsleben, P., & Herrmann, C. (2015). A simulation-based decision support for eco-efficiency improvements in production systems. *Journal of Cleaner Production, 105*, 389–405. doi:10.1016/j.jclepro.2014.12.082.

Talbi, E.-G. (2009a). Chapter 3: Population-based metaheuristics. In E.-G. Talbi (Ed.), *Metaheuristics from Design to Implementation* (pp. 190–307). John Wiley & Sons, Inc. doi:10.1002/9780470496916.ch3.

Talbi, E.-G. (2009b). Chapter 4: Metaheuristics for multiobjective optimisation. In *Metaheuristics from Design to Implementation* (pp. 308–384). doi:10.1007/s10288-010-0137-5.

Verfaillie, H. A., & Bidwell, R. (2000). Eco-efficiency: A guide to reporting company performance. *World Business Council for Sustainable Development*. Retrieved from www.gdrc.org/sustbiz/measuring.pdf. ISBN: 2-940240-14-0.

WBCSD. (2000a). Eco-efficiency. Creating more value with less impact. *World Business Council for Sustainable Development*. Retrieved from www.wbcsd.org.

WBCSD. (2000b). Measuring ecoefficiency: A guide to reporting company performance. Geneva, Switzerland. *World Business Council for Sustainable Development*, 40. Retrieved from http://oldwww.wbcsd.org/web/publications/measuring-eco-efficiency-portugese.pdf.

Yang, K. (2012). An improved particle swarm optimization for multi-objective discrete optimization. *International Conference on Information Management, Innovation Management and Industrial Engineering*, Sanya, China.

Yang, X. S., & Karamanoglu, M. (2013). Swarm intelligence and bio-inspired computation: An overview. *Swarm Intelligence and Bio-Inspired Computation*, 3–23. doi:10.1016/B978-0-12-405163-8.00001-6.

Zohal, M., & Soleimani, H. (2016). Developing an ant colony approach for green closed-loop supply chain network design: A case study in gold industry. *Journal of Cleaner Production, 133*, 314–337. doi:10.1016/j.jclepro.2016.05.091.

2

Distance-Based Function Approach for Optimal Design of Engineering Processes with Multiple Quality Characteristics

Taha-Hossein Hejazi and Shahin Behboudi
Amirkabir University of Technology

Hamed Baziyad
Tarbiat Modares University

CONTENTS

2.1 Introduction: Response Surface Methodology

The first introduction of response surface methodology (RSM) was made by Box and Wilson and dates back to 1951 (Box & Wilson, 1951, 1992). RSM brings some benefits to researchers, including reduction of experimental runs (Said & Amin, 2016), saving time (Moskowitz, 1994), and detecting the influence level of input variables of the model (Schutz, 1983). RSM provides conditions during which the number of experiments will be increased tremendously compared to the full factorial design (Box & Wilson, 1992). The laboratory test phase will be done better through the RSM approach. As a result, a short period of time will be needed for the test stage (Moskowitz, 1994). Different variables have various influences on the final quality of products, some of which have a low influence on the final quality and others largely affect the outputs. Schutz (1983) believed that RSM can identify the most effective variables using parameter estimation. In the next paragraph, the definition of RSM is presented.

RSM is based on experimental design and attempts to achieve its main aim that is known as "evaluating the optimal functioning of industrial facilities" (Gangil & Pradhan, 2017). Another remarkable issue considered in this paper is the minimum experimental effort among RSM methodology. RSM is a series of statistical and mathematical techniques used for improvement of an adequate functional relationship between an interest response and a number of related input variables (Khuri & Mukhopadhyay, 2010). Myer and Montgomery (2002) have introduced RSM as a collection of techniques based on statistics and mathematics, with some advantages such as developing, improving, and optimizing processes.

The quality characteristics called the response and independent variables are related to the input variables that influence the quality characteristic (Carley, Kamneva, & Reminga, 2004). In this chapter, $X(x_1, x_2, x_3..., x_n)$ represents independent variables and y represents the response variable. Thus, the general relationship between these two variable categories is presented in the format of the function as follows:

$$y = f(x_1, x_2, ..., x_n) + \varepsilon \tag{2.1}$$

The response variable follows two sources—namely independent variables and error. Error has been shown by ε symbol, and Carley et al. (2004) enumerated some of its effects such as measurement error on the response, background noise, and the effect of other variables. The symbol ε, as a statistical error, has a normal distribution with these features: $\varepsilon \sim N(0, \sigma^2)$. So, it is concluded that ε is identically and independently distributed (IID). The mean zero assumption of ε leads to the use of expected value for y instead of y. So, Eq. (1.1) was changed into Eq. (1.2) as

$$E(y) = \mu = f(x_1, x_2, \ldots, x_n) \tag{2.2}$$

Due to the different formats of true response function f, approximation tools should be used to detect the function from Carley et al. (2004). The choice of experimental design is dependent upon expert's abilities during the experiments. The linear function is considered as the simplest model used in RSM, and the general formulation of this function has been presented as

$$\frac{\partial \mu}{\partial x_2} = \beta_2 + 2\beta_{22}x_2 + \beta_{12}x_1 = 0 \tag{2.3}$$

$$\mu = \beta_0 + \sum_{i=1}^{k} \beta_i x_i + \varepsilon \tag{2.4}$$

The earlier model was appropriate only for the linear relationship between variables, and for the investigation of nonlinear relationships, other models should be used. Nonlinear models are divided into two categories: first-order and second-order models. The first one is suitable for small regions of independent variables with no effect of second order. When Eq. (2.4) is restricted to $k = 2$, only the main effects of the two variables, x_1 and x_2, are considered, which are known as the main effect model:

$$\mu = \beta_1 x_1 + \beta_2 x_2 + \beta_0 \tag{2.5}$$

With regard to interaction between independent variables, more complicated modes of Eq. (2.4) were represented in Eq. (2.6), which were part of the second-order models. The following terms represent a second-order interaction:

$$\mu = \beta_0 + \sum_{i=1}^{k} \beta_i x_i + \sum_{1<i<j}^{k} \beta_{ij} x_i x_j + \varepsilon \tag{2.6}$$

More developed second-order models consider another kind of relationship between variables, namely a quadratic relationship. Thus, the use of Eq. (2.7) rather than Eq. (2.6) leads to all kinds of second-order relationships between independent variables.

$$\mu = \beta_0 + \sum_{i=1}^{k} \beta_i x_i + \sum_{1<i<j}^{k} \beta_{ij} x_i x_j + \sum_{i=1}^{k} \beta_{ii} x_i^2 + \varepsilon \tag{2.7}$$

The explanation of all parameters used in Eqs. (2.1)–(2.6) is represented in Table 2.1. Also, it must be considered that the coefficient parameter β should be approximated via different techniques such as linear regression analysis.

TABLE 2.1

Explanation of Variables

Row	Variable	Explanation
1	β_0	Constant term
2	β_i	Coefficient of the linear parameters
3	x_i	Input variables
4	ε	Residual associated to the experiments
5	β_{ij}	Coefficients of interaction variables
6	β_{ii}	Coefficients of quadratic variables

2.2 RSM and Robust Design

Recently, robust design has been considered as a quality improvement tool in different industries (Shin, Hoang, Le, & Lee, 2016). Also, Shin et al. (2016) introduced another concept named RSM as a collection of mathematical and statistical tools that can be used in robust design issues with analyzing and modeling goals.

Insensitive design was the main motivation of robust design development, and external noises or tolerance effects will be minimized through robust design. Development of robust design theories has taken place in theories of other fields. According to such theories, robust design was categorized into three methods—namely the (i) Taguchi method, (ii) robust optimization, and (iii) robust design with the axiomatic approach (Park, Lee, Lee, & Hwang, 2006).

As shown earlier in formulas up to 2.6, all of them have a parameter called ε. From the point of RSM, this parameter has an uncontrolled nature, but control of noise parameters, ε, will be possible through a designed experiment.

2.3 RSM Phases

According to Bezerra, Santelli, Oliveira, Villar, and Escaleira (2008), RSM is applicable through six stages that are explained as follows:

1. Screening Experiment: In the initial stage, important independent parameters should be identified. At this step, this question must be answered: "which factors or parameters have the most important role in the response surface study?" The exact and correct answer to this question leads to reducing the number of candidate variables. Through the screening experiment step, fewer tests are needed.

2. Experimental Design: According to the selected parameters in the previous stage, the relevant experimental design should be selected. Details of this section have been proposed in Section 2.1.

3. Mathematical–Statistical Treatment of Data: In order to describe the response behavior, the mathematical equation that is selected from the experimental design stage is fitted in this phase.

4. Evaluation of Fitted Model: After fitting the mathematical model on the intended dataset, the model cannot be validated. So, the use of evaluation methods such as analysis of variance (ANOVA) seems necessary.

5. Determination of Optimal Conditions: Performing a displacement in the direction of optimal area is examined in this stage. In order to achieve this, optimal conditions must be defined. For example, we propose these conditions for Eq. (2.7) with two levels as follows:

$$\frac{\partial \mu}{\partial x_1} = \beta_1 + 2\beta_{11}x_1 + \beta_{12}x_2 = 0 \tag{2.8}$$

$$\frac{\partial \mu}{\partial x_2} = \beta_2 + 2\beta_{22}x_2 + \beta_{12}x_1 = 0 \tag{2.9}$$

6. Optimum Values: By solving Eqs. (2.7) and (2.8), critical points are gained. According to the type of objective function (minimum or maximum), the optimal point will be proposed.

2.4 Real Application

RSM has been used in many industries such as the welding and machining process. Bandyopadhyay, Panda, and Saha (2016) have used RSM to improve weld properties through the optimization process of Fiber Laser Welding of DP980 Steels. Box–Behnken design was the basis of RSM in this paper. Elatharasan and Kumar (2013) have developed a central composite design and mathematical model during RSM for friction stir welding (FSW) of the AA 6061-T6 aluminum alloy. The rotational speed of the tool, welding speed, and axial force have been considered as important welding parameters in this paper. In fact, these three variables have been found through the screening experiment phase of RSM.

RSM is useful for machining processes in regard to which many articles have been published. Rao et al. (2017) have investigated the cutting parameters effect—namely speed, feed, and depth of cut on the machinability of Niobium C-103 under dry machining conditions. This investigation has been done using RSM. Raj and Senthilvelan (2015) have used the Box–Behnken approach and multiparameter optimization for cutting conditions of Wire-EDM in order to achieve better surface roughness and material removal rate. Three variables have been selected in the study—namely pulse-on time, pulse-off time, and wire feed rate.

In Section 2.8, two real examples are presented in detail, both of which are related to welding and machining processes. Thus, the focus of the proposed real applications in the two last paragraphs is about these processes. Many papers have used RSM in other fields such as chemistry (Nyakundi & Padmanabhan, 2015), energy performance (Bliuc, Lepadatu, Iacob, Judele, & Bucur, 2017), nanocomposites (Ansari, Parsa, & Merati, 2017), etc.

2.5 Background and Definition of RSM

RSM is a collection of statistical and mathematical techniques useful for developing, improving, and optimizing processes (Myers, Montgomery, & Anderson-Cook, 2016). The method was introduced by Box[1] and Wilson in 1951. The main objective in response to surface methodology is to optimize the response variables by using design experiments. This method also has many applications in design and manufacturing.

2.5.1 Design of Experiments

One of the important aspects of RSM[2] is the design of experiments (Box & Draper, 1987), usually abbreviated as DOE. There are strategies to develop the model fitting of physical experiments, which can also be applied to numerical experiments. The main idea of DOE is to find the points to evaluate the response variables (Box & Draper, 1987).

Most of the criteria derived from the mathematical model of the process are used for the optimal design of experiments. Generally, polynomial models have different structures, so the corresponding experiments are used for specific issues. Due to the approximate accuracy of the model and the cost of constructing the response surface, the selection of DOEs is very important.

In traditional DOE, experiments need to be screened in the early steps of the process; in other words, the initial design variables were considered to not affect the responses. It is important to find significant design variables that have large effects on response variables.

Principles and definitions of the DOEs technique can be found in Box and Draper (1987), Atkinson & Donev (1992), and Myers et al., (2004), among the

---

[1] George Edward Pelham Box.
[2] Response surface methodology.

many others. Schoofs (1988) has reviewed the application of experimental design to structural optimization (Montgomery, 2017), and Unal et al. (1996) discussed the use of several designs for RSM and multidisciplinary design optimization and presented a complete review of using statistics in design (Box & Draper, 1987).

2.5.2 Methods of Experimental Design

Experiments are designed to achieve a variety of experimental objectives. A few of the more common types of experimental designs are as follows:

- The general $2k$ factorial design: It is a kind of design with k factors, each factor having two levels. The statistical model for this type of design includes k main effects, $\binom{k}{2}$ two-factor interaction,..., and one k-factor interaction; i.e., the regression model includes $2k - 1$ effect. Some beneficial guidelines for analyzing a $2k$ factorial design are as follows:

  a. Estimate the main effects

  b. Perform statistical experiments

  c. Refine the model by eliminating meaningless effects

  d. Analyze the normal probability plot of residuals

  e. Interpret the results

- The general 3^k factorial design: Like two factorial design, k factors are considered, each at three levels. They are usually referred to as low, intermediate, and high levels. Because of the model's possible curvature in the response function, the $3k$ designs were proposed. Two important tips should be noted as follows:

  a. These types of designs are not suitable for describing the model

  b. The $2k$ design augmented with center points is more efficient than the $3k$ design to obtain an indication of curvature.

- Completely randomized design: An experimental plan where the order in which the experiment is performed is completely random.

- Randomized-block design: In almost every experiment, increasing the variability of nuisance factors can affect the results. Generally, a nuisance factor is defined as a design factor that probably has an effect on the response, but these factors are not desirable. Sometimes, a nuisance factor is unknown and uncontrollable; i.e., we don't know about the existence of that factor, and it may even be changing levels during experimental design. When the resource of variability is known and controllable, its effect is systematically eliminated by a design technique called blocking. Blocking is one of

TABLE 2.2

An Example of Randomized-Block Design

Gender	Treatment	
	Pill	**Vaccine**
Male	250	250
Female	250	250

To access the details, see https://stattrek.com/statistics/dictionary.aspx?definition=randomized%20block%20design.

the important design techniques used extensively in various industrial experimentations (Ankenman & Dean, 2003). In the DOEs, the data obtained from the test are divided into "blocks" according to some criterion. The blocks are sequentially and randomly filled, e.g., Table 2.2 shows a randomized block design for a hypothetical medical experiment.

Gender is assigned to blocks. Each block represents the random allocation of individuals to treatment. For this design, 250 men get the pill, 250 men get the vaccine, 250 women get the pill, and 250 women get the vaccine.

It is obvious that the reaction of men and women to medications is different. This design ensures that the treatment conditions are the same for men and women. As a result, changing the treatment conditions cannot be related to gender. Therefore, as potential sources of variability and contradiction, gender is eliminated by random block design.

- Latin-square design: There are many test designs that use the block design principles. Suppose an experimenter is concerned with a single factor having p levels. The variability of the other two sources can be controlled in the experiment. If the effect of these two variables can be controlled by grouping them into blocks, then a Latin-square design may be appropriate (Montgomery, 2017). Consider a square with p rows and p columns; assign p treatments to the rows and columns so that each treatment will appear exactly once in a row and in a column, then we have $p*p$ Latin-square design. As an example, a 4*4 Latin-square design is shown in Figure 2.1.

2.5.3 Some Typical Applications of Experimental Design

Experimental design is considered as an important tool in science and engineering to improve the product of a process. In this manner, DOEs includes important activities such as new manufacturing process design, development, and process management (Myers et al., 2016). The application of experimental design techniques early in process development can result in

$$
\begin{array}{cccc}
A & B & C & D \\
D & A & B & C \\
C & D & A & B \\
B & C & D & A
\end{array}
$$

FIGURE 2.1
Example of Latin-square design.

1. Improved process yields
2. Reduced variability and closer conformance to nominal or target requirements
3. Reduced development time
4. Reduced overall cost

The experimental design method is one of the most important engineering design activities in which new products are developed and existing products are being improved (Myers et al., 2016). Some applications of experimental design in engineering design include

1. Evaluation and comparison of basic design configurations
2. Evaluation of material alternatives
3. Selection of design parameters so that the product will work well under a wide variety of field conditions, that is, so that the product is robust
4. Determination of key product design parameters that impact product performance
5. Formulation of new products

By using DOE, the manufacturing of products becomes more convenient. Also, DOE leads to increased field performance and reliability, lower product cost, and shorter product design and development time. It should be noted that the DOE has broad usages in marketing, market research, and general business operations (Myers et al., 2016).

2.6 Building Empirical Models

Basically, each approximate model is derived from the data obtained from the process or the system called empirical model. Hence, it can be manipulated by the experimenter or scientist. In general, building an empirical model can be written as

$$y = f(x_1, x_2, \ldots, x_n) + \varepsilon \qquad (2.10)$$

where ε represents the "random error" of the system.

2.6.1 Linear Regression Models

The first step in RSM is to find a suitable approximate model for the true functional relationship between responses and a set of independent variables. The approximate model is derived from the data collected from the process or system called as the empirical model. Multiple regression is one of the statistical techniques used to construct empirical models required in RSM.

As an example, suppose we want to develop an empirical model related to the effective average of loss and temperature and pressure, a first-order response surface model that might describe this relation is as follows:

$$y = \beta_0 + \beta_1 x_1 + \beta_2 x_2 + \varepsilon \qquad (2.11)$$

where y represents average of loss, x_1 represents temperature, and x_2 represents pressure. The earlier function is a multiple linear regression model with two independent variables. Typically, independent variables are called predictive variables. The term linear regression is used because Eq. (2.10) is a linear function of the parameters β_0, β_1, and β_2 (Myers et al., 2016). Also, ε is a random experimental error assumed to have zero mean and constant variance (Montgomery, 2017).

In general,

$$y = \beta_0 + \beta_1 x_1 + \beta_2 x_2 + \cdots + \beta_k x_k + \varepsilon \qquad (2.12)$$

is called a multiple linear regression model with k independent variables. The parameters β_j, where $j = 0, 1, 2, \ldots, k$, is called the regression coefficient. The parameter β_j represents the expected change in response y per unit change in x_j when all the remaining independent variables x_i $(i \neq j)$ are held constant (Myers et al., 2016).

$$y = \beta_0 + \beta_1 x_1 + \beta_2 x_2 + \beta_{12} x_1 x_2 + \varepsilon \qquad (2.13)$$

It is obvious that the more interaction we use, the more complex the model will be (Myers et al., 2016).

2.6.2 Other Approaches

In this section, some well-known approaches are to be discussed. In the following, we describe concepts such as polynomial regression, kriging, and neural network.

- Polynomial regression: There are many real-world problems that have no acceptable linear approximation. Hence, polynomial models have been used (Coelho & Neto, 2017). It is a technique like regression analysis, which describes the relationship between the value of independent variable and the corresponding conditional mean of response y. The basic form of polynomial regression can be written as follows:

$$y = \beta_0 + \sum_{j=1}^{m} \sum_{i=1}^{n} \beta_j x_i^{\,j} + \varepsilon \qquad (2.14)$$

Polynomial regression fits a nonlinear model to the data using the method of least squares error (LSE) (Fan & Gijbels, 1996). It has been shown in many papers that polynomial regression is more efficient than methods such as Monte-Carlo linear regression in predicting real-world phenomena (Roziqin, Basuki, & Harsono, 2016).

Polynomial regression approach is derived from geostatistics science, which is the best linear unbiased predictor for modeling spatially correlated random variation and estimating spatial autocorrelation; i.e. its prediction error variances are minimized; hence, it is often known as BULP.[3] The primary concept of this method was developed by Danie G. Krige for estimating the distribution of gold based on samples from a few boreholes. There are some applications of more forms of kriging to solve complex problems in the field of petroleum engineering, mining and geology, meteorology, soil science, precision agriculture, pollution control, and so on (Oliver & Webster, 2015). There are several versions of kriging method, but the ordinary kriging method is usually used, which is computationally practical and easier to implement. Also, ordinary kriging assumes a constant unknown mean only over the search neighborhood of X_1. To achieve the kriging system results, by considering the assumption of the model, its linear estimation equation is taken into account as follows:

$$z\left(x_0\right) = \begin{bmatrix} w_1 & w_2 & w_3 \cdots & w_n \end{bmatrix} \cdot \begin{bmatrix} z_1 \\ \vdots \\ z_n \end{bmatrix} = \sum_{i=1}^{n} w_i\left(x_0\right) \times z\left(x_i\right) \qquad (2.15)$$

---

[3] Best linear unbiased predictor.

- Neural network: In recent years, applications of neural network simulations are developed in many fields. This field was established before the advent of computers. Many important advances have been made by the use of inexpensive computer emulations. It is significant that neural networks have been applied to machine learning tasks (Behnke, 2003).

By considering their remarkable ability, neural networks can be used to extract patterns and detect trends by deriving the meaning from complicated or imprecise data. These trends are more complicated than being considered by humans or other computer techniques. A learning neural network is treated as an "expert" in the category of information it has been given to analyze. This expert is used for providing projections, given new situations of interest and answer "what if" questions. Other advantages include

1. Adaptive Learning: An ability to learn how to do tasks based on experience and according to the initial input data.
2. Self-Organization: An artificial neural network (ANN) can create its own organization or representation of information it receives during learning time.
3. Real-Time Operation: ANN calculations that are usually done with special hardware devices are being designed and manufactured, which take advantage of this capability.
4. Tolerance Fault through Coding Information Redundant: Certainly, partial degradation of a network leads to a reduction in the overall network performance. Also, some of its important features remain in line with the network vulnerability (Behnke, 2003).
5. An ANN is based on a collection of connected units called artificial neurons, each producing a sequence of real-valued activations. Each connection between neurons can transfer a signal to another neuron, and the receiving neuron can analyze the signal and then signal downstream neurons connected to it. Some neurons may affect the environment by doing activities. The real number of neurons is usually between 0 and 1. Neurons and synapses may also have weights that change as learning income that increases or decreases the signal strength. A simple sample of ANN is shown in Figure 2.2.

Based on the human brain algorithm, the main objective of the neural network approach is to solve problems. Over time, to pass information in the reverse direction and adjust the network to reflect that information, some attention must be focused on matching specific mental abilities.

Neural networks have some applications in different tasks, including computer vision, risk analysis, machine translation, error management, social network filtering, exploration of oil and gas, unmanned airplane control, medical diagnosis, and many other domains.

Hidden

Inputs   output

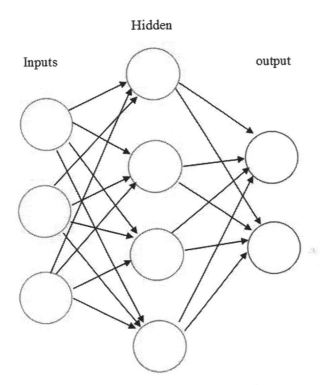

FIGURE 2.2
A simple example of ANN.

2.7 Optimization

The term "optimization" is usually used to improve the performance of many engineering processes and systems (Castilo, 2007). In many cases, optimization is an important stage to find the possible value of each factor for achieving the best response (Candioti, De Zan, Cámara, & Goicoechea, 2014). Here, we discuss single and multiple response optimizations (MROs), respectively.

2.7.1 Multiresponse Optimization

In many processes and systems, it is difficult to optimize multiple output variables, taking into account other factors that affect the function of the system (Hejazi, 2017; Hejazi & Seyyed-Esfahani, 2017). The MRO methodology is used for optimizing correlated multiple responses simultaneously. In fact, MRO is a good way to design an offline quality control that includes multiple input and output variables simultaneously (Hejazi & Seyyed-Esfahani, 2017).

The objective of this technique is to seek the best process setting that satisfies the individual response as much as possible by achieving their target with minimal variance, i.e., if quality characteristics of products are not in their desirable limit, products are completely unacceptable. It is necessary to say that response variables are to be optimized by methods such as multicriteria decision making and gray rational analysis (Espinoza-Escalante et al., 2008).

Some applications of MRO methodology have been used in many papers and textbooks. For instance, a mass manufacturing system includes different multiple stages. In these stages, the final stage may include a number of response variables that have multiple quality characteristics of interest. In fact, the characteristics of the response variable in the final stage are influenced by the operational conditions of the previous stages. So, it can be concluded that process performance is also affected by intermediate stages (Bera & Mukherjee, 2016). In this manner, there are some other aspects of MRO that have been studied in real-world problems; e.g. consider priority-based methods as one of the aspects of the MRO technique that discusses the most important response for optimization if the responses deviate from their desirable limit. In the following, if a unique solution is not found, then the steps are repeated so that an optimal solution can be obtained (Hejazi, Bashiri, Noghondarian, & Atkinson, 2011).

To find the best setting of parameters to optimize multiresponse models, a mathematical model, especially based on the distance-based approach and goal programming (GP), has been used (Hejazi, Salmasnia, & Bastan, 2013). It should be noted that some criterion such as distance to target and LSE can be considered so as to achieve an optimal value of responses.

2.8 Illustrative Examples

In recent years, using mathematical programming models in various industries has been taken into account to improve performance and to increase efficiency indices. Researches show that removing all variations is not possible, so manufacturers should focus on producing compatible products. This section presents applications of the proposed method using two different examples: (i) a typical machining process and (ii) FSW process.

2.8.1 MRO for Computer Numerical Control Machine Setting

Machining operation, as a very important process in the manufacturing industry, is mainly used as a mechanical process for cutting materials using special tools. The three main objectives that should be considered in the real machining applications are maximizing product rate, minimizing operational cost, and minimizing loss rate. For this purpose, factors such as depth of cut, feed rate, number of passes and spindle speed are considered as well as three response variables including output roundness, cycle time, and output diameter. It is necessary to note that the method presented in this section is a combination of RSM and a well-grounded distance-based approach called GP to cope with the optimization of several characteristics.

The primary information of designing experiments in the machining process is presented in Table 2.3.

The initial measurements are shown in Table 2.4. For this purpose, a full factorial design with four replications was performed (The source of data used for this purpose is available at www.cpkinfo.com-Quality Dimension's Standard statistical datasets).

The ANOVA procedure for primary experiments for the output diameter (in uncoded units) is as follows[4].

In the initial analysis, all factors were considered together with their interactions. For better validation of the regression model, the less significant terms of the initial full quadratic response surface were removed by considering the amount of p-value of each term. The final regression model can be written as follows:

$$\text{Output Diameter: } 11.9904 - 0.994381 \times \text{depth} + 0.00136174 \times \text{pass} \quad (2.16)$$

TABLE 2.3

Initial Experiments

| | CNC Machine Settings | | | | Outputs | | |
Run	Spindle Speed	Feed Rate	Depth of Cut	Number of Passes	Output Diameter	Output Roundness	Cycle Time
64	2,000	10	1	1	10.987	0.0008	13.5
63	1,000	10	1	1	10.988	0.0081	13.5
62	1,000	2	1	1	10.99	0.0115	63.5
61	2,000	10	1	2	10.998	0.0028	27
60	1,000	10	1	2	10.996	0.0034	27
34	2,000	10	0.01	1	11.988	0.0013	12.5
33	1,000	10	0.01	1	11.989	0.0048	12.5
32	1,000	10	0.01	2	11.988	0.0001	25
31	1,000	2	1	2	10.999	0.0028	127
4	2,000	10	1	2	10.999	0.0076	27
3	1,000	2	0.01	1	11.978	0.0019	62.5
2	2,000	10	0.01	1	11.987	0.0009	12.5
1	1,000	2	1	2	10.991	0.0043	127

---

[4] Calculations were done using Minitab V.16 statistical package.

TABLE 2.4

Analysis of Primary Experiments with
Siginificant Terms for Output Diameter

Term	Coefficients	P-Value
Constant	11.4907	0.000
Depth	−0.4961	0.000
Pass	0.0015	0.002
	R-square = 99%	

Also, the later plot shows that the residuals follow a normal distribution, and there are no undesirable trends in the process (Figure 2.3).

The results of ANOVA for the output roundness (in uncoded units) are also presented below (Table 2.5):

According to the results, the depth of cut and number of passes has a main effect on output roundness. The related regression model in terms of uncoded variables is

$$\text{Output roundness: } -6.88340 + 0.839778 \times \text{depth} - 0.597816 \times \text{pass} \quad (2.17)$$

Finally, the results of ANOVA for the output cycle time (for uncoded units) can be written as (Table 2.6)

$$\text{Output cycle time: } 112.485 - 9.375 \times \text{feed} + 1.5151 \times \text{depth}$$

$$+ 37.4949 \times \text{pass} - 3.125 \times \text{feed} \times \text{pass} + 0.505051 \times \text{depth} \times \text{pass} \quad (2.18)$$

It is important to mention that the factor spindle speed was not considered in all functions, because it has no effect on responses.

The method presented in this section is a combination of RSM and GP for optimization parameters in a machining process.

The goal is to minimize the adverse deviations of each response, so the model can be written as follows:

$$\text{Minw} = \frac{d_1 + d_{11}}{0.500048} + \frac{-d_2}{0.002972} + \frac{-d_3}{44.21161} \quad (2.19)$$

$$Z_1 = 11.9904 - 0.994381 \times \text{depth} + 0.00136174 \times (\text{pass} + 1) \quad (2.20)$$

$$Z_2 = -6.88340 + 0.839778 \times \text{depth} - 0.597816 \times (\text{pass} + 1) \quad (2.21)$$

$$Z_3 = 112.485 - 9.375 \times \text{feed} + 1.5151 \times \text{depth} + 37.4949 \times (\text{pass} + 1)$$

$$- 3.125 \times \text{feed} \times (\text{pass} + 1) + 0.505051 \times \text{depth} \times (\text{pass} + 1) \quad (2.22)$$

s.t.

$$11.81 \leq Z_1 \leq 11.83 \quad (2.23)$$

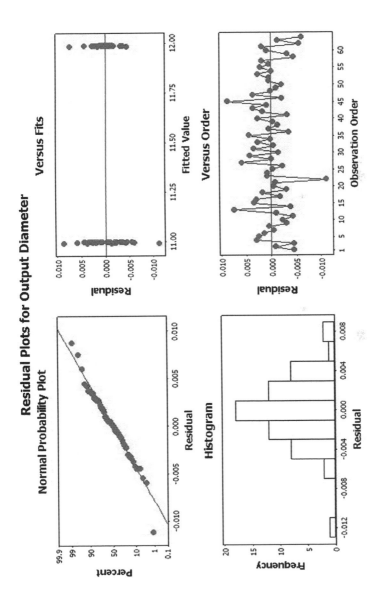

FIGURE 2.3
Residual plots for output diameter.

TABLE 2.5

Analysis of Primary Experiments for Output Roundness

Term	Coefficients	P-Value
Constant	0.002742	0.000
Depth	0.001183	0.001
Pass	−0.00107	0.002
R-square = 43.6%		

TABLE 2.6

Analysis of Primary Experiments
for Output Cycle Time

Term	Coefficients
Constant	57
Speed (rpm)	−0.00
Feed	−37.5
Depth	0.75
Pass	19
Speed * feed	−0.00
Speed * depth	−0.00
Speed * path	0.00
Feed * depth	0.00
Feed * pass	−12.5
Depth * path	0.25

$$Z_1 + d_1 - d_{11} = 11.82 \tag{2.24}$$

$$0 \le d_1 \le 0.01 \tag{2.25}$$

$$0 \le d_{11} \le 0.01 \tag{2.26}$$

$$Z_2 + d_2 = -4.60517 \tag{2.27}$$

$$d_2 \ge 0 \tag{2.28}$$

$$0 \le d_3 \le 28 \tag{2.29}$$

$$Z_3 + d_3 = 28 \tag{2.30}$$

$$2 \le \text{Feed rate} \le 10 \tag{2.31}$$

$$0.01 \leq \text{Depth of cut} \leq 1 \qquad (2.32)$$

$$\text{Pass} = 1 + y \qquad (2.33)$$

where
d_1, d_{11}:   undesirable deviation of output diameter
d_2:   undesirable deviation of output roundness
d_3:   undesirable deviation of output cycle time
Z_1:   objective value of output diameter
Z_2:   objective value of output roundness
Z_3:   objective value of output cycle time
y: binary variable

The optimization phase was performed, and the results are presented in Table 2.7.

According to the results, it is observed that the amount of acceptable product is 60%. Hence, the model has been revised by tightening the limit for d_1, d_{11}, d_2, and d_3; hence, the improved model reoptimized, and the results are tabulated in Table 2.8.

In order to compare optimal responses with initial responses, the results are summarized in Table 2.9.

The best value for diameter, roundness,[5] and cycle time is 11.82, 0, and 25, respectively. Due to the optimal solution compatibility with the optimal value of each response, the amount of loss has been considerably reduced by using the proposed approach.

TABLE 2.7

Optimal Solution of Model

Final Value	Name
−921.7840401	W
11.82998917	Z_1
−7.344595282	Z_2
25.30855172	Z_3
0.162686708	Depth of cut
10	Feed rate
0	y
0.01	d_{11}
2.739420677	d_2
2.691448259	d_3
0	d_1

---

[5] It should be noted that by taking antilog, the original value of the roundness would be available.

TABLE 2.8

Optimal Solution of the Revised Model

Final Value	Name
−919.2472636	W
11.821	Z_1
−7.337003719	Z_2
25.32681381	Z_3
0.171726672	Depth of cut
10	Feed rate
0	Y
0.001	d_{11}
2.731829114	d_2
2.673186167	d_3
0	d_1

TABLE 2.9

Comparing the Primary and Optimized Solutions

Responses	Primary Solution	Optimal Revised Model	Improvement Percentage
Z_1	11.49	11.821	10.85
Z_2	−7.65494411	−7.337003719	4.15
Z_3	95	25.32681381	74

2.8.2 Friction Stir Welding

FSW was invented by the Welding Institute in 1991 and is considered a process of solid-state welding (Thomas et al., 1995). In this method, heat generated by the friction and material flow is the reason for material binding. FSW does not use the melting of materials among the process of welding materials. The basis of welding is done based on the principle of severe material deformation (Krishnan, Maniraj, Deepak, & Anganan, 2016). Heidarzadeh, Khodaverdizadeh, Mahmoudi, and Nazari (2012) have introduced FSW as a solid-state joining process with some achievements in contrast to conventional fusion, such as avoiding big distortion, solidification cracking, porosity, oxidation, and other defects. This test design has been presented in Table 2.10.

This design has investigated the effects of five independent variables, namely rotational speed, linear speed, tilt angle degree, shoulder diameter, and pin diameter on two responses, namely ultimate tensile strength (UTS) and elongation. In the next stage, modeling of data in Table 2.11 has been represented in four parts: determining the response and control (independent) variables, determining the effective variables, modeling the target function, and modeling the constraints.

TABLE 2.10

DOE by Box–Behnken and the Results of UTS for Each Experiment

Test No.	Rotational Speed (rpm)	Linear Speed (mm/min)	Tilt Angle Degree (°)	Pin Diameter (mm)	Shoulder Diameter (mm)	UTS (MPa)	Elongation (%)
1	1,600	213	1	4	12	172	3.8
2	1,600	213	2.5	3	12	174	3.9
3	1,600	213	4	3	14	178	3.7
4	700	213	1	3	12	91	2.9
5	1,600	25	2.5	3	14	194	6.3
6	1,600	25	4	3	12	196	4.3
12	1,600	213	2.5	3	12	171	3.8
13	1,600	400	4	3	12	139	2.5
14	1,600	213	1	2	12	171	3.8
24	700	400	2.5	3	12	71	1
25	2,500	213	2.5	3	14	154	6.4
26	1,600	25	2.5	3	10	219	6.1
44	1,600	213	2.5	2	10	170	3.6
45	1,600	213	2.5	3	12	172	4
46	1,600	25	2.5	4	12	201	5.3

TABLE 2.11

Type of Variable Expression

Variable			Level		
Type	Name	Unit	1	2	3
Independent	Rotational speed	rpm	700	1,600	2,500
	Linear speed	mm/min	25	213	400
	Tilt angle degree	°	1	2.5	4
	Shoulder diameter	mm	2	3	4
	Pin diameter	mm	10	12	14
Response	UTS	MPa			
	Elongation	%			

2.8.2.1 Determining Responses and Control Variables

As previously mentioned, this model has five control variables and two response variables, explanations of which are shown in Table 2.12. The following information can be obtained from this table:

- Unit: This column presents different variable units. Be careful that elongation does not have any unit and is expressed in percentages

- Level: These columns are specific to the independent variables and show their different quantities. Note that every control variable has only three quantities (levels).

To better understand the FSW process and involved parameters in this process, Ghaffarpour, Aziz, and Hejazi (2017) have presented a schematic that is visible in Figure 2.4.

In fact, all independent variables in Table 2.12 have been identified among the screening experiment phase of RSM. The question that arises here is whether all these variables affect the final quality of welding. Which relations between these parameters are effective? To answer these questions, ANOVA has been used in this example. In order to use ANOVA, the random error should follow a normal distribution. If this condition is not met, normalization must be done. In our example, the normal test has been done by Minitab V.16 and shows that residuals follow normal distribution in both response parameters. In other words, residual quantities of UTS and elongation have a normal distribution in FSW illustration. Another condition that must be regarded is the independence of experiments that are met for both response variables.

TABLE 2.12

Response Surface Regression for UTS

Row	Term	Coef	SE Coef	T	P-Value
1	Constant	172.95	1.448	119.456	0.000
2	Rotational speed	32.77	1.983	16.531	0.000
3	Rotational speed × rotational speed	−31.69	1.982	15.986	0.000
4	Rotational speed × linear speed	−11.26	3.965	−2.840	0.007
6	R-square = 95.82%				

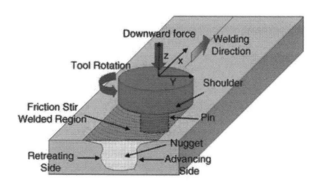

FIGURE 2.4
A schematic of FSW (Ghaffarpour et al., 2017).

2.8.2.2 Identifying Factors with Significant Effects

To identify significant effects in the model, the concept of p-value has been used. Thus, each parameter or relationship between them with p-value ≤ 0.05 has been considered as an effective factor. Table 2.13 has represented the output of Minitab V.16 for UTS. Modeling the target function for UTS has been done through information gained from Table 2.13. The term column shows the factors whose effects have been studied. The coef column shows the coefficient of each factor. Table 2.13 has been presented after elimination of meaningless variables.

As previously mentioned, factors with p-value ≤ 0.05 have been regarded as factors with significant effects. Such variables have been shown in Table 2.12. R-squared has been obtained as 95.82%, which represents high precession in the estimation of factor coefficients. Such tables have been shown for another response parameter, elongation, in Table 2.14. This table has only one factor with significant effect, namely linear speed. To gain more information about better modeling factors, significant effects have been eliminated. Modeling the target function of elongation has been done based on Table 2.14.

Effective factors for elongation have been presented in Table 2.14, namely rotational speed, linear speed, tilt angle degree, and shoulder diameter. Thus, these factors should be used for modeling the target function of elongation.

TABLE 2.13

Response Surface Regression for Elongation

Row	Term	Coef	SE Coef	T	P-Value
1	Constant	4.0219	0.05665	70.999	0.000
2	Rotational speed	1.4125	0.09605	14.706	0.000
3	Linear speed	−1.3315	0.09605	−13.863	0.000
4	Tilt angle degree	−0.2938	0.09605	−3.058	0.004
5	Shoulder diameter	0.4438	0.09605	4.620	0.000
6	R-square = 91.46%				

TABLE 2.14

Optimal Values among Independent Optimizations of Each Target Function

Rotational Speed	Linear Speed	Shoulder Diameter	Tilt Angle Degree	Value	Target
0.1169	1	0	0	214.48	E (UTS)
0	0	0	0	2.0736	Var (UTS)
1	−1	1	−1	7.5132	E (Elongation)
0	0	0	0	0.0032	Var (Elongation)

58 Mathematics in Engineering Sciences

2.8.2.3 Modeling the Target Function

This phase has been done under factors with significant effects (the term column of Tables 2.13 and 2.14) and their coefficient and standard error (SE) coefficient. Coef column has been used for the estimation of expected value of target functions. Also, to carry out robust optimization, SE coef columns have been used. In many real problems, there are several goals that should be optimized through related constraints. Problems with two or more responses are considered as multiresponses. In our FSW example, four responses have been modeled from Eqs. (2.34)–(2.37).

$$Z_1 = \max(E(UTS)) = 172.95 + 32.77 \times \text{rotational speed} - 31.69$$
$$\times \text{linear speed} - 49.26 \times \text{rotational speed}^2 - 11.26$$
$$\times \text{rotational speed} \times \text{linear speed} \tag{2.34}$$

$$Z_2 = \min(Var(UTS)) = 1.448^2 + 1.983^2 \times \text{rotational speed}^2 + 1.982^2$$
$$\times \text{linear speed}^2 + 2.455^2 \times \text{rotational speed}^2 + 3.965^2$$
$$\times \text{rotational speed}^2 \times \text{linear speed}^2 \tag{2.35}$$

$$Z_3 = \max(E(Elongation)) = 4.0219 + 1.41225$$
$$\times \text{rotational speed} - 1.3315 \times \text{linear speed} - 0.2938$$
$$\times \text{tilt angle degree} + 0.4438 \times \text{shoulder diameter} \tag{2.36}$$

$$Z_4 = \min(Var(Elongation)) = 0.05665^2 + 0.09605^2$$
$$\times \text{rotational speed}^2 + 0.09605^2 \times \text{linear speed}^2 + 0.09605^2$$
$$\times \text{shoulder diameter}^2 + 0.09605^2 \times \text{tilt angle degree}^2 \tag{2.37}$$

Target functions included four parts, two of which are related to maximizing expected value, Eqs. (2.34) and (2.36), and another one is related to minimizing variances, Eqs. (2.35) and (2.37). It should be noted that the goal of Eqs. (2.35) and (2.37) is robust optimization.

Constraint modeling has been presented based on the upper bound and lower bound of variables as

$$700 \le \text{rotational speed} \le 2,500 \tag{2.38}$$

$$25 \le \text{linear speed} \le 400 \tag{2.39}$$

$$1 \le \text{tilt angle degree} \le 4 \tag{2.40}$$

$$2 \leq \text{pin diameter} \leq 4 \tag{2.41}$$

$$10 \leq \text{shoulder diameter} \leq 14 \tag{2.42}$$

In this report, all calculations have been done based on coded parameters, among which each of the three variable levels changes into (−1, 0, +1). Thus, the form of equations from (2.38) to (2.42) has been shown as

$$-1 \leq \text{rotational speed} \leq +1 \tag{2.43}$$

$$-1 \leq \text{linear speed} \leq +1 \tag{2.44}$$

$$-1 \leq \text{tilt angle degree} \leq +1 \tag{2.45}$$

$$-1 \leq \text{pin diameter} \leq +1 \tag{2.46}$$

$$-1 \leq \text{shoulder diameter} \leq +1 \tag{2.47}$$

2.8.2.4 Optimization of FSW Model

To optimize the proposed model, multiple objective decision making (MODM) solutions have been used based on a distance function. Initially, each of the target functions should be optimized independently. Table 2.15 shows the optimized quantities of each target function.

In the next stage, the main target function must be estimated. To achieve this goal, the main target function has been estimated among formula (2.48) as

$$d_p = \min\left(\left(\sum_{i=1}^{m} w_i \cdot \left|\frac{Z_i^* - Z_i(x)}{Z_i^*}\right|^p\right)^{1/p}\right) \tag{2.48}$$

TABLE 2.15

Optimal Values among Main Target Functions

	Optimal Values	
Parameter	Coded	Real
Tilt angle degree	0.0823	2.3765
Shoulder diameter	0.1243	12.2487
Linear speed	0.174	162.525
Rotational speed	0.141	1,726.9356
F (UTS)	184.0656	
E (Elongation)	4.5322	
Var (UTS)	2.3057	
Var (Elongation)	0.0035	

If $p = 2$:

$$\min(Z_5) = \left(\left(\frac{Z_1 - 214.48}{214.48}\right)^2 + \left(\frac{Z_2 - 2.0736}{20.0736}\right)^2\right.$$

$$\left. + \left(\frac{Z_3 - 7.5132}{7.5132}\right)^2 + \left(\frac{Z_4 - 0.0032}{0.0032}\right)^2\right)^{\frac{1}{2}} \qquad (2.49)$$

Constraints have not changed. So, optimized values have been presented in Table 2.16 as

According to coded column, it is observed that optimal values of parameters are closer to the average range, zero, than the upper or lower bound, −1 or +1.

If the independent variables are set to proposed values in Table 2.15, target functions will achieve their best possible amounts. Be careful that this optimization is done with regard to robust optimization. Thus, if the independent variables are set according to Table 2.15, UTS and elongation of the optimized welded tool were equal to 184.06 MPa and 4.53%, respectively. Among robust optimization, variances of both response parameters were equal to 2.31 MPa^2 for UTS and 0.0032 for elongation. Due to the low variance of both response variables, the control of external noise variance is not difficult. In other words, if independent variables are set according to Table 2.15, the quantity of UTS and elongation are determined as

$$184.0656 - \sqrt{2.3057} \leq \text{UTS} \leq 184.0656 + \sqrt{2.3057} \qquad (2.50)$$

$$4.5322 - \sqrt{0.0035} \leq \text{Elongation} \leq 4.5322 + \sqrt{0.0035} \qquad (2.51)$$

The next issue to be considered is the amount of improvement in the target functions. Suppose that all independent variables are equal to zero. After optimization, the optimized parameters have been placed in target functions, and hence optimized values of target functions have been obtained. Improvement percentages of UTS and elongation were equal to 6.43% and 12.71%, respectively. Improvement in the variance of elongation is 98.91%, but

TABLE 2.16

Comparing Primary Solution and Optimal Solution of FSW

Response	Primary Solution	Optimal Revised Model	Improvement Percentage
Z_1	172.95	184.0656	6.43%
Z_2	2.0967	2.3057	−9.97%
Z_3	4.0219	4.5322	12.71%
Z_4	0.3209	0.0035	98.91%

variance of UTS has no improvement and increases by 9.97%. Such information has been presented in Table 2.16.

2.9 Conclusion

In recent years, using mathematical programming models in various industries have been taken into account to improve performance and increase efficiency indices. In order to achieve this target, two concepts namely quality engineering and MRO are presented in the following. The quality engineering discipline deals with the analysis of a manufacturing system at all stages to improve the quality of the production process and its output. The quality of products is commonly characterized by some variables called responses, which should be optimized with respect to existing controllable factors. Increasing the quality of products and reducing production wastes are achieved by using statistical methods, which will reduce the cost of the product. Another technique is MRO, which is considered by many scientists and engineers. The main idea of MRO is to optimize the response variables simultaneously by determining the best setting of variables. In this chapter, a numerical procedure for the computer numerical control (CNC) machine and FSW process has been applied, respectively. The CNC machine process includes four design factors, namely spindle speed, feed rate, depth of cut, and number of passes, that were optimized by modeling based on the GP method. The advantage of the GP approach demonstrated that the amount of loss has been considerably reduced. Hence, the model that has been achieved by the GP method can be used as an effective tool so as to help decision makers cope with the problems they face. Following the previous technique, FSW was introduced as a welding method. To carry out modeling, five parameters have been selected through the screening experiment phase. These variables were rotational speed, linear speed, tilt angle degree, pin diameter, and shoulder diameter. By determining effective variables, the pin diameter was eliminated, and modeling was done based on the other four parameters. In this example, UTS and elongation were response variables. The robust design was considered as an important part of target modeling to minimize external noise.

2.10 Funding

This work was supported by the Iran's National Elites Foundation [15/96595].

References

Ankenman, B. E., & Dean, A. M. (2003). Chapter 8. Quality improvement and robustness via design of experiments. In R. Khattree & C. R. Rao (Eds.), Handbook of Statistics (Vol. 22, pp. 263–317). Amesterdam, Boxton: Elsevier.

Ansari, M. H., Parsa, J. B., & Merati, Z. (2017). Removal of fluoride from water by nanocomposites of POPOA/Fe_3O_4, POPOA/TiO_2, POPOT/Fe_3O_4 and POPOT/TiO_2: Modelling and optimization via RSM. *Chemical Engineering Research and Design, 126*, pp. 1–18.

Bandyopadhyay, K., Panda, S. K., & Saha, P. (2016). Optimization of fiber laser welding of dp980 steels using rsm to improve weld properties for formability. *Journal of Materials Engineering and Performance, 25*(6), pp. 2462–2477.

Behnke, S. (2003). *Hierarchical Neural Networks for Image Interpretation* (Vol. 2766). Heidelberg: Springer.

Bera, S., & Mukherjee, I. (2016). A multistage and multiple response optimization approach for serial manufacturing system. *European Journal of Operational Research, 248*(2), pp. 444–452.

Bezerra, M. A., Santelli, R. E., Oliveira, E. P., Villar, L. S., & Escaleira, L. A. (2008). Response surface methodology (RSM) as a tool for optimization in analytical chemistry. *Talanta, 76*(5), pp. 965–977.

Bliuc, I., Lepadatu, D., Iacob, A., Judele, L., & Bucur, R. D. (2017). Assessment of thermal bridges effect on energy performance and condensation risk in buildings using DoE and RSM methods. *European Journal of Environmental and Civil Engineering, 21*(12), pp. 1466–1484.

Atkinson, A. C., & Donev, A. N. (1992). *Optimum Experimental Designs*. Oxford: Clarendon.

Box, G. E. P., & Draper, N. R. (1987). *Empirical Model-Building and Response Surfaces* (Vol. 424). New York: John Wiley & Sons.

Box, G. E., & Wilson, K. B. (1951). On the experimental attainment of optimum conditions. *Journal of the Royal Statistical Society: Series B (Methodological), 13*(1), pp. 1–38.

Box, G. E. P., & Wilson, K. B. (1992). On the experimental attainment of optimum conditions. In S. Kotz, N. L. Johnson (Eds.), *Breakthroughs in Statistics. Springer Series in Statistics (Perspectives in Statistics)*. New York: Springer.

Candioti, L. V., De Zan, M. M., Cámara, M. S., & Goicoechea, H. C. (2014). Experimental design and multiple response optimization. Using the desirability function in analytical methods development. *Talanta, 124*, pp. 123–138.

Carley, K. M., Kamneva, N. Y., & Reminga, J. (2004). Response surface methodology (No. CMU-ISRI-04–136). Carnegie-Mellon University, Pittsburgh, PA, School of Computer Science.

Castilo, E. D. (2007). *Process Optimization: A Statistical Approach*. Uniontown, PA: Springer.

Coelho, F., & Neto, J. P. (2017). A method for regularization of evolutionary polynomial regression. *Applied Soft Computing, 59*, pp. 223–228.

Elatharasan, G., & Kumar, V. S. S. (2013). An experimental analysis and optimization of process parameter on friction stir welding of AA 6061-T6 aluminum alloy using RSM. *Procedia Engineering, 64*, pp. 1227–1234.

Espinoza-Escalante, F. M., Pelayo-Ortiz, C., Gutiérrez-Pulido, H., González-Álvarez, V., Alcaraz-González, V., & Bories, A. (2008). Multiple response optimization analysis for pretreatments of Tequila's stillages for VFAs and hydrogen production. *Bioresource Technology, 99*(13), pp. 5822–5829.

Fan, J., & Gijbels, I. (1996). *Local Polynomial Modelling and Its Applications: Monographs on Statistics and Applied Probability 66* (Vol. 66, 0-412-98321-4). London: Chapman and Hall.

Gangil, M., & Pradhan, M. K. (2017). Modeling and optimization of electrical discharge machining process using RSM: A review. *Materials Today: Proceedings, 4*(2), pp. 1752–1761.

Ghaffarpour, M., Aziz, A., & Hejazi, T.-H. (2017). Optimization of friction stir welding parameters using multiple response surface methodology. *Proceedings of the Institution of Mechanical Engineers, Part L: Journal of Materials: Design and Applications, 231*(7), pp. 571–583.

Heidarzadeh, A., Khodaverdizadeh, H., Mahmoudi, A., & Nazari, a. E. (2012). Tensile behavior of friction stir welded AA 6061-T4 aluminum alloy joints. *Materials & Design, 37*, pp. 166–173.

Hejazi, T. H. (2017). A multiresponse model for reliability-based simulation optimization in systems subjected to random external stresses. *Quality and Reliability Engineering International, 33*(6), pp. 1225–1233.

Hejazi, T. H., Bashiri, M., Noghondarian, K., & Atkinson, A. C. (2011). Multiresponse optimization with consideration of probabilistic covariates. *Quality and Reliability Engineering International, 27*(4), pp. 437–449.

Hejazi, T. H., Salmasnia, A., & Bastan, M. (2013). Optimization of correlated multiple response surfaces with stochastic covariate. *International Journal of Computer Theory and Engineering, 5*(2), p. 341.

Hejazi, T. H., & Seyyed-Esfahani, M. (2017). Multistage-multiresponse models for dynamic quality chain design problems. *Quality and Reliability Engineering International, 33*(6), pp. 1263–1279.

Khuri, A. I., & Mukhopadhyay, S. (2010). Response surface methodology. *Wiley Interdisciplinary Reviews: Computational Statistics, 2*(2), 128–149, pp. 1939–0068.

Krishnan, M. M., Maniraj, J., Deepak, R., & Anganan, K. (2016). Prediction of optimum welding parameters for FSW of aluminium alloys AA6063 and A319 using RSM and ANN.

Montgomery, D. C. (2017). *Design and Analysis of Experiments.* New York: John Wiley & Sons.

Moskowitz, H. R. (1994). Chapter 5. Product optimization: Approaches and applications. In H. J. H. MacFie & D. M. H. Thomson (Eds.), Measurement of Food Preferences (pp. 97–136). London: Blackie Academic & Professional.

Myer, R. H., & Montgomery, D. C. (2002). *Response Surface Methodology: Process and Product Optimization Using Designed Experiment* (pp. 343–350). New York: John Wiley & Sons.

Myers, R. H., Montgomery, D. C., Vining, G. G., Borror, C. M., & Kowalski, S. M. (2004). Response surface methodology: A retrospective and literature survey. *Journal of Quality Technology, 36*(1), pp. 53–77.

Myers, R. H., Montgomery, D. C., & Anderson-Cook, C. M. (2016). *Response Surface Methodology: Process and Product Optimization Using Designed Experiments.* Hoboken, NJ: John Wiley & Sons.

Nyakundi, E. O., & Padmanabhan, M. N. (2015). Green chemistry focus on optimization of silver nanoparticles using response surface methodology (RSM) and mosquitocidal activity: Anopheles stephensi (Diptera: Culicidae). *Spectrochimica Acta Part A: Molecular and Biomolecular Spectroscopy, 149*, pp. 978–984.

Oliver, M. A., & Webster, R. (2015). *Basic Steps in Geostatistics: The Variogram and Kriging*. New York: Springer.

Park, G.-J., Lee, T.-H., Lee, K. H., & Hwang, K.-H. (2006). Robust design: An overview. *AIAA Journal, 44*(1), pp. 181–191.

Raj, D. A., & Senthilvelan, T. (2015). Empirical modelling and optimization of process parameters of machining titanium alloy by wire-EDM using RSM. *Materials Today: Proceedings, 2*(4–5), pp. 1682–1690.

Rao, K. S., Rao, C. S. P., Bose, P. S. C., Rao, B. B., Ali, A., & Kumar, K. K. (2017). Optimization of machining parameters for surface roughness on turning of niobium alloy C-103 by using RSM. *Materials Today: Proceedings, 4*(2), pp. 2248–2254.

Roziqin, M. C., Basuki, A., & Harsono, T. (2016). A comparison of Montecarlo linear and dynamic polynomial regression in predicting dengue fever case.

Said, K. A. M., & Amin, M. A. M. (2016). Overview on the response surface methodology (RSM) in extraction processes. *Journal of Applied Science & Process Engineering, 2*(1), pp. 2289–7771.

Schoofs, A. J. G. (1988). Experimental design and structural optimization. *In G. I. N. Rozvany & B. L. Karihaloo (Eds.), Structural Optimization* (pp. 307–314). Dordrecht: Springer.

Schutz, H. G. (1983). Multiple regression approach to optimization. *Food Technology, 37*(11), pp. 46–62.

Shin, S., Hoang, T.-T., Le, T.-H., & Lee, M.-Y. (2016). A new robust design method using neural network. *Journal of Nanoelectronics and Optoelectronics, 11*(1), pp. 68–78.

Thomas, W. M., Nicholas, E. D., Needham, J. C., Murch, M. G., Temple-Smith, P., & Dawes, C. J. (1995). U.S. Patent No. 5460317. U.S. Patent and Trademark Office: Washington, DC.

Unal, R., Lepsch, R., Engelund, W., & Stanley, D. (1996). Approximation model building and multidisciplinary design optimization using response surface methods. In *6th Symposium on Multidisciplinary Analysis and Optimization*, Bellevue, WA. p. 4044.

3

Application of Finite Element Method in Enhancing the Performance of Biomedical Implants

Mohit Pant
National Institute of Technology

Mohit Kumar
Auxein Medical Private Limited

Sahil Garg
Apeejay Institute of Management Technical Campus

CONTENTS

3.1 Introduction

Biomechanics can be comprehended as an amalgamation of the concepts of biological entities, such as tissues or bones, and mathematical formulas of mechanical and physical aspects. Understanding of biomechanics is paramount in the development of drugs, comprehension of wound healing procedures, and can help scientists in the estimation of peak physical prowess of human body. The first graphical representation of human anatomy was sketched by Leonardo Da Vinci in the 16th century [1]. Since the dawn of the 18th century, mathematical approaches were implemented on the study of kinesis and ambulation of animals [2]. Groundbreaking work on soft tissues [3] and bones [4] later acted as missing pieces of puzzles in bridging the fields of biology and mechanics. Various experimental testing procedures and their improvements to approximate mechanical properties of tissues such as tensile strength, hardness, etc. were reported [5]. The quest of perfection led to the application of computational techniques in the field of biomechanics.

Finite element (FE) method (FEM) based analysis is one of the most extensively used computational methods used for numerical simulation of a variety of engineering problems [6–8]. The complex field of biomedical engineering utilized the applications of FEM for modeling dental implants and bone implants [9–11].

In 1972, Brekelmans *et al.* [12] highlighted the advantages FEM over experimental studies of the mechanical behavior of skeletal part. Their studies were based on 2D model of femur because of its simple design and unproblematic replacement options. A theory for remodeling of internal and external bone was applied and tested using FEM to imitate internal bone in a proximal femur by Beaupre *et al.* [13]. Computational fluid dynamics (CFD) tool based on FEM has proved to be a base for the design of rotary blood pumps [14], and this tool has reduced costs tremendously, and its efficacy had been proved when the pump was tested experimentally using particle flow visualization. Dopico-González [15] worked on a heuristic approach to analyze an FE model on an uncemented hip replacement that considered variability in bone–implant version angle. El-Anwar and El-Zawahry [16] used a number of dimensional FE models to study dental implants based on the diameter and length of implants and concluded that an increase in the length and implant diameter provides better stress distribution on the bone. Zhang *et al.* [17] provided a new 3D FEM-based model of femur in partial with volume rendering, which proved to be more apt for FE analysis. Finite element analysis (FEA) based free vibration mode analysis of femur bone fracture was performed by Kumar *et al.* [18]. Chen *et al.* [19] used nondestructive 3D microcomputed tomography (microCT) based FE (microFE) models to estimate bone mechanical properties at tissue level.

The future scope of computational methods lies with the use of meshfree methods (MMs) in biomechanics. MMs provide a myriad of applications [20], and Belinha *et al.* [21] used Natural Neighbor Radial Point Interpolation Method for bone tissue remodeling analysis. More work on implant analysis, e.g. fracture analysis, using MMs is currently under progress in our labs as well.

This chapter discusses the applicability of FEM in complete analysis of implants manufactured in an industry. The novelty of work lies in the design of implant based on the requirements and physical attributes of Indian patients. This chapter provides the analysis of 3.5-mm reconstruction bone plate and expert femur nail.

3.2 Results and Discussions

The static and dynamic analysis of implants shown in this section is carried out for two cases of implants: 3.5-mm reconstruction bone plate and expert femur nail. The objectives of the study are as follows:

- To analyze stresses and deformation on the bone–plate interface as per surgical planning using FEA.
- Static and dynamic testing of 3.5-mm reconstruction plate according to ASTM standard F-382.
- To analyze stresses and deformation on the bone–nail interface using FEA.
- Static and dynamic testing of expert femur nail according to ASTM F1264.

The rationale behind the sample selection is to select extreme size of plate from different lengths to have sufficient plate length available for analysis. ABAQUS 6.13 software is used for pre- and postprocessing of static testing, and FE-safe 6.4 software is used for dynamic testing. The process of test follows static FEA according to surgical approach. In vitro static FEA testing setup was created using assembly Computer aided design (CAD) models of clavicle bone and 3.5-mm reconstruction plate. Loading and boundary conditions were applied in ABAQUS. Both normal and worst-case compression loading were applied. Stress and deformation analysis locate the critical areas. Static four-point bending test as per ASTM F382 is used to determine the performance characteristics of bone plates. Dynamic testing using FEA is performed in the framework of FE-safe for fatigue analysis of FE models. It is used alongside commercial FEA software to estimate the location of fatigue cracks and the probability of survival at different service lives. For static FEA, if the values of stresses are in safer limits as according to applicable standards to that material, then the product is in safer limit. For dynamic FEA, fatigue life up to 1,000,000 cycles was predicted using different loadings.

3.2.1 Literature Survey on 3.5-mm Reconstruction Plate and Expert Femur Nail

The purpose of this FEA study is to check the safety performance of 3.5-mm reconstruction plate (refer Table 3.1) and expert femur nail (refer Table 3.2) made of titanium by calculating the deformation and stresses on plate and its performance after surgery. So the focus of the literature research was to evaluate the various types of FEA techniques and experimental techniques that researchers have utilized for their studies to evaluate the safety of 3.5-mm reconstruction plate and expert femur nail.

TABLE 3.1

Brief Literature on 3.5-mm Reconstruction Plate

S. No.	Literature Source	Type of Study	Reason for Selection/Rejection
1	Kemper *et al.* [22]	Experimental study	The purpose of this study was to determine the influence of loading direction on the structural response of the human clavicle subjected to three-point bending.
2	Cronskär and Bäckström [23]	FEA study	The aim of the study was to develop a method for the realistic simulation of stresses and displacements in the bone and fixation device and to use this method to make comparisons between a conventional reconstruction plate and a customized plate, designed from patient-specific CT data.
3	Rusovici *et al.* [24]	FEA study	The research investigated the use and prediction accuracy of advanced material models within an FE framework to predict dynamic stresses and strains within the clavicle body, which would be subjected to a dynamic load.
4	Cronskär *et al.* [25]	FEA study	This paper addresses the evaluation of clavicle fixation devices by means of computational models.
5	Ni *et al.* [26]	FEA study	This study aims to compare the construct stability, stress distribution, and fracture micromotion of three fixations based on FE method.
6	Santos *et al.* [27].	Experimental study	This experimental in vitro study evaluated the influence of screw length on the mechanical properties of a locking reconstruction plate designed with locking rings inserted into plate holes.
7	Berber *et al.* [28]	Experimental study	The purpose of this study was to assess the stability of a developmental pelvic reconstruction system that extends the concept of triangular osteosynthesis with fixation anterior to the lumbosacral pivot point.

3.2.1.1 Conclusion of Literature Research

Based on the literature research, it was concluded that FEA techniques are the most efficient methods to reduce the design iterations of implants. Various design and evaluation techniques to check the safety performance of implants were found. It is also concluded that virtual surgical planning techniques are a powerful tool for optimal placement of expert femur nail and 3.5-mm reconstruction plate, which is very important for the proper functioning of bone–implant assembly postsurgery. Material of the implant, design of the implant according to Indian population, and safety performance of these implants are the keys areas of research nowadays.

TABLE 3.2

Brief Literature on Expert Femur Nail

S. No.	Literature Title	Type of Study	Reason for Selection/Rejection
1	Apel *et al.* [29]	FEA study	This study aims to find the stability of introchanteric fractures of femur under axial loading.
2	Ivanov *et al.* [30]	FEA study	This article presents new intramedullary (IM) nail design for femur diaphyseal fracture osteosynthesis. Biomechanical comparison of the two IM nails (standard locking and new developed) was conducted.
3	Hawi *et al.* [31]	Retrospective Study	This retrospective cohort study aimed to determine whether the use of computer navigation during femoral nailing procedures reduced postoperative femoral malrotation and leg length discrepancy as well as the number of revision cases.
4	Wan *et al.* [32]	Retrospective study	The purpose of this study was to identify the underlying cause by simulating the forces involved in a controlled laboratory setting, and then to illustrate some intraoperative tips on how to detect this misalignment and suggest solutions prevent this intraoperative complication.
5	Hyung *et al.* [33]	Experimental study	The aim was to use preoperative templating and 3D printed model to characterize the technical difficulties associated with the use of current commercially available IM nail systems for the management of atypical femur fractures (AFFs) with severe bowing.

3.2.2 Case 1: 3.5-mm Reconstruction Bone Plate

The test sample/specimen for case 1, i.e. 3.5-mm reconstruction bone plate is shown in Figure 3.1. The 3D model of clavicle bone is shown in Figure 3.2a. Material properties were assigned using the ABAQUS software. Material properties according to ISO 5832-3 are used for plate and screws. Table 3.3 of material properties are given later:

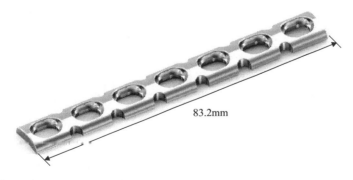

83.2mm

FIGURE 3.1
3.5-mm reconstruction plate.

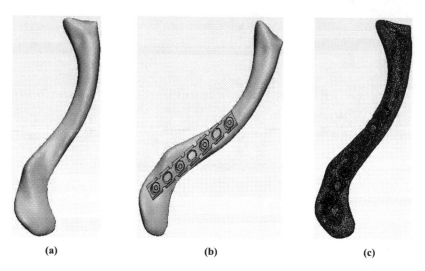

(a)   (b)   (c)

FIGURE 3.2
(a) 3D model of clavicle bone, (b) bone and plate assembly, and (c) meshed bone and plate model.

TABLE 3.3

Properties of Components of Model for Case 1

Component	Material Properties		
	Young's Modulus	Poisson Ratio	Density
Cortical bone	20 GPa	0.3	2,000 kg/m^3
Titanium	106 GPa	0.33	4,700 kg/m^3

The placement of the plate and screws in the assembly were done as per the surgical planning shown in Figure 3.2b. The volume mesh assembly of the bone and plate is constructed using tetrahedral C3D10M element. The total number of elements in the model is 132,295, as shown in Figure 3.2c. The loading and boundary conditions were determined on literature works by Cronskär *et al.* [25] on loading of the clavicle bone. Results are presented as contour plots, which can be plotted using standard FE viewers in Figure 3.3a and b.

Maximum stress of 258 MPa was observed in the 3.5-mm reconstruction plate under the specified loading and boundary conditions. The maximum stress in the result is less than the yield strength of titanium, as given in ISO 5832-3. Henceforth, the stresses on 3.5-mm reconstruction plate are under safer limits.

Under the applied load, there was a maximum deformation of 0.39 mm, as shown in Figure 3.4, which is elastic in nature. Maximum deflection is at the distal end of the clavicle bone and at the head portion of the 3.5-mm reconstruction plate.

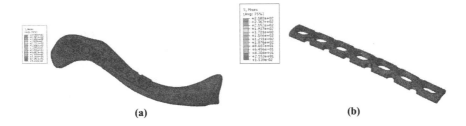

FIGURE 3.3
Stress distribution of (a) bone–plate assembly and (b) plate.

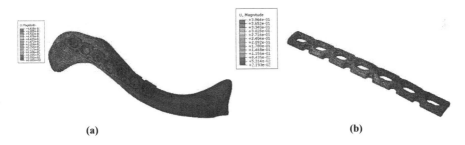

FIGURE 3.4
Displacement of (a) bone–plate assembly and (b) plate.

There is a negligible deformation in the 3.5-mm reconstruction plate. So, "3.5-mm reconstruction plate" is safe for use as the elastic deformation is less.

For the worst-case scenario, a five time load than the normal load was applied to the bone. Maximum stress of 1,291 MPa was observed, as shown in Figure 3.5, in the 3.5-mm reconstruction plate under the same boundary conditions as used in the normal loading scenario. The maximum stress in the result is little more than the yield strength of titanium, as given in ISO 5832-3.

FIGURE 3.5
Stress distribution in the bone plate assembly for worst-case scenario.

3.2.2.1 Four-Point Bend Test of the Plate

The test setup for four-point bending test is designed according to **ASTM F382**. The goal of this experiment is to understand bending stiffness, bending structural stiffness, and bending strength from a single cycle (static) bend test on metallic bone plates using cylindrical rollers. Following are the configurations of FEA for four-point bending test in accordance with the setup shown in Figure 3.6, and the parameters are $a = 12\,mm$, $h = 12\,mm$, and diameter of rollers used = 6 mm.

A maximum displacement load of 5 mm was applied on the plate, and its deflection is shown in Figure 3.7a and b. The load vs. displacement graph is shown in Figure 3.8.

Based upon the load vs. displacement graph, the values of proof load, bending stiffness, bending strength, and structural bending stiffness were calculated as shown in Table 3.4.

3.2.2.2 Dynamic Testing for Fatigue Test Case 1

To perform fatigue analysis, FE-safe requires three inputs:

- The stresses in each point in the model: FE-safe can use elastic stresses from an elastic FE analysis.
- A description of loading: Load histories can be imported from industry-standard file formats or entered at the keyboard.
- Material Data: Fatigue properties of the component material are required.

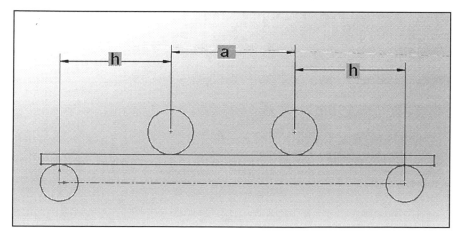

FIGURE 3.6
Four-point bend test setup for 3.5-mm reconstruction bone plate.

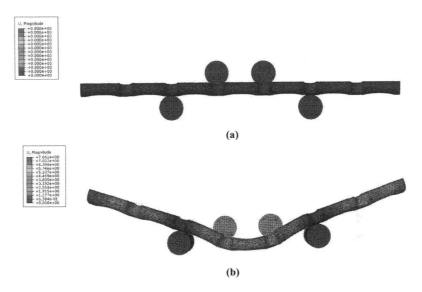

FIGURE 3.7
Four-point bend test of titanium plate. Case 1: (a) initial step and (b) final step.

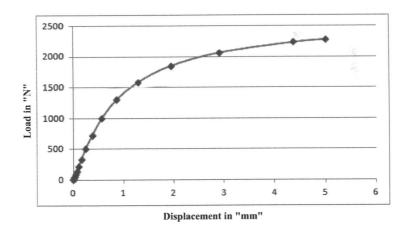

FIGURE 3.8
Four-point bend test: load vs. displacement graph, case 1.

TABLE 3.4

Results from Four-Point Bend Test for Case 1

Material	Proof Load (N)	Displacement at Proof Load (mm)	Bending Stiffness (N/mm)	Bending Strength (N-m)	Structural Bending Stiffness
Titanium	900	5	1,778	5.4	1.28

TABLE 3.5

Results of Fatigue Test for Case 1

S. No.	Min. Load (N)	Max. Load (N)	No. of Cycles	Result
1	8	80	1,000,000	No fracture
2	9	90	1,000,000	No fracture
3	10	100	1,000,000	No fracture
4	20	200	1,563	Fracture
5	30	3,000	139	Fracture

For the purpose of analysis, the surface finish defined is $1.6 < Ra \leq 4\,\mu m$, and the material is set to titanium. Brown Miller Algorithm is used for the analysis. The results of fatigue tests are shown in Table 3.5. The test parameters are as follows:

- Loading waveform: Sinusoidal
- Frequency: 5 Hz
- Default endurance limit: E7 cycles

3.2.3 Case 2: Expert Femur Nail and Femur Bone

The test sample/specimen for case 2, i.e., expert femur nail and femur bone is shown in Figure 3.9.

The 3D model of clavicle bone is shown in Figure 3.10a. Material properties were assigned using the ABAQUS software. Material properties according to ISO 5832-3 are used for plate and screws. Table 3.6 of material properties is given later:

The placement of the nail in the assembly was done as per the surgical planning shown in Figure 3.10b. The volume mesh assembly of the bone and nail is constructed using tetrahedral C3D10M element. The total number of elements in the model is 380,0242, as shown in Figure 3.10c. An average human body weight is around 80 kg, so under normal conditions, the top of the femur bone would experience a load of around 500 N. Hence, a load of 500 N was applied at the top of femur bone and was constrained from the bottom. Results are presented as contour plots, which can be plotted using standard FE viewers in Figure 3.11a and b.

400mm

FIGURE 3.9
Expert femur nail.

TABLE 3.6

Material Properties of Components of Model for Case 2

Component	Material Properties		
	Young's Modulus	Poisson Ratio	Density
Cortical bone	20 GPa	0.3	2,000 kg/m³
Titanium	106 GPa	0.33	4,700 kg/m³

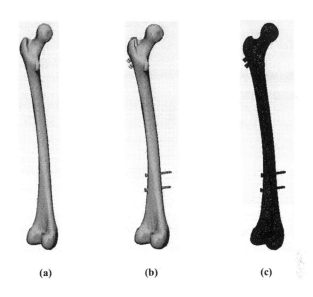

FIGURE 3.10

(a) 3D model of femur bone, (b) bone, nail, and screws assembly, and (c) meshed bone, nail, and screws model.

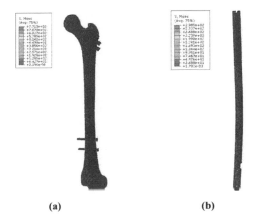

FIGURE 3.11

Stress distribution: (a) bone, nail, and screws assembly and (b) plate.

A maximum stress of 298.5 MPa was observed in the expert femur nail under the specified loading and boundary conditions. The maximum stress in the result is less than the yield strength of titanium, as given in ISO 5832-3. And hence, the stresses on the expert femur nail are under safer limits.

Under the applied load of 500 N, there was a maximum deformation of 3.7 mm in femur bone and 3.3 mm in the nail, as shown in Figure 3.12a and b, which is elastic in nature. Maximum deflection is at the proximal end of the femur bone and at the head portion of the expert femur nail.

There is a negligible deformation in the expert femur nail. So "expert femur nail" is safer for fixation in femur fractures.

For the worst-case scenario, a five time load than the normal load was applied to the bone. A 2,500 N load was applied at the proximal end of the femur bone. Maximum stress of 1,493 MPa was observed in the expert femur nail, as shown in Figure 3.13 under the same boundary conditions as used in the normal loading scenario.

3.2.3.1 Four-Point Bend Test of the Plate

The test setup for four-point bending test is designed according to **ASTM F382**. The goal of this experiment is to understand bending stiffness, bending structural stiffness, and bending strength from a single cycle (static) bend test on metallic bone plates using cylindrical rollers. Following are the configurations of FEA for four-point bending test in accordance with the setup shown in Figure 3.14, and the parameters are $a = 12$ mm, $h = 12$ mm, and diameter of rollers used $= 6$ mm.

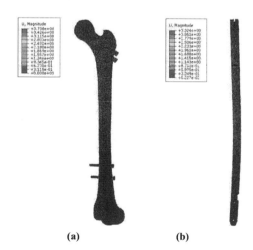

(a)   (b)

FIGURE 3.12
Displacement: (a) bone, nail, and screws assembly and (b) plate.

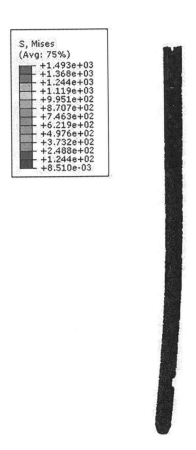

FIGURE 3.13
Stress distribution for femur plate worst-case scenario.

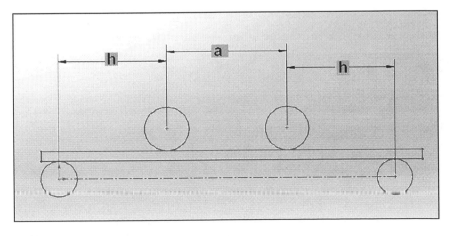

FIGURE 3.14
Four-point bend test setup for expert femur nail.

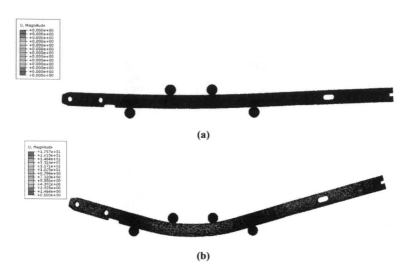

(a)

(b)

FIGURE 3.15
Four-point bend test of titanium plate. Case 2: (a) initial step and (b) final step.

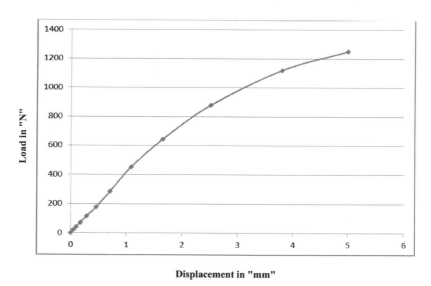

Displacement in "mm"

FIGURE 3.16
Four-point bend test: load vs. displacement graph, case 2.

A maximum displacement load of 5 mm was applied on the nail, and its deflection is shown in Figure 3.15a and b. The load vs. displacement graph is shown in Figure 3.16.

Based on the load vs. displacement graph, the values of proof load, bending stiffness, bending strength, and structural bending stiffness were calculated as shown in Table 3.7.

TABLE 3.7

Results of Four-Point Bend Test for Case 2

Material	Proof Load (N)	Displacement at Proof Load (mm)	Bending Stiffness (N/mm)	Bending Strength (N-m)	Structural Bending Stiffness
Titanium	700	1.8	363	14	9.6

TABLE 3.8

Results of Fatigue Test for Case 2

S. No.	Min. Load (N)	Max. Load (N)	No. of Cycles	Result
1	10	100	1,000,000	No fracture
2	20	200	1,000,000	No fracture
3	30	300	298,966	No fracture
4	40	400	17,014	Fracture
5	50	500	2,583	Fracture

3.2.3.2 Dynamic Testing for Fatigue Test Case 2

The procedure for this test is similar to the one adopted for the earlier Case 1. For the purpose of analysis, the surface finish defined is $1.6 < Ra \leq 4$ µm, and the material is set as titanium. Brown Miller Algorithm is used for the analysis. The results of fatigue tests are shown in Table 3.8. The test parameters are as follows:

- Loading waveform: Sinusoidal
- Frequency: 5 Hz
- Default endurance limit: E7 cycles

3.3 Conclusions

The work establishes the applicability of engineering-based analysis in biomedical sector by conjugating many spheres of science into one. The FEM analysis enables industry to save cost and provide insight into the points where the product can fail.

The conclusion is based on the results of various FEA results that "3.5 mm reconstruction plate" and "expert femur nail" have good strength and high load-bearing capability. During realistic loading, maximum stress on plate and nail was 258 and 298.5 MPa, respectively, which is quite lower to the yield strength of titanium, as given in ISO 5832-3. The focus of this study was to evaluate the safety performance of these implants using FEA. FEA results show that the mechanical strength of these implants is good because all the stresses and deformations are within safety limits. Still, improvements in case of applications with respect to worst-case scenario and fatigue tests, are in process. These implants have so far been successful, and postsurgeries and future improvisations are on the way.

References

1. C.D. O'Malley, J.B. Saunders, C.M. Saunders, L. da Vinci, Leonardo da Vinci on the human body: the anatomical, physiological, and embryological drawings of Leonardo da Vinci, (1952) 506. file://catalog.hathitrust.org/Record/001575669.
2. G. Borelli Alfonso, *Borelli's on the Movement of Animals - On the Force of Percussion*, 1st ed., Switzerland: Springer, 2015.
3. K. Langer, On the anatomy and physiology of the skin: III. The elasticity of the cutis, *Br. J. Plast. Surg.* 31 (1978) 185–199. doi: 10.1016/S0007-1226(78)90081-4.
4. J. Wolff, The classic: On the inner architecture of bones and its importance for bone growth, *Clin. Orthop. Relat. Res.* 468 (2010) 1056–1065. doi: 10.1007/s11999-010-1239-2.
5. A.A. Glaser, R.D. Marangoni, J.S. Must, T.G. Beckwith, G.S. Brody, G.R. Walker, W.L. White, Refinements in the methods for the measurement of the mechanical properties of unwounded and wounded skin, *Med. Electron. Biol. Eng.* 3 (1965) 411–419. doi: 10.1007/BF02476136.
6. V.M. Puri, R.C. Anantheswaranb, The finite-element method in food processing: A review, *J. Food Eng.* 19 (1993) 247–274.
7. T.Y. Chao, W.K. Chow, H. Kong, A review on the applications of finite element method to heat transfer and fluid flow, *Int. J. Archit. Sci.* 3 (2002) 1–19.
8. S.L. Ho, W.N. Fu, Review and future application of finite element methods in induction motors, *Electr. Mach. Power Syst.* 26 (2007) 111–125. doi: 10.1080/07313569808955811.
9. J.-P. Geng, K.B.C. Tan, G.-R. Liu, Application of finite element analysis in implant dentistry: A review of the literature, *J. Prosthet. Dent.* 85 (2001) 585–598. doi: 10.1067/mpr.2001.115251.
10. R.C. van Staden, H. Guan, Y.C. Loo, Application of the finite element method in dental implant research, *Comput. Methods Biomech. Biomed. Eng.* 9 (2006) 257–270. doi: 10.1080/10255840600837074.
11. S.K. Parashar, J.K. Sharma, A review on application of finite element modelling in bone biomechanics, *Perspect. Sci.* 8 (2016) 696–698. doi: 10.1016/j.pisc.2016.06.062.
12. W.A.M. Brekelmans, H.W. Poort, T.J.J.H. Slooff, A new method to analyse the mechanical behaviour of skeletal parts, *Acta Orthop.* 43 (1972) 301–317. doi: 10.3109/17453677208998949.
13. G.S. Beaupre, T.E. Orr, D.R. Carter, An approach for time-dependent bone modeling and remodeling-application a preliminary remodeling simulation, *J. Orthop. Res.* 8 (1990) 662–670.

14. G.W. Burgreen, J.F. Antaki, Z.J. Wu, A.J. Holmes, Computational fluid dynamics as a development tool for rotary blood pumps, *Artif. Organs.* 25 (2001) 336–340. doi: 10.1046/j.1525-1594.2001.025005336.x.

15. C. Dopico-González, A.M. New, M. Browne, A computational tool for the probabilistic finite element analysis of an uncemented total hip replacement considering variability in bone-implant version angle, *Comput. Methods Biomech. Biomed. Eng.* 13 (2010) 1–9. doi: 10.1080/10255840902911536.

16. M.I. El-Anwar, M.M. El-Zawahry, A three dimensional finite element study on dental implant design, *J. Genet. Eng. Biotechnol.* 9 (2011) 77–82. doi: 10.1016/j.jgeb.2011.05.007.

17. Y. Zhang, W. Zhong, H. Zhu, Y. Chen, L. Xu, J. Zhu, Establishing the 3-D finite element solid model of femurs in partial by volume rendering, *Int. J. Surg.* 11 (2013) 930–934. doi: 10.1016/j.ijsu.2013.06.843.

18. A. Kumar, H. Jaiswal, T. Garg, P.P. Patil, Free vibration modes analysis of femur bone fracture using varying boundary conditions based on FEA, *Procedia Mater. Sci.* 6 (2014) 1593–1599. doi: 10.1016/j.mspro.2014.07.142.

19. Y. Chen, E. Dall'Ara, E. Sales, K. Manda, R. Wallace, P. Pankaj, M. Viceconti, Micro-CT based finite element models of cancellous bone predict accurately displacement once the boundary condition is well replicated: A validation study, *J. Mech. Behav. Biomed. Mater.* 65 (2017) 644–651. doi: 10.1016/j.jmbbm.2016.09.014.

20. S. Garg, M. Pant, Meshfree methods: A comprehensive review of applications, *Int. J. Comput. Methods.* 15 (2018) 1830001-1–1830001-85. doi: 10.1142/S0219876218300015.

21. J. Belinha, L.M.J.S. Dinis, R.M. Natal Jorge, The meshless methods in the bone tissue remodelling analysis, *Procedia Eng.* 110 (2015) 51–58. doi: 10.1016/j.proeng.2015.07.009.

22. A.R. Kemper, J.D. Stitzel, C. Mcnally, H.C. Gabler, S.M. Duma, Biomechanical response of the human clavicle : The effects of loading direction on bending properties, *J. Appl. Biomech.* 25 (2009) 165–174.

23. M. Cronskär, M. Bäckström, Modeling of fractured clavicles and reconstruction plates using CAD, finite element analysis and real musculoskeletal forces input, *WIT Trans. Biomed. Health.* 17 (2013) 235–243. doi: 10.2495/BIO130211.

24. R. Rusovici, J.T. O'Brien, I. Ghita, Finite element modeling of human clavicle under dynamic loading, *Biomed. Eng. (NY).* 10 (2013) 531–537. doi: 10.2316/P.2013.791-165.

25. M. Cronskär, J. Rasmussen, M. Tinnsten, Combined finite element and multibody musculoskeletal investigation of a fractured clavicle with reconstruction plate, *Comput. Methods Biomech. Biomed. Eng.* 18 (2013) 1–10. doi: 10.1080/10255842.2013.845175.

26. M. Ni, W. Niu, D.W. Wong, W. Zeng, J. Mei, M. Zhang, Finite element analysis of locking plate and two types of intramedullary nails for treating mid-shaft clavicle fractures, *Injury.* 47 (2016) 1918–1623. doi: 10.1016/j.injury.2016.06.004.

27. R.R. Santos, S.C. Rahal, C.M. Neto, C.R. Ribeiro, A.S.C. Edson, C.R. Foschini, F.S. Agostinho, L.D.R. Mesquita, Biomechanical analysis of locking reconstruction plate using mono- or bicortical screws, *Mater. Res.* 19 (2016) 588–593.

28. O. Berber, A. Amis, A.C. Day, Biomechanical testing of a concept of posterior pelvic reconstruction in rotationally and vertically unstable fractures, *J. Bone Joint Surg. Br.* 93 (2011) 237–244. doi: 10.1302/0301-620X.93B2.24567.

29. D. Apel, A. Patwardhan, M. Pinzur, W. Dobozi, Axial loading studies of unstable intertrochanteric fractures of the femur, *Clin. Orthop. Relat. Res.* 246 (1989) 156–164.

30. D. Ivanov, A. Barabash, Y. Barabash, Preclinical biomechanics of a new intra-medullary nail for femoral diaphyseal fractures, *Russ. Open Med. J.* 4 (2015) 1–7. doi: 10.15275/rusomj.2015.0205.
31. N. Hawi, E. Liodakis, E.M. Suero, T. Stuebig, M. Citak, C. Krettek, Radiological outcome and intraoperative evaluation of a computer-navigation system for femoral nailing: A retrospective cohort study, *Injury.* 45 (2014) 1632–1636. doi: 10.1016/j.injury.2014.05.039.
32. J. Wan, K. Derly, O.C. Jiandong, H. Benoit, C. Mauffrey, Prevention of inac-curate targeting of proximal screws during reconstruction femoral nailing, *Eur. J. Orthop. Surg. Traumatol.* 26 (2016) 391–396. doi: 10.1007/s00590-016-1769-8.
33. J. Hyung, Y. Lee, O. Shon, H. Chul, J. Wan, Surgical tips of intramedullary nailing in severely bowed femurs in atypical femur fractures: Simulation with 3D printed model, *Injury.* 47 (2016) 1318–1324. doi: 10.1016/j.injury.2016.02.026.

4

Magnetic Nanofluids—A Novel Concept of Smart Fluids

Vimal Kumar Joshi
Dronacharya Group of Institutions

Paras Ram
National Institute of Technology Kurukshetra

CONTENTS

4.1 Brief Introduction of the Magnetic Nanofluids (Ferrofluids)

Magnetic nanofluid (MNF), often known as ferrofluid, is a stable colloidal suspension of magnetic nanoparticles having surfactant coatings in a carrier liquid. Stephen Pappell at The National Aeronautics and Space Administration (NASA) (1960) first developed ferrofluids as an alternative for controlling fluids in the space. These fluids are responsive to the externally applied magnetic field, even in the absence of gravitational force, and have significant effects in heat transfer phenomena comparative to the conventional fluids. These unique features made them vital for application in space and more useful in heat transfer problems. The study of rheological properties, especially the thermal property of such fluids, has an immense importance in many areas of engineering and industries, such as rotating machinery, thermal power generator system, nuclear reactors, storage device, gas turbine rotors, high-speed computer disk drives, medical equipment, air cleaning machines, crystal growth process, etc.

4.2 Composition of MNFs

MNFs are composed of synthesized magnetic particles Fe_3O_4, γ-Fe_2O_3, or $CoFe_2O_4$, having an average size of about 10nm dispersed in a carrier liquid like water, kerosene, hydrocarbon, toluene, fluorocarbon, glycol, and lubricants. To prevent the clumping of nanoparticles due to the magnetic field, they are provided the necessary coating of dispersing agents known as surfactant.

There are three major constituents of MNFs:

1. Magnetic nanoparticles
2. Surfactant coating or dispersive agents
3. Carrier/base fluids

TABLE 4.1

The Physical Properties of Some Carrier Liquids Surfactants or Dispersive Agents

Surfactant Name	Molecular Formula	Density	Viscosity
Oleic acid	$C_{18}H_{34}O_2$	0.895 g/mL	27.64 mPa s (25°C)
Tetramethyl ammonium hydroxide	$C_4H_{13}NO$	1.015 g/cm³ (20%–25% aqueous solution)	3.13 cP (19°C)
Citric acid	$C_6H_8O_7$	1.665 g/cm³ (anhydrous) 1.542 g/cm³ (18°C, monohydrate)	6.5 cP (50% aq. sol.)
Lecithin	$C_{35}H_{66}NO_7P$	1.0305 g/cm³ (20°C)	

The composition of a typical MNF consists of about 5% magnetic particles, 10% surfactant coating materials, and 85% carrier liquids, by volume.

4.2.1 Magnetic Nanoparticles

Iron (Fe), Cobalt (Co), Magnetite (Fe_3O_4), Hematite (Fe_2O_3), Maghemite (γ-Fe_2O_3), $CoFe_2O_4$, or Fe-C.

4.2.2 Surfactants or Dispersive Agents

In an MNF, there is a possibility of agglomeration due to the influence of van Der Waals forces that decay/decelerate the flow and may reduce its life. To avoid this difficulty, there is a need of a surfactant coating, usually a hydrocarbon, on each metallic particle. Some of the surfactant materials with their rheological properties are given in Table 4.1.

4.2.3 Carrier/Base Fluids (Table 4.2)

TABLE 4.2

The Physical Properties of Some Carrier Liquids at 20°C and 50°C [1]

Fluid	Density (ρ_{20}/ρ_{50})	Specific Heat Capacity (Cp_{20}/Cp_{50})	Thermal Conductivity (k_{20}/k_{50})	Dynamic Viscosity (μ_{20}/μ_{50})
Water	1.011	1.000	0.940	1.816
Ethylene glycol	1.021	0.945	0.961	3.263
Mineral oil (10. NF)	1.022	0.933	1.027	3.592
Dielectric liquid (FC-77)	1.041	0.954	1.041	1.616
Glycerine	1.015	0.923	0.997	8.809

4.2.4 MNFs Considered under the Study (Table 4.3)

TABLE 4.3

Various MNFs with Their Physical Properties

Properties	FC-72 [2]	Taiho W-40 [3]	C1-20B [4]	EMG-901 [5]	90 G [6]
Base fluids	Fluorocarbon (Perfluorohexane C_6F_{14})	Water (H_2O)	Kerosene	Hydrocarbon	Hydrocarbon
Density (kg/m^3)	1.68×10^3	1.4×10^3	1.25×10^3	1.53×10^3	0.972×10^3
Coefficient of thermal expansion (1/K)	1.6×10^{-3}	0.026×10^{-4}	0.86×10^{-3}	6×10^{-4}	9×10^{-4}
Pyromagnetic coefficient (A/mk)	–	240	80	30	–
Dynamical viscosity in zero magnetic field (kg/ms)	6.4×10^{-4}	3.99×10^{-2}	6×10^{-3}	9.95×10^{-3}	15×10^{-3}
Thermal diffusivity (m^2/s)	3.084×10^{-8}	64.3×10^{-8}	5×10^{-8}	8.2×10^{-8}	8.7×10^{-8}
Prandtl number	12.3	44.3	128	79.3	176.4

4.3 Boundary Layer Flow

The theory of boundary layer is developed by a famous German Scientist Ludwig Prandtl (1904). According to Prandtl, there exist two regions in a viscous flow over a solid surface:

i. The first region is a thin layer of a viscous fluid in contact with the solid surface in which the viscosity affects significantly, known as the boundary layer.

ii. Another region lies outside the boundary layer, where the viscous effects are negligible or the fluid is considered as nonviscous.

Boundary layer is observed in the various phenomena of nature such as planetary boundary layer and aircraft wing boundary layer due to wing's airflow. Thus, the layer of fluid where the free stream velocity becomes zero is known as the hydrodynamic boundary layer. But in the theory of heat transfer, besides momentum boundary layer, a thermal boundary layer also occurs.

4.3.1 Rotating Disk Problems

Due to a vast range of applications, the rotating disk has become a most common geometry for investigating different kind of flows. The study of rotating disk has numerous applications in gas and steam turbines, pumps, and other rotating machines [7] under the classical problem of fluid dynamics. The rotating disk has become a significant geometry of study, which allows an analytical and theoretical idea in solution to the equation of motion in three-dimensional flows (Figure 4.1).

There are several reasons for considering various problems of MNF flow in a prototype system of rotating disk. The classical problem of an ordinary viscous flow over a semi-infinite rotating disk was first discussed by Kármán [8]. Cochran [9] and Benton [10] further extended the problem of Kármán for the unsteady case and improved the results to the higher degree of accuracy. After the pioneering study of Kármán, the rotating disk problems again came into limelight. Many researchers devoted their study on studying the behavior of boundary layer flow (BLF) over the rotating disk and related solutions. Mithal [11] investigated the ordinary viscous steady flow due to a rotating disk under the influence of uniform suction. Attia [12] studied the time-dependent boundary layer state due to the external uniform magnetic field. Attia and Aboul-Hassan [13] analyzed the rotating

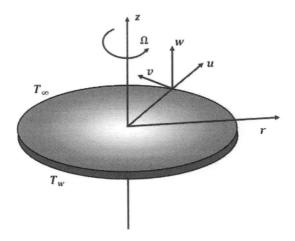

FIGURE 4.1
Schematic diagram of the rotating disk problems.

surface flow, considering the effects of Hall effect and uniform axial magnetic field. Turkyilmazoglu [14] employed homotopy analysis method to express the analytical solution of the BLF in the presence of a rotating disk and uniform suction or injection. Ram et al. [15] described the magnetoviscous effects of an MNF on flow profiles and obtained a solution applying the Neuringer–Rosensweig model.

4.3.2 Stretchable Rotating Disk Problems

In this type of problem, the disk rotates with radial stretching at a constant rate. The fluid layer in contact with the disk surface thus rotates with the same angular velocity, and due to viscosity, a boundary layer is generated. In such flows, the fluid is thrown radially outward as an action of centrifugal force supplemented by the radial stretching of the disk. The boundary layer that flows with such kind of geometries has applications in the processes of extrusion in metal and plastic industries (Figure 4.2).

4.3.3 Stationary Disk Problems (Bodewadt Flow)

The three-dimensional flow of rotating viscous fluids over a stationary surface have importance in many natural phenomena like predicting the behavior of tectonic plates due to the geological stretching by studying the dynamics of hurricanes/tornadoes/cyclone, and also in studying the mechanical rotor–stator system. The revolving flow of an incompressible viscous fluid over a stationary disk is known as Bodewadt flow, recognized by Bodewadt [16]. In this type of flow, the outward flow of the fluid due to fluid rotation is balanced by the radial pressure gradient (Figure 4.3).

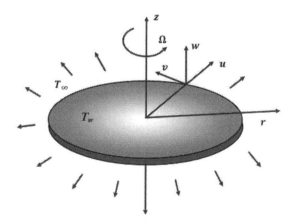

FIGURE 4.2
Schematic diagram of the stretchable rotating disk problems.

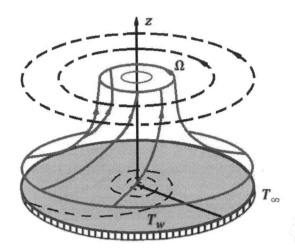

FIGURE 4.3
Schematic diagram of the stationary disk problems.

4.4 Variable Viscosity

Viscosity is the rheological property of a fluid that plays an influential role in the generation of boundary layers over the surface of a solid body moving through fluid or fluid flowing past the solid body. In the case of ordinary viscous fluid flows, the viscosity is not constant, but depends on other properties of the fluid such as temperature, pressure, depth, etc. Thus, the viscosity of a fluid is generally a variable quantity.

4.4.1 Temperature-Dependent Viscosity

Temperature is one of the common factors affecting the viscosity of a fluid. For an ordinary viscous fluid, the viscosity decreases with an increase in the temperature of liquids, whereas for gases, it increases.

4.4.2 Depth-Dependent Viscosity

To study the behavior of mantle flow with convective mixing, Gurnis and Davies [17] have considered depth-dependent viscous effects. Stegman et al. [18] explained the depth dependency of the viscosity on mantle flow while maintaining a uniform midocean ridge basalt reservoir. Cserepes [19] performed a numerical experiment on three-dimensional structures of thermal convection in a rectangular domain, taking the joint effect of heating mode and depth-dependent viscosity. Roos and Schuttelaars [20] described the effects of both time- and depth-dependent eddy viscosity on sand wave formation.

4.4.3 Geothermal Viscosity (Depth- and Temperature-Dependent Viscosity)

In the current study of magnetic nanoliquids, we have taken the viscosity of MNFs as variable depending on both depth and temperature connected through the following relation:

$$\mu(z,T) = \frac{\mu_\infty(1-\alpha z)}{\{1+\alpha(T-T_\infty)\}}, \quad (\alpha \geq 0).$$

Similar types of exponential relations are already being used by many other researchers (Lai and Kulacki [21], Ling and Dybbs [22]). Torrance and Turcotte [23] demonstrated the influence of a depth-dependent and temperature-dependent viscous fluid on BLF with finite-amplitude convection. Korenaga and Jordan [24] explained the onset of convection in an incompressible fluid under the influence of geothermal viscosity using the two-dimensional numerical simulation. Such problems of flows are of interest in the study of geophysics as well as in many technical processes.

4.5 Porous Medium

A porous medium is a solid matrix material containing pores that may or may not be interconnected. Many applications of porous medium are seen in civil engineering (e.g., the flow of water in aquifers, the propagation of stresses under foundation of structures, and transport of pollutants in aquifers), agricultural engineering (e.g., the movement of solutes and water in the root zone in the soil), and reservoir engineering (e.g., the flow of oil, water, and gases in petroleum reservoir). Examples of porous materials are soils, beach sand, glass beads, catalyst pellets, sandstone, limestone, concrete, cement, bricks, lungs, and kidneys, aquifers from which ground water is pumped, packed-bed in chemical engineering industry, the root zone in agricultural, paper, cloth, rye bread, wood, and cork.

4.6 Transport Phenomena

In the fluid flow, transport phenomenon generally refers to the movement of various entities (such as momentum, energy, or mass) through a fluid due to diffusion or convection. Due to the variations of concentration in a medium, a relative motion of various chemical species occurs, which is generally referred as diffusion. While the velocity variations lead to the transport of momentum, the temperature difference results in the transport of energy, usually known as heat conduction. The transport phenomena have a variety of applications like surface coating of magnetic inks in computer hard disks, detection of toxic chemicals in soils and streambeds, separation of proteins in downstream bioprocessing, etc.

4.6.1 Heat Transfer

The exchange of heat is the flow of thermal energy from a hotter or a colder object according to the law of thermodynamics. Whenever there is a temperature difference, heat naturally flows from hotter to colder areas until the temperature is in equilibrium across the whole area. The transfer of heat cannot be stopped, but resisted or reflected. Heat transfer is commonly associated with fluid dynamics.

The analytical study of heat flow and temperature distribution is of great interest in a broad area of applied sciences and engineering. In heat transfer problems, we often use the terms heat and temperature; however, there is a difference between the two. Heat is a form of energy, and generally, its direction of flow is from hotter objects to the colder one. It is usually represented by the symbol Q with units as calorie in the CGS (Centimetre Gram Second) system and joule in the SI (International System) system. On the other hand, the temperature is the unit to measure the degree of hotness with the unit kelvin in SI system, which is used to predict the direction of heat flow.

Heat transfer occurs primarily through the following three modes:

Conduction

Convection

Radiation

In real-life problems, the temperature is distributed in a medium under the influence of all the earlier three modes. Therefore, the interaction of one mode with other modes cannot be ignored. However, for an analysis, one of the heat transfer modes is considered, taking negligible effects of the other two modes. Here, a brief description of these three heat transfer modes is given later.

4.6.1.1 Conduction

Heat conduction is the transfer of internal energy among molecules/atoms. The atoms of a solid body vibrate faster at a higher temperature than at a lower temperature in the same or other system in contact and transfer some of their internal energy to the neighboring molecules/atoms. Heat conduction has a dominating role in heat transfer within a solid body/system while fluids are comparatively less conductive.

4.6.1.2 Convection

The heat transfer through convection mode is only found in a fluid medium and occurs in the form of bulk mass transfer. When there is a temperature difference of the fluid and the adjacent solid surface, the transfer of heat takes place from the heated surface to the fluid by means of the movement of fluid molecules. If heat transport happens with the motion of particles by the effects of buoyancy resulting from the variation in density, the transfer of heat is termed as free/natural convection, while the process is called forced convection if the external agencies like fan and blower are applied for enhancing the heat transfer.

4.6.1.3 Radiation

All substances including solid, liquid, and gaseous matters emit heat at elevated temperatures. The transfer of heat energy as photon and electromagnetic waves is termed as thermal radiation (*also called as infrared radiation*). Radiation does not require a medium to propagate unlike conduction and convection, since it does not rely on any contact between the source and the object. It is a form of transport of energy without any medium and exchange of mass, in which electromagnetic waves travel at the speed of light.

4.6.2 Mass Transfer

The movement of fluid particles from high concentration to low concentration area is referred as mass transfer. It occurs due to the concentration difference by means of pressure, temperature, and density gradients. The mass transfer occurs in many fields of engineering and applied sciences (physical, chemical, and biological). Few examples of mass transfer are osmosis, respiratory mechanisms, gas absorption, distillation, crystallization adsorption, air humidification, ion exchange, etc. The most common real life examples are the evaporation of water to the atmosphere, evaporation of clouds, smoke diffusion from a tall chimney, the dissolution of salt and sugar in water, dispersion of fog, etc.

Mass transfer can be classified into two categories.

    i. Diffusive mass transfer
    ii. Convective mass transfer

4.6.2.1 Diffusive Mass Transfer

The diffusive mass transfer occurs due to the movement of fluid particles of one component to another. It happens either by the temperature, pressure, or concentration gradients and is calculated as diffusion flux, $J = -D\dfrac{\partial C}{\partial x}$.

where D is the diffusion constant and $\dfrac{\partial C}{\partial x}$ is the concentration gradient per unit length.

4.6.2.2 Convective Mass Transfer

The mechanism includes the bulk motion of mass between two immiscible fluids or boundary surface, and the moving fluid is referred to as convective mass transfer. Again, it is categorized into free/natural and forced convective mass transfer. In natural convection, the movement of the species happens due to the difference in density resulting from the concentration or the temperature difference or by mixing of varying compositions, whereas in forced convection, the forced movement of mass occurs due to external agencies.

4.7 Nondimensional Physical Parameters

The dimensionless parameters play an important role in observing the qualitative behavior of a physical problem. These dimensionless parameters provide different physical phenomena for fluids:

4.7.1 Reynolds Number

A British scientist Osboene Reynolds (1842–1912) introduced a nondimensional number to distinguish the nature of flow as laminar and turbulent, which was later on known as Reynolds number. He defined a critical Reynolds number Re_{crit} such that the flow with Reynold number $Re < Re_{crit}$ is laminar, otherwise turbulent. Mathematically, it is defined as the ratio of the inertial to viscous force and represented as

$$Re = \frac{F_i}{F_v} = \frac{\rho \overline{V} \, d\overline{V}/dx}{\mu d^2 \, \overline{V}/dx^2}$$

$$= \frac{\rho \overline{V}\overline{V}/L}{\mu \overline{V}} = \frac{\rho \overline{V}L}{\mu} = \frac{\overline{V}L}{v},$$

where $\overline{V}$ is the characteristic velocity of the flow.

4.7.1.1 Physical significance

The large value of Re indicates that the fluid is lightly viscous. Water and air are fluids of small viscosity, and in motion, they usually give very high Reynolds number. Also, it is a known fact that if the value of Re exceeds a certain critical value, the flow ceases to be laminar near the boundaries. In these circumstances, the flow becomes irregular in that eddies or just that of the boundaries, resulting in a turbulent flow. The flow of fluid is characterized as

Laminar if $Re < 2{,}300$,

Transient if $2{,}300 < Re < 4{,}000$,

Turbulent if $Re > 4{,}000$.

4.7.2 Prandtl Number

Prandtl number is an important dimensionless number associated with the properties of fluids. It is directly related with the thermal and velocity boundary layer thickness. Mathematically, it is defined as the ratio of momentum diffusivity to the thermal diffusivity of the fluid.

$$Pr = \frac{\text{Kinematic viscosity}}{\text{Thermal diffusivity}} = \frac{\nu}{\alpha} = \frac{\mu/\rho}{k/\rho c_p} = \frac{\mu c_p}{k}.$$

In the problems of heat transfer, Prandtl number, specifically, controls the relative speed with which the momentum and heat energies of fluids are transmitted.

4.7.3 Eckert Number

To characterize dissipation in dimensionless analysis, the Eckert number is introduced as a ratio of kinetic energy and enthalpy.

$$Ec = \frac{\text{Kinetic energy}}{\text{Enthalpy}} = \frac{\overline{V}^2}{c_p \Delta \theta},$$

where $\overline{V}$: Characteristic velocity of the flow and
    $\Delta\theta$: Temperature difference of the flow.

4.7.4 Schmidt Number

The ratio of diffusivity of momentum and mass is known as the Schmidt number, which plays a major role in the problems of convective mass transfer.

$$Sc = \frac{\text{Viscous diffusion rate}}{\text{Mass diffusion rate}} = \frac{\nu}{D} = \frac{\mu}{\rho D}.$$

4.7.5 Nusselt Number (Rate of Heat Transfer)

The dimensionless number defined as the ratio of convective and conductive thermal resistance of the fluid is called the Nusselt number and is given as

$$Nu = \frac{-h}{(T_w - T_\infty)}\left(\frac{\partial \theta}{\partial y}\right)_{y=0},$$

where h is the characteristic length and $(T_w - T_\infty)$ is the difference of temperature between the wall and the fluid. The Nusselt number for convection is always greater than 1, since the flow motion enhances heat transfer.

4.7.6 Sherwood Number (Rate of Mass Transfer)

The dimensionless quantity Sh is known as Sherwood number and is defined as

$$Sh = \frac{-h}{(C_w - C_\infty)}\left(\frac{\partial C}{\partial y}\right)_{y=0}, \text{ where } \frac{\partial C}{\partial y} \text{ is the mass concentration gradient.}$$

4.7.7 Skin Friction Coefficients

The skin friction coefficient is used to represent the dimensionless shearing stress at the surface of the body and is defined as

$$C_f = \frac{\tau_w}{\frac{1}{2}\rho U_\infty^2} = \frac{1}{\frac{1}{2}\rho U_\infty^2}\left(\frac{\partial u}{\partial y}\right)_{y=0}, \text{ where } \left(\frac{\partial u}{\partial y}\right) \text{ is the velocity gradient.}$$

In case of fluid flow past a solid body (or when a solid body is moved through fluid), the viscous property of the fluid layer just in contact with the body sets up a shear stress, which resists the flow. This shear stress is also called skin friction. Generation of skin friction at the boundary is a natural consequence of viscous flow and no-slip condition at the wall.

For the laminar boundary flow of an ordinary viscous fluid along a flat body, the local wall shear stress is defined as

$$\tau_w = \mu \frac{\partial u}{\partial y}\bigg|_{y=0}$$

where μ is the fluid viscosity and u is the boundary layer velocity along the direction of the plate (y-axis).

In nondimensional form, the shear stress can also be expressed as skin friction coefficient as

$$C_f = \frac{\tau_w}{\frac{1}{2}\rho U_\infty^2},$$

where ρ is the fluid density and U_∞ is the free stream velocity.

The cylindrical coordinates and Newtonian formulae, the radial and tangential stress at the disk surface are given as

$$\tau_r = \mu\left(\frac{\partial u}{\partial z} + \frac{\partial w}{\partial r}\right)\bigg|_{z=0} = \mu Re^{1/2}\Omega U''(0),$$

$$\tau_t = \mu\left(\frac{\partial v}{\partial z} + \frac{1}{r}\frac{\partial w}{\partial \varphi}\right)\bigg|_{z=0} = \mu Re^{1/2}\Omega V'(0).$$

And, the radial and tangential skin coefficients are, respectively, given as

$$C_{f_r} = Re^{-\frac{1}{2}} U''(0)$$

$$C_{f_\phi} = Re^{-\frac{1}{2}} V'(0)$$

where $Re = \dfrac{r^2 \omega}{v}$ is the rotational Reynolds number, an U' and V are the dimensionless radial and tangential velocities.

4.8 Ferrohydrodynamic Model for Problem Formulation

The Neuringer–Rosensweing (NR) model (1964) for the unsteady flow of MNFs and convective heat transfer (in the absence of the magnetocaloric effects) over a semi-infinite disk is applied. The magnetization $\vec{M}$ is considered to be aligned in the direction of applied magnetic field. The governing equations are a slightly modified form of Navier–Stokes equations in addition to the Maxwell equations of magnetization. The model consists of the following governing equations:

The continuity equation

$$\nabla \cdot \vec{q} = 0 \tag{4.1}$$

The equation of motion

$$\rho \left[\frac{\partial \vec{q}}{\partial t} + (\vec{q} \cdot \nabla) \vec{q} \right] = -\nabla p + \mu_0 \left(\vec{M} \cdot \nabla \right) \vec{H} + \nabla^2 \left(\mu \vec{q} \right) \tag{4.2}$$

In case of an unsteady flow of MNF embedded in the porous medium, the equation of motion (4.2) becomes

$$\rho \left[\frac{\partial \vec{q}}{\partial t} + (\vec{q} \cdot \nabla) \vec{q} \right] = -\nabla p + \mu_0 \left(\vec{M} \cdot \nabla \right) \vec{H} + \nabla^2 \left(\mu \vec{q} \right) - \frac{\mu}{K} \vec{q} \tag{4.2a}$$

The equation of energy

$$\rho C_p \left[\frac{\partial T}{\partial t} + (\vec{q} \cdot \nabla) T \right] = k \nabla^2 T \tag{4.3}$$

If the viscous dissipation is taken into consideration, the equation of energy (4.3) becomes

$$\rho C_p \left[\frac{\partial T}{\partial t} + (\vec{q} \cdot \nabla) T \right] = k \nabla^2 T + \varphi_D. \tag{4.3a}$$

The equation of energy considering the viscous dissipation besides the thermal radiation is given as

$$\rho C_p \left[\frac{\partial T}{\partial t} + (\vec{q} \cdot \nabla) T \right] = k \nabla^2 T + \varphi_D - \frac{\partial q_r}{\partial z}. \tag{4.3b}$$

The equation of mass transfer

$$\frac{\partial C}{\partial t} + (\vec{q} \cdot \nabla) C = D \nabla^2 C. \tag{4.4}$$

Maxwell equations

$$\vec{B} = \mu_0 \left(\vec{H} + \vec{M} \right), \nabla \times \vec{H} = \vec{0}. \tag{4.5}$$

The NR model has been popularly adopted by many researchers since last few decades. Using NR model, Verma and Vedan [25,26] and Verma and Singh [27] investigated the paramagnetic Couette and helical flow with heat transfer and flow through a porous annulus, respectively.

4.9 Boundary Conditions

In disk-related problems, it is assumed that the fluid does not move relative to the disk surface in the tangential direction. Rather, it sticks to the surface, and this phenomenon is termed as the no-slip condition. Two point boundary conditions are necessary to be specified along with the modeled governing equations.

In a BLF due to a stationary disk, the fluid rotates with a constant angular velocity, but the fluid layer in contact with the disk remains stationary due to viscosity, and a boundary layer is generated over the disk surface. The boundary conditions of flow near a stationary disk are as follows:

$$\left. \begin{array}{l} u = 0, v = 0, w = 0, p = 0, T = T_w \quad \text{at } z = 0 \\[2mm] u \to 0, v \to r\Omega, p \to 0, T \to T_\infty \quad \text{as } z \to \infty \text{ and} \\[2mm] w \text{ tends to some finite positive value as } z \to \infty \end{array} \right\} \tag{4.6}$$

While in rotating disk problems, the layer of fluid in contact with the disk also rotates with the same angular velocity of the disk and is thrown radially outward due to the action of centrifugal force. This radial outward flow is balanced by the axial downward fluid flow. Thus, the boundary conditions specifying the flow are as follows:

$$u = 0, v = r\Omega, w = 0, p = 0, T = T_w \quad \text{at } z = 0$$

$$u \to 0, v \to 0, p \to 0, T \to T_\infty \quad \text{as } z \to \infty \text{ and}$$

$$w \text{ tends to some finite negative value as } z \to \infty$$

(4.7)

The boundary conditions for the radially stretchable disk along with the angular rotation are as follows:

$$u = sr, v = r\Omega, w = 0, p = 0, T = T_w \quad \text{at } z = 0$$

$$u \to 0, v \to 0, p \to 0, T \to T_\infty \quad \text{as } z \to \infty \text{ and}$$

$$w \text{ tends to some finite negative value as } z \to \infty$$

(4.8)

4.10 Similarity Transformations

The need of the similarity transformation arises to reduce the system of partial differential equations (PDEs) having n-independent variable into the system of $(n-1)$ independent variables. If $n = 2$, the system of equations will be transformed into a set of ordinary differential equations (ODEs). In 1921, Kármán introduced these famous transformations, serving the purpose of solving the BLF over an infinite disk, which is later known as Von Kármán transformations. For the BLF over a rotating/stationary disk, the transformations used are as follows:

$$\eta = \frac{z}{\delta}, u = r\Omega U'(\eta), v = r\Omega V(\eta),$$

$$w = \frac{v}{\delta} W(\eta) \text{ and } T - T_\infty = (T_w - T_\infty)\theta(\eta)$$

(4.9)

For the BLF over a stretchable rotating disk, the transformations used are as follows:

$$\eta = \frac{zRe}{\delta}, u = rs\, U'(\eta), v = rs\, V(\eta), w = \frac{s}{\delta Re} W(\eta),$$

$$p = \frac{s^2}{Re^2} P(\eta) \text{ and } T - T_\infty = \Delta T\, \theta(\eta)$$

(4.10)

where s is the rate of radial stretching of disk and $Re = \sqrt{\dfrac{s}{v}}$ is the Reynolds number.

4.11 Numerical Scheme

The three-dimensional BLF over a stationary/rotating/stretchable rotating semi-infinite plates are mathematically represented by the system of coupled nonlinear PDEs (4.11–4.14) together with their respective boundary conditions (4.15) specified in the semi-infinite flow domain. In this problem, we have found the solution for BLF problems using the numerical techniques that combine the Runge–Kutta method along with the shooting technique. First of all, the governing equations are converted into a system of coupled nonlinear ODEs using suitable transformations. Then, the numerical solution of the transformed model system is obtained in MATLAB® tool ODE45 with shooting technique for the initial guesses.

The dimensionless set of model equations governing the time-dependent Bodewadt flow of MNFs in the presence of a porous medium is as follows:

$$\left(1-\frac{\varepsilon_1\eta}{Re}\right)U''' - \frac{\varepsilon_1}{Re}U'' - \left(1-\frac{\varepsilon_1\eta}{Re}\right)(1+\varepsilon\theta)^{-1}\varepsilon\theta'U''$$

$$+\frac{(1+\varepsilon\theta)}{Re^2}\left[2\eta U'' + R\left(V^2 - U'^2\right) + 2RUU'' - R - \frac{2BR}{Re} - R\beta U'\right] = 0 \quad (4.11)$$

$$\left(1-\frac{\varepsilon_1\eta}{Re}\right)V'' - \frac{\varepsilon_1}{Re}V' - \left(1-\frac{\varepsilon_1\eta}{Re}\right)(1+\varepsilon\theta)^{-1}\varepsilon\theta'V'$$

$$+\frac{(1+\varepsilon\theta)}{Re^2}\left[2\eta V' + 2R(UV' - VU') - R\beta V\right] = 0 \quad (4.12)$$

$$\theta'' + \frac{2Pr}{Re^2}(\eta + RU)\theta' = 0 \quad (4.13)$$

$$\varphi'' + \frac{2Sc}{Re^2}(\eta + RU)\varphi' = 0 \quad (4.14)$$

where the prime denotes derivative with respect to dimensionless number η.
The dimensionless boundary conditions are

$$\left.\begin{array}{l} U'(0) - 0, \ V(0) - 0, \ W(0) - 0, \ \theta(0) = 1, \ \varphi(0) - 1 \\[4pt] U'(\infty) \to 0, \ V(\infty) \to 1, \ \theta(\infty) \to 0, \ \varphi(\infty) \to 0 \\[4pt] W(\infty) \text{ tends to some positive value.} \end{array}\right\} \quad (4.15)$$

Initially, we reduce the modeled problem into initial value problem by reducing the higher order differential equations into first order using the following process:

$$U = y_1, U' = y_2, U'' = y_3, V = y_4, V' = y_5, \theta = y_6, \theta' = y_7, \phi = y_8, \phi' = y_9 \quad (4.16)$$

$$y_3' = \left(\frac{1}{1 - \dfrac{\varepsilon_1 \eta}{Re}}\right)\left(\frac{\varepsilon_1}{Re} y_3 + \left(1 - \frac{\varepsilon_1 \eta}{Re}\right)(1 + \varepsilon y_6)^{-1} \varepsilon y_7 y_3 - \frac{(1+\varepsilon y_6)}{Re^2}\left(2\eta y_3 + R\left(y_4{}^2 - y_2{}^2\right)\right.\right.$$

$$\left.\left. - 2R y_1 y_3 + R + \frac{2BR}{Re} + R\beta y_2\right)\right), \tag{4.17}$$

$$y_5' = \left(\frac{1}{1 - \dfrac{\varepsilon_1 \eta}{Re}}\right)\left(\frac{\varepsilon_1}{Re} y_5 + \left(1 - \frac{\varepsilon_1 \eta}{Re}\right)(1 + \varepsilon y_6)^{-1} \varepsilon y_7 y_5\right.$$

$$\left. - \frac{(1+\varepsilon y_6)}{Re^2}\left(2\eta y_5 + 2R\left(y_1 y_5 - y_4 y_2\right) - R\beta y_4\right)\right), \tag{4.18}$$

$$y_7' = \frac{-2Pr}{Re^2}\left(\eta + R y_1\right) y_7, \tag{4.19}$$

$$y_9' = \frac{-2Sc}{Re^2}\left(\eta + R y_1\right) y_9, \tag{4.20}$$

with boundary conditions as

$$y_1 = y_2 = 0, y_4 = 0, y_6 = 1, y_8 = 1, \quad \text{at } \eta = 0$$

$$y_2 = 0, y_4 = 0, y_6 = 0, y_8 = 0 \quad \text{as } \eta \to \infty. \tag{4.21}$$

The values of $y_3(0)$, $y_5(0)$, and $y_7(0)$, which are not given at the initial conditions, are guessed by taking help of the shooting technique. The semi-infinite domain may be replaced with the finite domain $[0, \eta_\infty)$ in such a way that the asymptotic behavior of the flow can be closely approximated at the boundaries. The accuracy of the guessed values is validated by equating the calculated values of y_2, y_4, y_6, and y_8 with their actual boundary values.

Table 4.4 demonstrates the effects of Sc and Pr on skin friction factors, Nusselt number and Sherwood number. Since the Schmidt number is not directly appearing in dimensionless momentum equations and temperature equation; therefore, no influence on the distribution of velocity and temperature is prevailing. On the other hand, a significant rise in reactant Sherwood number magnitude has been noted for higher values of Sc. A marginal decrease in $-\varphi'(0)$ and a tremendous increase in $-\theta'(0)$ have been observed

TABLE 4.4

Variations of Friction Factor Coefficients ($U''(0)$, $V''(0)$), Local Nusselt Number ($-\theta'(0)$) and Sherwood Number ($-\varphi'(0)$) for Various Values of Sc and Pr

Sc	Pr	$U''(0)$	$V''(0)$	$-\theta'(0)$	$-\varphi'(0)$
1	50	0.220115	2.758014	100.000391	2.314975
3	50	0.220115	2.758014	100.000391	6.018304
5	50	0.220115	2.758014	100.000391	10.003186
7	50	0.220115	2.758014	100.000391	14.002168
9	50	0.220115	2.758014	100.000391	18.001772
4	50	0.212423	1.322091	20.013690	2.019347
4	75	0.212554	1.323894	30.009475	2.019265
4	100	0.212624	1.324799	40.007236	2.019219
4	125	0.212667	1.325343	50.005851	2.019188
4	150	0.212697	1.325706	60.004910	2.019167

due to increasing values of *Pr*, whereas skin friction factors show nominal enhancement in their magnitude with *Pr*.

Schmidt number is found to have a decreasing influence on concentration profiles (Figure 4.4). Physically, Schmidt number is classified as the ratio of kinematic viscosity to mass diffusivity; therefore, as *Sc* increases, mass diffusivity decreases, which is further responsible for the reduction in concentration profiles. Also, when fluids of higher Prandtl number are used, the system cooled up rapidly as the heat dissipation capacity increases (Figure 4.5).

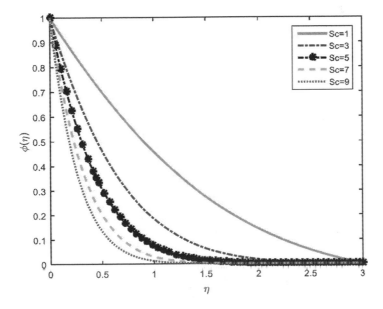

FIGURE 4.4
The effects of Schmidt number on concentration profile.

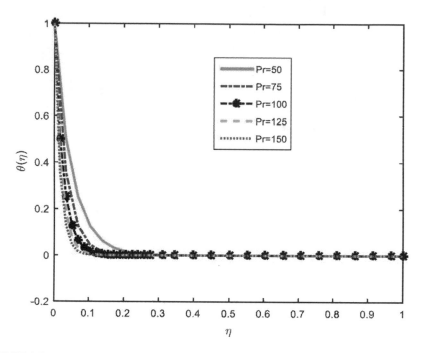

FIGURE 4.5
The effects of Prandtl number on temperature profile.

4.12 Applications of MNFs

The suspension of nanoparticles on conventional fluids sometimes enhanced the thermal conductivity of fluids even by 100 times than that of the carrier fluids. Many experimental investigations have been performed to demonstrate the enhancing and lowering of heat transfer behavior of MNF so that they can be utilized in improving the process performance of thermal devices. The study of the rheological behavior of magnetic nanosuspensions have numerous commercialized applications in engineering and applied sciences (physical and medical sciences), such as various industrial thermal devices, computer storage devices, magnetic sealing, energy conversion system, thermal power generating system, nuclear reactors, transportation, biomedicine, etc. These fluids are also used in lubrication process in dampers and bearings. In electronics, ferrofluids are greatly applied to several devices, for example, sensors, accelerometer, densimeters, pressure transducers, electromechanical, energy converters etc. One special application of MNFs is their use as magnetic ink for high-speed, inexpensive and silent printers. Due to the effect of the Curie's law, these fluids become less magnetic at high temperature, and the heated fluid is replaced from its position with the cold fluid, attracted by the externally applied magnetic field. This property of ferrofluids manifests in its heat reducing capabilities, which finds its heat controller applications in electric motors, high rotating surface devices, and hi-fi loudspeakers.

They are also found to be useful in the field of biomedicine due to magnetically targeted drug delivery (anticancer agents such as radionuclides, cancer-specific antibodies, genes, etc.) to a certain area of human body. In this process, the ferrofluids carry the drugs and travel to the selective area using the applied magnetic field. It has also been used in nanosurgery to separate a cancer tumor and in magnetic hyperthermia to generate heat by electromagnetic energy. Several ferrofluids are marketed as contrast agents for monitoring the activity of tissues in the brain and also for toxin removal process in cancer treatment. Magnetic fluids are also used for repairing detachment of the retinal in eye surgery and as a contrast medium for X-ray examinations.

References

1. Shin, S., & Cho, Y. I. (1996). Forced convection behavior of a dielectric fluid (FC-77) in a 2: 1 Rectangular duct. *International Communications in Heat and Mass Transfer, 23*(5), 731–744.
2. Rini, D. P., Chen, R. H., & Chow, L. C. (2002). Bubble behavior and nucleate boiling heat transfer in saturated FC-72 spray cooling. *Transactions-American Society of Mechanical Engineers Journal of Heat Transfer, 124*(1), 63–72.
3. Snyder, S. M., Cader, T., & Finlayson, B. A. (2003). Finite element model of magnetoconvection of a ferrofluid. *Journal of Magnetism and Magnetic Materials, 262*(2), 269–279.
4. Hong, C. Y., Jang, I. J., Horng, H. E., Hsu, C. J., Yao, Y. D., & Yang, H. C. (1997). Ordered structures in Fe_3O_4 kerosene-based ferrofluids. *Journal of Applied Physics, 81*(8), 4275–4277.
5. Weilepp, J., & Brand, H. R. (1996). Competition between the Bénard-Marangoni and the Rosensweig instability in magnetic fluids. *Journal de Physique II, 6*(3), 419–441.
6. Tangthieng, C., Finlayson, B. A., Maulbetsch, J., & Cader, T. (1999). Heat transfer enhancement in ferrofluids subjected to steady magnetic fields. *Journal of Magnetism and Magnetic Materials, 201*(1), 252–255.
7. Owen, J. M., & Roger, R. H. (1989). Flow and heat transfer in rotating-disc systems. Volume I-Rotor-stator systems. NASA STI/Recon Technical Report A, 90.
8. Kármán, T. V. (1921). Über laminare und turbulente Reibung. *ZAMM-Journal of Applied Mathematics and Mechanics/Zeitschrift für Angewandte Mathematik und Mechanik, 1*(4), 233–252.
9. Cochran, W. G. (1934, July). The flow due to a rotating disc. *Mathematical Proceedings of the Cambridge Philosophical Society,* 30, No. 3, 365–375. Cambridge University Press.
10. Benton, E. R. (1966). On the flow due to a rotating disk. *Journal of Fluid Mechanics, 24*(4), 781–800.

11. Mithal, K. G. (1961). On the effects of uniform high suction on the steady flow of a non-Newtonian liquid due to a rotating disk. *The Quarterly Journal of Mechanics and Applied Mathematics, 14*(4), 403–410.
12. Attia, H. A. (1998). Unsteady MHD flow near a rotating porous disk with uniform suction or injection. *Fluid Dynamics Research, 23*(5), 283.
13. Attia, H. A., & Aboul-Hassan, A. L. (2004). On hydromagnetic flow due to a rotating disk. *Applied Mathematical Modelling, 28*(12), 1007–1014.
14. Turkyilmazoglu, M. (2010). Analytic approximate solutions of rotating disk boundary layer flow subject to a uniform suction or injection. *International Journal of Mechanical Sciences, 52*(12), 1735–1744.
15. Ram, P., Bhandari, A., & Sharma, K. (2010). Effect of magnetic field-dependent viscosity on revolving ferrofluid. *Journal of Magnetism and Magnetic Materials, 322*(21), 3476–3480.
16. Bodewadt, V. U. (1940). Die drehströmung über festem grunde. *ZAMM-Journal of Applied Mathematics and Mechanics/Zeitschrift für Angewandte Mathematik und Mechanik, 20*(5), 241–253.
17. Gurnis, M., & Davies, G. F. (1986). The effect of depth-dependent viscosity on convective mixing in the mantle and the possible survival of primitive mantle. *Geophysical Research Letters, 13*(6), 541–544.
18. Stegman, D. R., Richards, M. A., & Baumgardner, J. R. (2002). Effects of depth-dependent viscosity and plate motions on maintaining a relatively uniform mid-ocean ridge basalt reservoir in whole mantle flow. *Journal of Geophysical Research: Solid Earth, 107*(B6), 1–10.
19. Cserepes, L. (1993). Effect of depth-dependent viscosity on the pattern of mantle convection. *Geophysical Research Letters, 20*(19), 2091–2094.
20. Roos, P. C., & Schuttelaars, H. M. (2013). Influence of time-and depth-dependent eddy viscosity on the formation of tidal sandwaves. VLIZ Special Publication.
21. Lai, F. C., & Kulacki, F. A. (1990). The effect of variable viscosity on convective heat transfer along a vertical surface in a saturated porous medium. *International Journal of Heat and Mass Transfer, 33*(5), 1028–1031.
22. Ling, J. X., & Dybbs, A. (1987). *Forced convection over a flat plate submersed in a porous medium: Variable viscosity case (No. CONF-871234-).* American Society of Mechanical Engineers, New York.
23. Torrance, K. E., & Turcotte, D. L. (1971). Thermal convection with large viscosity variations. *Journal of Fluid Mechanics, 47*(1), 113–125.
24. Korenaga, J., & Jordan, T. H. (2002). Onset of convection with temperature-and depth-dependent viscosity. *Geophysical Research Letters, 29*(19), 1–4.
25. Verma, P. D. S., & Vedan, M. J. (1978). Helical flow of ferrofluid with heat conduction. *Journal of Mathematical and Physical Sciences, 12*, 377–389.
26. Verma, P. D. S., & Vedan, M. J. (1979). Steady rotation of a sphere in a paramagnetic fluid. *Wear, 52*(2), 201–218.
27. Verma, P. D. S., & Singh, M. (1981). Magnetic fluid flow through porous annulus. *International Journal of Non-linear Mechanics, 16*(3–4), 371–378.

5

An Exposition on Mathematical Models Involving Various Types of Differential Equations

Jonnalagedda Vasundhara Devi
Gayatri Vidya Parishad College of Engineering (Autonomous)

Farzana A. McRae
Catholic University of America

Zahia Drici
Illinois Wesleyan University

CONTENTS

5.1 Introduction

Understanding physical phenomena and trying to predict or estimate their future behavior has been an ongoing endeavor. Many such phenomena can be studied using differential equations as mathematical models. Results relating to the existence of solutions of such equations are very important, for otherwise, one would be searching for a solution that does not exist. Knowing the existence of solution is one thing, but finding it is another issue. But if it is known that there is a unique solution, then further investigation can be considered. If it can be shown theoretically that the solutions continuously depend on the initial data and that they are differentiable with respect to initial values, then the qualitative behavior of the solutions can be found without having any prior information about solution. This pioneering work was done by Lyapunov, wherein he constructed a function whose behavior under certain conditions gave information regarding the stability of the solution. Thus, a systematic qualitative study of a nonlinear differential equation deals with the existence and uniqueness of solutions along with their continuous dependence on initial conditions, differentiability with respect to initial values, and stability.

Various physical phenomena demanded a better mathematical model than an ordinary differential equation (ODE). For example, the prey–predator model developed by Volterra evolved as an integrodifferential equation (InDE). In some other situations, it was observed that a time lag that needed to be considered and this gave rise to a delay differential equation (DDE). Similarly, modeling nondeterministic phenomena gave rise to a random differential equation (RDE). In many physical phenomena, there are disturbances, which are considered as impulses in time giving way to an impulsive differential equation (IDE). A fuzzy differential equation (FDE) is used to model propagation of uncertainty in dynamical systems. Multivalued functions gave rise to a set differential equation (SDE). Derivatives of arbitrary order have provided a framework to model problems in a variety of disciplines, giving rise to a fractional differential equation (FrDE). Differential equations with anticipation were introduced to model systems whose behavior at a given time is influenced by their anticipated future potential states. Systems with dependence on both past history and a desired future state gave rise to differential equation with retardation and anticipation (DE with R&A). Matrix differential equations (MDE) occur frequently in applications, such as mathematical physics, optimal control, and dynamic programming. On the other hand, graphs were used naturally to study organizational structures, and time-varying graphs are used to understand structures that vary with time in this setup. Once a derivative of a graph was introduced, a graph differential equation (GDE) could be envisaged. Owing to the isomorphism that exists between a graph and a matrix, GDE was studied alongside MDE.

The differential equations mentioned above are some of the models developed to study physical phenomena. These models have been extensively studied, and there is a vast volume of related information available. We give a brief description of the approaches used to study the initial value problem (IVP) of an ODE. We present results on the existence and uniqueness of solutions, and their continuous dependence on initial data and their differentiability with respect to initial values as well as stability.

Under a continuity condition, Peano's theorem guarantees the existence of a solution to an ODE, whereas Picard's theorem guarantees uniqueness under a Lipschitz condition [1]. More general uniqueness results can be found in Ref. [2]. Both existence and uniqueness are also obtained by using fixed point theorems. The more widely known theorems are Banach, Tychonoff, and Krasnoselskii fixed point theorems, the last one being a combination of the first two theorems [3]. All the aforementioned results work in the setup of integral equations, that is, the given ODE is transformed into an integral equation where the integral is treated as an operator to which the fixed point theory is applied. The two concepts of continuous dependence of solutions on initial data and differentiability with respect to initial values are necessary to develop the variation of parameters formula and are also useful in the study of stability and perturbation theories.

In 1892, Lyapunov established two theorems, now known as Lyapunov theorems, which predict uniform and uniform asymptotic stability of a system, without having any knowledge of the solution. He constructed a function called a Lyapunov function, which was essential in obtaining his results [4]. To accommodate different types of physical systems, new notions such as eventual stability and partial stability were introduced. In case of systems that undergo disturbances but still remain stable, for example, an aeroplane under turbulence, the idea of practical stability was introduced. Corresponding to various stability definitions, practical stability concepts were also introduced [5]. Subsequently, all these stability definitions were unified under one setup, giving rise to the concept of stability in terms of two measures [6].

After Professor V. Lakshmikantham introduced the idea of using differential inequalities and the technique of comparison, there has been a dynamic change in the study of qualitative theory [7]. Using the concept of lower and upper functions (solutions), the fundamental differential inequality result and the comparison theorem were developed. This approach was used to systematically develop the theory of differential and integral inequalities [1, 8]. These two volumes contain the foundational work on differential, delay differential, and integral equations.

Using the method of lower and upper solutions, two iterative methods the monotone iterative technique (MIT) [9] and the quasilinearization (QL) technique were developed [10]. Both these techniques are constructive in nature and guarantee the existence of solutions in a closed sector. The MIT gives the existence of extremal solutions in a closed sector with a linear rate of convergence, and the QL gives the existence and uniqueness of solutions in a closed sector with quadratic convergence.

The Lyapunov function has been exploited along with the comparison principle to study many other qualitative properties of the solutions, such as various types of boundedness, Lagrange stability, and nonuniform stability. Also, results dealing with Lyapunov-like functions and vector Lyapunov functions were established. Lyapunov-like functions together with the theory of differential and integral inequalities were used to further study nonlinear differential equations [11].

Lyapunov-like functions are used as a vehicle to transform a complicated differential system into a relatively simpler system. It is then sufficient to study the simpler system. Both Lyapunov technique and the nonlinear variation of parameters formula use the concept of norm to study perturbation theory, and as a consequence, any advantage the perturbation term offers in terms of stability is lost in the process. To address this problem, a new technique, called the variational Lyapunov method was developed. An example given in Ref. [12] showcases the advantage that this technique offers. The usual assumption in studying stability and other similar concepts is that two solutions start at the same time with different initial values. But in practical situations, this is not always the case, as the initial time may also differ. Taking this into account, differential equation having initial time difference was considered [13–15].

A periodic boundary value problem (PBVP) is another important problem associated with many mathematical models developed to study physical phenomena exhibiting systematic oscillatory behavior. The study of these problems is relatively difficult compared with that of IVPs of the corresponding mathematical models. Recently in Ref. [16], the authors used the MIT for IVPs to obtain a unique solution of the PBVP. This approach is very advantageous and has been developed for models involving ODEs, and graph and matrix differential equations. For details, see Refs. [17–19].

As stated earlier a systematic study involving the concepts mentioned above and many more notions, such as global solutions, has been done for most of the mathematical models described earlier. In the following sections, for each of these models, some of the results are presented. Since the differential inequality result is fundamental to the comparison principle, this result is given for every mathematical model.

5.2 Ordinary Differential Equations

We focus on the theory of nonlinear ODEs, as ODEs are extensively used to model physical phenomena. We present some basic existence and uniqueness results and proceed to introduce some iterative techniques used to find solutions of ODEs. The concepts of solutions depending continuously on initial values and differentiablitiy of solutions with reference to initial conditions are given, and results pertaining to Lyapunov stability and practical stability are also presented.

We begin with nonlinear IVP

$$x' = p(t, x)$$
$$x(t_0) = x_0,$$

(5.1)

where $x \in C^1 \left[\mathbb{R}, \mathbb{R}^n \right]$ and $p \in C \left[J \times \mathbb{R}^n, \mathbb{R}^n \right]$, $J = \left[t_0, t_0 + a \right)$

A widely used existence theorem is Peano's theorem, which gives existence of a solution of Eq. (5.1) in a small neighborhood of the initial value, (t_0, x_0) [1].

Theorem 5.2.1

Let $p \in C[R_0, \mathbb{R}^n]$, where $R_0 = \{(t,x) : |t - t_0| \le a$ and $\|x - x_0\| \le b, \|p(t,x)\| \le M\}$ on R_0. Then, Eq. (5.1) has at least one solution on $t_0 \le t \le t + \alpha, \alpha = \min\left(a, \dfrac{b}{M}\right)$.

A frequently used theorem, which not only gives existence but also uniqueness is Picard's Theorem [20] and, is given below.

Theorem 5.2.2

Let p in $C[R_0, \mathbb{R}^n]$, where $R = \{(t,x) : |t - t_0| \le a, \|x\| - x_0 \le b\}$ and satisfy

$$\|p(t, x_1) - p(t, x_2)\| \le L(\|x_1 - x_2\|, (t, x_1) \text{ and } (t, x_2)) \in R_0, L > 0.$$

Then, Eq. (5.1) has a unique solution on $(t_0 - h, t_0 + h) \subseteq [t_0 - a, t_0 + a], \quad h > 0$.

The two theorems above involve transforming the differential equation into an integral equation, whereas the introduction of the concept of differential inequalities makes it possible to work directly with the differential equation. A fundamental result on scalar differential inequalities is the following Ref. [1].

Theorem 5.2.3

Assume that

$$\rho' \le p(t, \rho),$$

$$\theta' > p(t, \theta),$$

with $p \in C[J \times \mathbb{R}^n, \mathbb{R}^n]$, and $\rho, \theta \in C^1[\mathbb{R}, \mathbb{R}^n]$. If $p(t_0) < \theta(t_0)$, then $\rho(t) < \theta(t)$, $t \in J$.

Using the concept of differential inequalities, the notions of upper and lower solutions and iterative techniques were introduced. The following concept is from Ref. [9].

Definition 5.2.4

A function $v \in C^1[J, \mathbb{R}]$ is said to be an upper solution of Eq. (5.1) if $v' \ge p(t, v)$, and a lower solution of Eq. (5.1) if the inequalities are reversed.

Then, using differential inequalities and the notion of upper and lower solutions, it was shown that the solution of the IVP (Eq. 5.1) is sandwiched between the lower and upper solutions. Thus, the existence of solutions is guaranteed by the lower and upper solutions, which also act as bounds for the solutions. In other words, the existence of solutions in a closed sector is obtained with this approach [9].

In both Peano's and Picard's theorems, the method of successive approximations is used to find the solution of an integral equation corresponding to the IVP (Eq. 5.1). But the sequences in these theorems are not necessarily monotone, and the rate of convergence is not known. Using the method of lower and upper solutions, a constructive method called the MIT was developed to obtain the solution of the IVP (Eq. 5.1). In this method, two monotone sequences of functions, say ρ_n and θ_n, which are solutions of linear differential equations obtained from the given problem, are constructed. These sequences converge to the minimal and maximal solutions of the IVP (Eq. 5.1) under certain conditions.

Definition 5.2.5

Let $\psi(t)$ be a solution of IVP of the scalar differential equation (Eq. 5.1) on $[t_0, t_0 + a)$. Then, $\psi(t)$ is said to be a maximal solution of Eq. (5.1) if, for every solution $x(t)$ of Eq. (5.1) existing on $[t_0, t_0 + a)$, the inequality

$$x(t) \le \psi(t), \quad t \in [t_0, t_0 + a)$$

holds. A minimal solution $\sigma(t)$ may be defined similarly by reversing the above inequality.

Next, a theorem involving MIT is as follows [9]:

Theorem 5.2.6

Let $p \in C[J \times \mathbb{R}, \mathbb{R}]$, ρ_0, θ_0 be lower and upper solutions of Eq. (5.1) such that $\rho_0 < \theta_0$ on J. Suppose further that $p(t, u_1) - p(t, u_2) \ge -M(u_1 - u_2)$ for $\rho_0 \le u_1 \le u_2 \le \theta_0$ and $M > 0$. Then, there exist monotone sequences $\{\rho_n\}, \{\theta_n\}$ such that $\rho_n \to \sigma$ and $\theta_n \to \psi$ as $n \to \infty$, uniformly and monotonically on J, and σ and ψ are minimal and maximal solutions of Eq. (5.1), respectively.

In the MIT, though the sequences converge monotonically, the rate of convergence is linear. The method of QL, initially developed by Bellman and Kalaba [9], guarantees quadratic convergence. In this method, assuming that $p(t, x)$ in Eq. (5.1) is convex, a sequence of iterates that converges quadratically to the solution of Eq. (5.1) is constructed. Starting in 1992, this method was generalized to apply to problems where $p(t, x)$ is not convex but satisfies a convexity-like condition, resulting in what is now called the generalized QL method (GQL) [21].

We present below a theorem using this method [22]. The sequence of iterates in this theorem are solutions of linear differential equations obtained from Eq. (5.1).

Theorem 5.2.7

Assume that

(A1) ρ_0 and $\theta_0 \in C^1[J, \mathbb{R}]$ are lower and upper solutions of Eq. (5.1) satisfying $\rho_0 \leq \theta_0$ on J;

(A2) $p \in C[\Omega, \mathbb{R}]$, $p_x(t, x)$, $p_{xx}(t, x)$ exist and are continuous, satisfying $p_{xx}(t, x) + \phi_{xx}(t, x) > 0$ on $\Omega = \{(t, x) : \rho(t) \leq x \leq \theta_0(t), t \in J\}$, where $\phi \in C[\Omega, \mathbb{R}]$ and $\phi_x(t, x), \phi_{xx}(t, x)$ exist, are continuous and $\phi_{xx}(t, x) > 0$ on Ω.

Then, there exist monotone sequences $\{\rho_n(t)\}, \{\theta_n(t)\}$ which converge uniformly to the unique solution of Eq. (5.1) and the convergence is quadratic.

An IVP is said to be well posed if it has a unique solution, and if the solutions are continuous and differentiable with respect to initial conditions, and satisfy condition (ii) in Theorem 5.2.9 (given below). The next two results deal with the continuous dependence and differentiability of a solution of Eq. (5.1) with respect to initial conditions Ref. [1].

Theorem 5.2.8

Let $p \in C[J \times \mathbb{R}^n, \mathbb{R}^n]$ and, for $(t, x), (t, y) \in J \times \mathbb{R}^n$,

$$\|p(t, x) - p(t, y)\| \leq g(t, \|x - y\|),$$

where $g \in C[J \times \mathbb{R}_+, \mathbb{R}_+]$. Assume that $u(t) \equiv 0$ is the unique solution of the differential equation (DE)

$$u' = g(t, u), \tag{5.2}$$

such that $u(t_0) = u_0$. Then, if the solutions $u(t, t_0, u_0)$ of DE (5.2) through every point (t_0, u_0) are continuous with respect to initial values (t_0, u_0), the solutions $x(t, t_0, x_0)$ of Eq. (5.1) are unique and continuous with respect to the initial values (t_0, x_0).

Theorem 5.2.9

Assume that $p \in C[J \times \mathbb{R}^n, \mathbb{R}^n]$ and possesses continuous partial derivatives $\partial p / \partial x$ on $J \times \mathbb{R}^n$. Let the solution $x(t, t_0, x_0)$ of Eq. (5.1) exist for $t \geq t_0$, and let

$$H(t,t_0,x_0) = \frac{\partial p(t, x(t,t_0,x_0))}{\partial x}.$$

Then,

(i) $\Phi(t,t_0,x_0) = \dfrac{\partial x(t, x(t,t_0,x_0))}{\partial x_0}$ exists and is the solution of the DE

$$y' = H(t,t_0,x_0)y, \tag{5.3}$$

such that $\Phi(t,t_0,x_0)$ is the identity matrix;

(ii) $\dfrac{\partial x(t,t_0,x_0)}{\partial t_0}$ exists, is the solution of DE (5.3), and satisfies the relation

$$\frac{\partial x(t, x(t,t_0,x_0))}{\partial t_0} = -\Phi(t,t_0,x_0)\cdot p(t_0,x_0), \quad t \geq t_0. \tag{5.4}$$

Next, we present some results pertaining to the stability of the solution of a well-posed problem from Ref. [10].

Let $x(t,t_0,x_0)$ be any solution of the differential system

$$x' = p(t,x) \quad x(t_0) = x_0, \quad t_0 \geq 0 \tag{5.5}$$

where $p \in C[J \times S_\rho, \mathbb{R}^n]$, and $S_\rho = [x \in \mathbb{R}^n : \|x\| < \rho]$.

Set $p(t,0) = 0, \quad t \in J$, so that zero solution of Eq. (5.5) through $(t_0,0)$ exists. We now list some stability concepts.

Definition 5.2.10

The trivial solution $x \equiv 0$ of Eq. (5.5) is

(S_1) equistable if "for each $\epsilon > 0, t_0 \in J$, there exists a positive function $\delta = \delta(t_0,\epsilon)$ that is continuous in t_0 for each ϵ such that the inequality

$$\|x_0\| \leq \delta$$

implies

$$\|x(t,t_0,x_0)\| < \epsilon, \quad t \geq t_0";$$

(S_2) quasi-equiasymptotically stable if "for each $\epsilon > 0, t_0 \in J$, there exist positive numbers $\delta_0 = \delta_0(t_0)$ and $T = T(t_0,\epsilon)$, such that, for $t \geq t_0 + T$ and $\|x_0\| < \delta_0$,

$$\|x(t,t_0,x_0)\| < \epsilon";$$

(S_3) equiasymptotically stable if "(S_1) and (S_2) hold simultaneously."

It is convenient to introduce certain classes of monotone functions in order to characterize Lyapunov functions from Ref. [1].

Definition 5.2.11

A function $\phi(r)$ is said to belong to the class $\mathcal{K}$ if "$\phi \in C[[0,\rho), \mathbb{R}_+]$, $\phi(0) = 0$, and $\phi(r)$ is strictly monotone increasing in r."

Definition 5.2.12

A function $\sigma(t)$ is in $\mathcal{L}$ if "$\sigma \in C[J, \mathbb{R}_+]$, $\sigma(t)$ is strictly decreasing in t, and $\sigma(t) \to 0$ as $t \to \infty$."

Definition 5.2.13

A function $\phi(t,r)$ is in class $\mathcal{KK}$ if "$\phi \in C[J \times [0,\rho), \mathbb{R}_+]$, $\phi \in \mathcal{K}$ for each $t \in J$, and ϕ is monotone increasing in t for each $r > 0$ and $\phi(t,r) \to \infty$ as $t \to \infty$ for each $r > 0$."

Definition 5.2.14

A function $V(t,x)$ with $V(t,0) = 0$ is said to be positive definite (negative definite) if "there exists a function $\phi(r) \in \mathcal{K}$ such that the relation

$$V(t,x) \ge \phi(\|x\|), \quad \left(\le -\phi(\|x\|)\right)$$

is satisfied for $(t,x) \in J \times S_\rho$."

Definition 5.2.15

A function $V(t,x) \ge 0$ is said to be decrescent if "a function $\phi(r) \in \mathcal{K}$ exists, such that

$$V(t,x) \le \phi(\|x\|), \quad (t,x) \in J \times S_\rho."$$

In order to use Lyapunov's second method along with the comparison principle to study the stability properties of Eq. (5.5), we need the scalar differential equation

$$u' = g(t,u), \quad u(t_0) = u_0 \ge 0, \quad t_0 \ge 0 \qquad (5.6)$$

where $g \in [J \times \mathbb{R}_+, \mathbb{R}]$. We suppose that $g(t,0) \equiv 0$ so that $u = 0$ is a solution of DE (5.6) through $(t_0, 0)$. Parallel to the definitions (S_1)–(S_3), we designate by (S_1^*)–(S_3^*) the stability definitions relative to the zero solution of DE (5.6).

Definition 5.2.16

The zero solution $u = 0$ of DE (5.6) is said to be $\left(S_1^*\right)$ equistable if "for each $\epsilon > 0, t_0 \in J$, there exists a positive function $\delta = \delta(t_0, \epsilon)$ that is continuous in t_0 for each ϵ, such that

$$u(t, t_0, u_0) < \epsilon, \quad t \geq t_0,$$

provided

$$u_0 \leq \delta".$$

The definitions $\left(S_2^*\right)$ and $\left(S_3^*\right)$ can be formulated similarly.

Below criteria for the asymptotic stability of the solution $x = 0$ of Eq. (5.5) is given from Ref. [1].

Theorem 5.2.17

Suppose $V(t, x)$ and $g(t, u)$ exist and satisfy the following assumptions:

(i) $g \in C[J \times \mathbb{R}_+, \mathbb{R}]$ and $g(t, 0) \equiv 0$.
(ii) $V \in C[J \times S_\rho, \mathbb{R}_+]$, $V(t, 0) \equiv 0$, and $V(t, x)$ is positive definite and locally Lipschitzian in x.
(iii) $D^+ V(t, x) \leq g(t, V(t, x)), (t, x) \in J \times S_\rho,$

where $D^+ V(t, x) = \lim_{h \in 0^+} \sup \dfrac{1}{h}\left[V(t + h, x + hf(t, x)) - V(t, x)\right]$

Then, the equiasymptotic stability of the solution $u = 0$ of DE (5.6) assures the equiasymptotic stability of the trivial solution of Eq. (5.5).

In the study of stability of nonlinear systems, an interesting set of problems deals with bringing states close to certain sets rather than to the state $x = 0$. The desired state of a system may be mathematically unstable, but the system may oscillate sufficiently near this state so that its performance is acceptable. Such is the case, for example, of the trajectory of some aircraft and missiles. Thus, a notion of stability that is neither weaker nor stronger than Lyapunov stability is desired. LaSalle and Lefschetz suggested a name for such a concept and called it practical stability [23].

Definition 5.2.18

The IVP (Eq. 5.1) is said to be practically stable if, given (λ, A) with $0 < \lambda < A$, we have $\|x_0\| < \lambda$ implies $\|x(t)\| < A, \quad t \geq t_0$.

Corresponding to Definition 5.2.18, we define practical stability notions for the scalar differential equation

$$u' = g(t, u), \quad u(t_0) \geq u_0, \tag{5.7}$$

where $g \in C[\mathbb{R}_+ \times \mathbb{R}_+, \mathbb{R}]$.

Definition 5.2.19

DE (5.7) is said to be practically stable if, given $0 < \lambda < A$, we have $u_0 < \lambda$ implies $u(t) < A, \quad t \geq t_0$ for some $t_0 \in \mathbb{R}_+$.

The following is a basic practical stability result.

Theorem 5.2.20

Assume that for $(t, x) \in \mathbb{R}_+ \times \mathbb{R}^n$, $g \in C\left[\mathbb{R}_+^2, \mathbb{R}\right]$

$$\left[x, p(t, x)\right]_+ \leq g\left(t, \|x\|\right)$$

where $[x, y]_+ = \lim_{h \to 0^+}\left(\frac{1}{h}\|x + hy\| - \|x\|\right), \quad x, y \in \mathbb{R}^n$.

Then, the practical stability properties of DE (5.7) imply the corresponding practical stability properties of Eq. (5.1).

There are many applications of ODEs, such as in analysis of electrical networks, in the study of motion of coupled systems of oscillating springs, in harmonic analysis, projectile motions, security and terrorism related models, and heat transfer, to name a few.

5.3 Integrodifferential Equations

Among many types of mathematical models developed for studying physical phenomena involving past history, InDEs play an important role. Abel, Lotka, Volterra, and a few others initiated the study of integral as well as InDEs. Beginning with the prey–predator model [24] studied by Volterra, InDEs have been used as mathematical models to study phenomena in various disciplines, such as elasticity, viscosity, epidemics, population dynamics, image processing, finance, neural networks, biomedicine, and many more.

Some important results pertaining to the qualitative study of nonlinear InDEs that were established by researchers over the years and were consolidated in Ref. [25] are given below.

Consider the IVP of InDE equation given by

$$
\left.
\begin{aligned}
x'(t) &= p\big(t, x(t)\big) + \int_{t_0}^{t} G\big(t, s, x(s)\big)\, ds \\[2mm]
x(t_0) &= x_0.
\end{aligned}
\right\}
\tag{5.8}
$$

Of the many types of existence results proved for IVP (Eq. 5.8), we present a result that uses Tychonoff's fixed point theorem and the concept of comparison. Using the solution of a scalar InDE, we obtain the existence of a solution of the IVP (Eq. 5.8). The scalar InDE is obtained by using estimates of $p(t, x)$ and $G(t, s, x)$.

Theorem 5.3.1

Assume that

(i) $p \in C\big[\mathbb{R}_+ \times \mathbb{R}^n, \mathbb{R}^n\big]$, $f \in C\big[\mathbb{R}_+{}^2, \mathbb{R}_+\big]$, $f(t, u)$ is monotone nondecreasing in u for each $t \in \mathbb{R}$ and

$$
\big|p(t, x)\big| \le f\big(t, |x|\big), \quad (t, x) \in \mathbb{R}_+ \times \mathbb{R}^n;
$$

(ii) $G \in C\big[\mathbb{R}_+{}^2 \times \mathbb{R}^n, \mathbb{R}^n\big]$, $H \in C\big[\mathbb{R}_+{}^3, \mathbb{R}_+\big]$, $H(t, s, u)$ is monotone nondecreasing in u for each $(t, s) \in \mathbb{R}_+{}^2$ and

$$
\big|G(t, s, x)\big| \le H\big(t, s, |x|\big), \quad (t, s, x) \in \mathbb{R}_+{}^2 \times \mathbb{R}^n;
$$

(iii) for every $w_0 > 0$, the scalar integrodifferential IVP

$$
w'(t) = f\big(t, w(t)\big) + \int_{t_0}^{t} H\big(t, s, w(s)\big)\, ds, \quad w(t_0) = w_0
\tag{5.9}
$$

has a solution $w(t)$ existing for $t \ge t_0$;

(iv) $\displaystyle \int_{s}^{t} \big|G(\sigma, s, x(s))\big|\, d\sigma \le N$ for $t, s \in \mathbb{R}_+, x \in C\big[\mathbb{R}_+, \mathbb{R}^n\big]$.

Then, for every $x_0 \in \mathbb{R}^n$ such that $|x_0| \le w_0$, there exists a solution $x(t)$ of IVP (Eq. 5.8) for $t \ge t_0$ satisfying $|x(t)| \le w(t)$, $t \ge t_0$.

Next, parallel to the setup of differential inequalities, many types of integrodifferential inequalities have been developed. We give the following implicit integrodifferential inequality result, where the integral is introduced as an operator S satisfying certain properties.

Theorem 5.3.2

Assume that

(i) $P \in C[\mathbb{R}_+ \times \mathbb{R}^3, \mathbb{R}]$ and $P(t,x,y,z)$ is nondecreasing in x for each (t,y,z) and nonincreasing in z for each (t,x,y);

(ii) S maps $C[\mathbb{R}_+, \mathbb{R}]$ into $C[\mathbb{R}_+, \mathbb{R}]$ and for $u_1, u_2 \in C[\mathbb{R}_+, \mathbb{R}]$, the inequality $u_1(t) \le u_2(t)$, $t_0 \le t \le t_1, t_0 \ge 0$ implies $Su_1 \le Su_2$ at $t = t_1$;

(iii) $v, w \in C^1[\mathbb{R}_+, \mathbb{R}]$ and

$$P(t, v', v, Sv) < 0, P(t, w', w, Sw) \ge 0, \quad t \ge t_0.$$

Then, $v(t_0) < w(t_0)$ implies $v(t) < w(t)$, $t \ge t_0$.

We next need the following definition.

Definition 5.3.3

The maximal solution $r(t)$ and the minimal solution $\gamma(t)$ of an IVP of the InDE are solutions of Eq. (5.8) and satisfy the relations $\gamma(t) \le x(t) \le r(t)$, $t \in \mathbb{R}_+$, where $x(t)$ is any solution of Eq. (5.8).

The existence of maximal and minimal solutions of the scalar integrodifferential IVP is required when using the concept of comparison.

Theorem 5.3.4

Assume that
$f \in C[J \times \mathbb{R}, \mathbb{R}]$, $H \in C[J \times J \times \mathbb{R}, \mathbb{R}]$, $H(t, s, w)$ is nondecreasing in w for each

(t, s) and $\int_s^t |H(\sigma, s, w(s))| d\sigma \le N$ for $t_0 \le s \le t \le t_0 + a$, $w \in \Omega_0$, where $J = [t_0, t_0 + a]$

and $\Omega_0 = \{w \in C[J, \mathbb{R}] : |w(t) - w_0| \le b\}$. Then, there exist maximal and minimal solutions for the scalar IVP (Eq. 5.9) on $(t_0, t_0 + \alpha]$ for some $0 < \alpha < a$.

As integrodifferential inequalities play an important role in the study of InDEs, it is useful to estimate a function, say $m(t)$, satisfying an integrodifferential inequality by the maximal solution, say $r(t)$, of the corresponding InDE. The following is a result in this direction.

Theorem 5.3.5

Assume that $f \in C[\mathbb{R}_+{}^2, \mathbb{R}]$, $H \in C[\mathbb{R}^3, \mathbb{R}]$, $H(t, s, w)$ is nondecreasing in w for each (t, s) and for $t \ge t_0$,

$$D_- m(t) \le f\left(t, m(t)\right) + \int_{t_0}^{t} H\left(t, s, m(s)\right) ds,$$

where $m \in C[\mathbb{R}_+, \mathbb{R}]$ and $D_- m(t) = \liminf_{h \to 0^-} h^{-1}[m(t+h) - m(t)]$. Suppose that $r(t)$ is the maximal solution of InDE Eq. (5.9) existing on $[t_0, \infty)$, then,

$$m(t) \le r(t), \quad t \ge t_0,$$

provided $m(t_0) \le u_0$.

In the above theorem the comparison equation is a scalar InDE and that finding solutions of even a simple InDE is difficult compared with an ODE. Thus, having the comparison equation as an ODE would be more fruitful.

To establish a comparison theorem of the aforementioned type requires the following lemma [25].

Lemma 5.3.6

Let $f_0, f \in C[\mathbb{R}_+{}^2, \mathbb{R}]$ satisfy

$$f_0(t, u) \le f(t, u), \quad (t, u) \in \mathbb{R}_+{}^2.$$

Then, the right maximal solution $r(t, t_0, w_0)$ of

$$w' = f(t, w), \quad w(t_0) = w_0 \ge 0 \tag{5.10}$$

and the left maximal solution $\eta(t, T, v_0)$ of

$$u' = f_0(t, u), \quad u(T) = v_0 \ge 0 \tag{5.11}$$

satisfy the relation

$$r(t, t_0, w_0) \le \eta(t, T, v_0), \quad t \in [t_0, T],$$

whenever $r(T, t_0, w_0) \le v_0$.

The above lemma has some interesting features. It introduces the concept of left and right maximal solutions. The left maximal solution has its initial value at the point (T, v_0). Note that $[t_0, T]$ is the time interval considered. On the other hand, the right maximal solution is the usual one starting at (t_0, u_0).

A comparison theorem involving the maximal solution of an ODE to estimate a function satisfying an integrodifferential inequality is as follows.

Theorem 5.3.7

Let $m \in C[\mathbb{R}_+, \mathbb{R}_+]$, $f \in C\left[\mathbb{R}_+^2, \mathbb{R}\right]$, $H \in C\left[\mathbb{R}_+^3, \mathbb{R}\right]$ and

$$D_- m(t) \leq f(t, m(t)) + \int_{t_0}^t H(t, s, m(s)) ds, \quad t \in I_0,$$

where $I_0 = \{t \geq t_0 : m(s) \leq \eta(s, t, m(t)), t_0 \leq s \leq t\}$, $\eta(t, T, v_0)$ is the left maximal solution of DE (Eq. 5.11) existing on $[t_0, T]$. Assume that

$$f_0(t, u) \leq F(t, u; t_0),$$

where

$$F(t, u; t_0) = f(t, u) + \int_{t_0}^t H(t, s, u) ds$$

and $r(t)$ is the maximal solution of

$$u' = F(t, u; t_0), u(t_0) = u_0$$

existing on $[t_0, \infty)$. Then,

$$m(t_0) \leq u_0 \text{ implies } m(t) \leq r(t), \quad t \geq t_0.$$

Observe that in the above theorem though the left maximal solution exists on $[t_0, T]$, the right maximal solution exists on $[t_0, \infty)$ and the result holds on $[t_0, \infty)$.

Once the notion of inequalities is in place, the lower and upper solutions of the InDE IVP (Eq. 5.8) are presented. If the lower and upper solutions of Eq. (5.8) are known, then the existence of a solution in a closed sector $\Omega = \{(t, x): \rho(t) \leq x \leq \theta(t), \quad t \in [t_0, T]\}$ can be guaranteed.

We present here MIT given in Refs. [9,25] developed for a PBVP of a nonlinear InDE given by

$$u'(t) = p(t, u(t), (Su)(t)), \quad u(0) = u(2\pi), \tag{5.12}$$

where $p \in C[[0, 2\pi] \times \mathbb{R} \times \mathbb{R}, \mathbb{R}]$, $Su(t) = \int_{t_0}^t G(t, s) u(s) ds$ and $G \in C[[0, 2\pi] \times [0, 2\pi], \mathbb{R}_+]$.

Lemma 5.3.8

Let $m \in C^1[[0, 2\pi], \mathbb{R}]$ be such that

$$m' \leq -Mm - NSm, \quad m(0) \leq m(2\pi), \tag{5.13}$$

where $M > 0$, $N \geq 0$. Then, $m(t) \leq 0$, for $0 \leq t \leq 2\pi$, provided one of the following conditions hold:

(a) $2Nk_0\pi\left(e^{2M\pi} - 1\right) \leq M$;
   or
(b) $2\pi[M + 2\pi Nk_0] \leq 1$, where $0 \leq k_0 = \max G(t,s)$ for $(t,s) \in [0, 2\pi] \times [0, 2\pi]$ and $G(t,s) \geq 0$.

The following result is the MIT developed for PBVP (Eq. 5.12).

Theorem 5.3.9

Assume that $\rho, \theta \in C^1[[0, 2\pi], \mathbb{R}]$ are lower and upper solutions relative to PBVP (Eq. 5.12) and ρ, θ are such that
   (H_0) $\rho, \theta \in C^1[[0, 2\pi], \mathbb{R}]$, such that $\rho(t) \leq \theta(t)$,

$$\rho' \leq p(t, \rho, S\rho), \quad \rho(0) \leq \rho(2\pi),$$

and

$$\theta' \geq p(t, \theta, S\theta), \quad \theta(0) \geq \theta(2\pi);$$

(H_1) whenever $\rho(t) \leq \bar{u} \leq u \leq \theta(t)$ and $\rho(t) \leq \bar{\phi}(t) \leq \phi(t) \leq \theta(t)$, $p(t, u, S\phi) - p(t, \bar{u}, S\bar{\phi}) \geq -M(u - \bar{u}) - NS(\phi - \bar{\phi})$, $t \in [0, 2\pi]$, where M and N are positive constants satisfying $2Nk_0\pi e^{2M\pi} < M$, $k_0 = \max G(t,s)$, on $[0, 2\pi] \times [0, 2\pi]$.
   Then, there exist monotone sequences $\{\rho_n(t)\}, \{\theta_n(t)\}$ with $\rho_0 = \rho$, $\theta_0 = \theta$ such that $\lim_{n \to \infty} \rho_n(t) = \gamma(t)$ and $\lim_{n \to \infty} \theta_n(t) = r(t)$ uniformly on $[0, 2\pi]$, and γ and r are minimal and maximal solutions of PBVP (Eq. 5.12), respectively, satisfying $\rho(t) \leq \gamma(t) \leq r(t) \leq \theta(t)$, on $[0, 2\pi]$.

   In order to study perturbation theory and stability theory, the following results are essential. The notion of continuous dependence of solutions on initial values is given in Theorem 1.6.1, differentiability of solutions with respect to initial values in Theorem 1.8.2, and nonlinear variation of parameters in Theorem 1.8.3, all in Ref. [25].
   We next proceed to give some stability results. The definitions of various types of stability run parallel to the definitions given in Section 5.2. Consider the InDE IVP

$$x'(t) = p\big(t, x(t), (Sx)(t)\big), t \geq t_0, x(t_0) = x_0, \tag{5.14}$$

where $(Sx)(t) = \int_{t_0}^{t} G(t, s, x(s))ds$, $p \in C[\mathbb{R}_+ \times B(\epsilon) \times \mathbb{R}^n, \mathbb{R}^n]$ and $G \in C[\mathbb{R}_+ \times \mathbb{R}_+ \times S(\rho), \mathbb{R}_+]$, and $B(\epsilon) = \{x : |x| < \epsilon\}$, where this system admits the zero solution if $p(t, 0, 0) \equiv 0$ and $G(t, s, 0) \equiv 0$.
   The following theorem deals with various types of stability concepts.

Theorem 5.3.10

Assume that

(i) $V \in C[\mathbb{R}_+ \times S(\rho), \mathbb{R}_+]$, $V(t, x)$ is locally Lipschitzian in x, and $b(\|x\|) \leq V(t, x) \leq a(\|x\|)$, $a, b \in \mathcal{K}(t, x) \in \mathbb{R}_+ \times S(\rho)$;

(ii) w_0 and $w \in C[\mathbb{R}_+ \times \mathbb{R}_+, \mathbb{R}]$, $w_0(t, u) \leq w(t, u)$, $\eta\left(t, t^0, v_0\right)$ is the left maximal solution of

$$v' = w_0(t, v), \quad v\left(t^0\right) = v_0 \geq 0,$$

existing on $t_0 \leq t \leq t^0$, and $r(t, t_0, u_0)$ is the right maximal solution of

$$u' = w(t, u), \quad u(t_0) = u_0 \tag{5.15}$$

existing on $[t_0, \infty)$;

(iii) $D\_V(t, x(t)) \leq w(t, V(t, x(t)))$ on Ω, where $\Omega = \left\{x \in C[\mathbb{R}_+, \mathbb{R}^n]:\right.$ $\left. V(s, x(s)) \leq \eta(s, t, V(t, x(t))), \quad t_0 \leq s \leq t\right\}$.

Then, the stability properties of the zero solution of DE (Eq. 5.15) imply the corresponding stability properties of the zero solution of DE (Eq. 5.14).

In order to prove the above theorem, it is enough to obtain the estimate

$$V(t, x(t, t_0, x_0)) \leq r(t, t_0, V(t_0, x_0)), \quad t > t_0,$$

where $x(t) = x(t, t_0, x_0)$ is any solution of Eq. (5.14) and $r(t, t_0, u_0)$ is the maximal solution of the DE (Eq. 5.15). Then, using the condition (i) and the stability properties of the zero solution of the DE (Eq. 5.15), one can prove the corresponding stability properties of the zero solution IVP of InDE (Eq. 5.14) by the standard arguments.

In Ref. [25], the motion of an unbounded one-dimensional nonlinear viscoelastic body is modeled by an InDE. Further, the presence of InDE is showcased in population models.

5.4 Delay Differential Equations

DDEs appear in models of physical systems whose future states depend not only on the present states but also on their past history. In this section, we consider DDEs of the form

$$x'(t) = f(t, x_t) \tag{5.16}$$

where f is a suitable functional. Given any $\tau > 0$, let $x \in C\big[[-\tau,\infty),\mathbb{R}\big]$ for any $\tau > 0$ and let x_t be a translation of the restriction of x to the interval $[t-\tau,t]$. Then,

$$x_t(s) = x(t+s), \quad -\tau \le s \le 0.$$

The graph of x_t is the graph of x on $[t-\tau,t]$ shifted to the interval $[-\tau,0]$.
Let

$$C_\rho = \big\{\phi \in C\big[[-\tau,0],\mathbb{R}^n\big]: \phi_0 < \rho\big\}$$

where $\|\phi\|_0 = \max\limits_{\tau \le s \le 0}\|\phi(s)\|$, and $\rho > 0$.

A solution of the DDE (Eq. 5.16) is defined as follows [8].

Definition 5.4.1

A function $x(t_0,\phi_0)$ is said to be a solution of the DDE (Eq. 5.16) with the given initial function $\phi_0 \in C_\rho$ at $t = t_0 \ge 0$, if there exists a number $A > 0$ such that

(i) $x(t_0,\phi_0)$ is defined and continuous on $[t_0 - \tau, t_0 + A]$ and $x_t(t_0,\phi_0) \in C_\rho$, for $t_0 \le t \le t_0 + A$;
(ii) $x_{t_0}(t_0,\phi_0) = \phi_0$;
(iii) The derivative of $x(t_0,\phi_0)$ at t, $x'(t_0,\phi_0)(t)$ exists for $t \in [t_0,t_0 + A)$ and satisfies (Eq. 5.16) for $t \in [t_0,t_0 + A)$.

The following results are from Ref. [8]. We begin with a local existence theorem that is proved using Schauder's fixed point theorem.

Theorem 5.4.2

Let $f \in C\big[I \times C_\rho, \mathbb{R}^n\big]$ where $I = [t_0,t_0 + a]$. Then, given an initial function $\phi_0 \in C_\rho$ at $t = t_0 \ge 0$, there exists an $\alpha > 0$ such that there is a solution $x(t_0,\phi_0)$ of Eq. (5.16) on $[t_0 - \tau, t_0 + \alpha)$.

The following lemma is needed to prove the next theorem.

Lemma 5.4.3

Let $q \in C\big[[t_0 - \tau,\infty),\mathbb{R}_+\big]$ satisfy

$$D_-q(t) \le g\big(t,\|q_t\|_0\big), \quad t > t_0,$$

where $q \in C[I \times \mathbb{R}_+, \mathbb{R}_+]$ and $D_-q(t) = \liminf\limits_{h\to 0^-}\big[q(t+h)-q(t)\big]h^{-1}$. If $r(t)$ is the maximal solution of the following IVP

$$u' = g(t, u), \quad u(t_0) = u_0 \geq 0, \tag{5.17}$$

existing for $t \geq t_0$. Then, $q(t) \leq r(t)$, $t \geq t_0$, provided $\|q_{t_0}\|_0 \leq u_0$.

We now state a global existence theorem.

Theorem 5.4.4

Let $f \in C[I \times C^n, \mathbb{R}^n]$, and for $(t, \phi) \in I \times C^n$,

$$\|f(t, \phi)\| \leq g\left(t, \|\phi\|_0\right),$$

where $g \in C[I \times \mathbb{R}_+, \mathbb{R}_+]$ and is nondecreasing in u for each $t \in I$, and $C^n = C[[-\tau, 0], \mathbb{R}^n]$. Assume that the solutions $u(t) = u(t, t_0, u_0)$ of the DE (Eq. 5.17) exist for all $t \geq t_0$. Then, the largest interval of existence of any solution $x(t_0, \phi_0)$ of the DDE (Eq. 5.16) is $[t_0, \infty)$.

Next, we consider continuous dependence of solutions on initial conditions.

Lemma 5.4.5

Let $f \in C[I \times C_\rho, \mathbb{R}^n]$ and, for $t \in I$, $\phi \in C_\rho$, let

$$G(t, r) = \max_{\|\phi\|_0 < r} \|f(t, \phi)\|.$$

Assume $r^*(t, t_0, 0)$ is the maximal solution of

$$u' = G(t, u)$$

through $(t_0, 0)$. Then, if $x(t_0, \phi_0)$ is any solution of the DDE (Eq. 5.16) with the initial function ϕ_0 at $t = t_0$, we have

$$\|x_t(t_0, \phi_0) - \phi_0\|_0 \leq r^*(t, t_0, 0),$$

on the common interval of existence of $x(t_0, \phi_0)$ and $r^*(t, t_0, 0)$.

Theorem 5.4.6

Let $f \in C[I \times C_\rho, \mathbb{R}^n]$ and, for $t \in I$, $\phi, \psi \in C_1$,

$$\|f(t, \phi) - f(t, \psi)\| \leq g\left(t, \|\phi(0) - \psi(0)\|\right),$$

where $g \in C[I \times [0, 2\rho), \mathbb{R}_+]$. Assume that $u(t) \equiv 0$ is the only solution of the scalar differential equation (5.17) through $(t_0, 0)$. I if the solutions $u(t, t_0, u_0)$

of DE (Eq. 5.17) through every point (t_0, u_0) exist for $t \geq t_0$ and are continuous with respect to the initial values (t_0, u_0), then the solutions $x(t_0, \phi_0)$ of the DDE (Eq. 5.16) are unique and continuous with respect to the initial values (t_0, ϕ_0).

Next, two results pertaining to the stability of the zero solution of Eq. (5.16) are given. Set $f(t, 0) \equiv 0$ so that the zero solution exists. Also, let the solution $x(t_0, \phi_0)$ of DDE (Eq. 5.16) exist.

First, we state the definition of stability and asymptotic stability.

Definition 5.4.7

The zero solution of DDE (Eq. 5.16) is said to be stable if, for any $\epsilon > 0$ and $t_0 \in I$, there exists $\delta > 0$ such that $\|\phi_0\|_0 \leq \delta$ implies $\|x_t(t_0, \phi_0)\|_0 < \epsilon$, $t \geq t_0$.

Definition 5.4.8

The trivial solution of DDE (Eq. 5.16) is said to be asymptotically stable if it is stable, and in addition, for any $\epsilon > 0$, $t_0 \in I$, there exist positive numbers δ_0 and T such that $\|\phi_0\|_0 \leq \delta_0$ implies $\|x_t(t_0, \phi_0)\|_0 < \epsilon$, $t \geq t_0 + T$.

Next, criteria for stability and asymptotic stability of the trivial solution of DDE (Eq. 5.16) are given in the following theorems.

Theorem 5.4.9

Let $f \in C[I \times C_\rho, \mathbb{R}^n]$, $g \in C[I \times [0, \rho), \mathbb{R}_+]$, $g(t, 0) \equiv 0$, and   for $t \in I$, $\phi \in C_\rho$ such that

$$\|\phi_0\|_0 = \|\phi_0(0)\|,$$

the following inequality holds

$$\|f(t, \phi)\| \leq f(t, \|\phi(0)\|).$$

Then, if the trivial solution of the scalar differential equation (5.17) is stable, the trivial solution of DDE (Eq. 5.16) is stable.

Theorem 5.4.10

Let $f \in C[I \times C_\rho, \mathbb{R}^n]$, $g \in C[I \times [0, \rho), \mathbb{R}_+]$ and $g(t, 0) \equiv 0$.
   Assume that

$$A(t) \liminf_{h \to 0^-} h^{-1}\left[\|\phi(0) + hf(t, \phi)\| - \|\phi(0)\|\right] + \|\phi(0)\| D_- A(t) \leq g(t, \|\phi(0)\| A(t)),$$

for $t > t_0$ and $\phi \in C_\rho$ satisfying

$$\|\phi\|_0 |A_t|_0 = \|\phi(0)\| A(t),$$

where $A(t) \geq 1$ is continuous on $[t_0 - \tau, \infty)$ and $A(t) \to \infty$ as $t \to \infty$. Then, the stability of the trivial solution of the DE (Eq. 5.17) implies the asymptotic stability of the trivial solution of DDE (Eq. 5.16).

We now present a uniqueness result obtained using the method of GQL [21]. First, we state a lemma needed to prove the uniqueness result.

Lemma 5.4.11

Suppose that

(i) $u, v \in C[\tilde{I}, \mathbb{R}] \cap C^1[I, \mathbb{R}]$, $f \in C[I \times C, \mathbb{R}]$, $\tilde{I} = [-\tau, T]$ and

$$v' \geq f(t, v_t), \quad u' \leq f(t, u_t), \quad t \in I;$$

(ii) $f(t, \phi) - f(t, \psi) \leq L \int_\tau^0 [\phi(s) - \psi(s)] ds, \quad t \in I \quad$ whenever $\quad \phi(s) \geq \psi(s),$

$-\tau \leq s \leq 0$, $L > 0$ is a constant, and $f(t, \phi)$ is quasinon decreasing in ϕ for each $t \in I$, that is, whenever $\psi(s) \leq \phi(s)$, $-\tau \leq s < 0$, and $\psi(0) = \phi(0)$, $f(t, \psi) \leq f(t, \phi)$, $t \in I$.

Then, $u(s) \leq v(s)$, $-\tau \leq s \leq 0$ implies $u(t) \leq v(t)$, $t \in I$, provided that $L + e^{-Lr} > 1$.

We list the following hypotheses:
(H$_1$) $f \in C[I \times C, \mathbb{R}]$;
(H$_2$) $u_o, v_0 \in C[I, \mathbb{R}] \cap C^1[I, \mathbb{R}]$ and $v_0'(t) \geq f(t, v_{0t})$, $u_0' \leq f(t, v_{0t})$, $t \in I$, $u_0(s) \leq v_0(s)$, $\tau \leq s \leq 0$;
(H$_3$) The Frechet derivative $f_\phi(t, \phi)$ exists, and is a continuous linear operator satisfying

(i) $f_\phi(t, \phi)\psi \leq L \int_{-\tau}^0 \psi(s) ds$, $\phi, \psi \in C$, $u_{0t} \leq v_{0t}, t \in J$;
(ii) $f(t, \phi) \geq f(t, \psi) + f_\phi(t, \psi)(\phi - \psi)$ whenever $u_{0t} \leq \psi \leq \phi \leq v_{0t}$, $t \in I$; and
(iii) $\psi_1(s) \leq \psi_2(s), -\tau \leq s \leq 0, \psi_1(0) = \psi_2(0)$ implies that $f_\phi(t, \psi)\psi_1 \leq f_\phi(t, \psi)\psi_2$, $\psi_1, \psi_2 \in C, t \in I$ and $u_{0t} \leq \psi_1 \leq \psi_2 \leq v_{0t}$.

(H$_4$) $\|f_\phi(t, \psi_1) - f_\phi(t, \psi_2)\| \leq L_2 \|\psi_1 - \psi_2\|_0^\beta$ where $t \in I$, $\psi_1, \psi_2 \in C_p$, $L_2 > 0$, $\|\phi\|_0 = \max_{-\tau \leq s \leq 0} |\phi(s)|$ and $0 \leq \beta \leq 1$.

An existence result is as follows.

Theorem 5.4.12

Let the hypotheses (H$_1$) to (H$_4$) hold. Then, there exist monotone sequences $\{u_n(t)\}, \{v_n(t)\}$ which converge uniformly to the unique solution of DDE (Eq. 5.16) on I and the convergence is superlinear.

We conclude this section by noting that DDEs are useful as mathematical models for studying problems in the life sciences, finance, and in engineering. For example, in Ref. [26], a selection dynamic model with time delay is proposed to quantitatively study cancer stem cells. Additional examples of applications in biological systems can be found in the special issue of the journal *Complexity* [27] devoted to such applications. Time-delay systems also appear in engineering. For example, when feedback controls are introduced to stabilize a system, some inevitable delays are introduced [28]. In finance, differential equations with time delay have been used to account for delayed responses in security markets [29].

5.5 Stochastic and Random Differential Equations

The mathematical modeling of several real-world problem leads to differential systems that involve some inherent randomness that arise in a system due to various types of unforeseen external factors. The study of nonlinear random differential systems is a very important area in modern applied mathematics.

We give in this section some results pertaining to the fundamental concepts of RDEs and inequalities in the framework of sample calculus. Random differential inequalities and Lyapunov functions are crucial to the study of a variety of qualitative aspects of solutions, including stability of RDEs, similar to the role that differential inequalities play in the study of deterministic differential equations.

We start by defining a sample solution of an RDE and then give a basic existence result.

Let $\Omega \equiv (\Omega, A, P)$ be a complete metric space. Set $B(z, \rho) = \{x \in \mathbb{R} : |z - x| < \rho,$ for a fixed $z \in \mathbb{R}\}$.

For $\rho > 0$, let $\bar{B}(z, \rho)$ be the closure of $B(z, \rho)$.

We provide below the classes of functions that are required for our study.

$C[\mathbb{R}_+ \times B(z, \rho), R[\Omega, \mathbb{R}]]$ is the class of sample continuous $\mathbb{R}$-valued random functions $f(t, x)$ whose realizations are denoted by $f(t, x, \omega)$.

$M[\mathbb{R}_+ \times B(z, \rho), R[\Omega, \mathbb{R}]]$ is the class of $\mathbb{R}$-valued random functions $f(t, x)$ such that $f(t, x(t))$ is product-measurable whenever $x(t)$ is product-measurable.

$IB[I, R[\Omega, \mathbb{R}_+]]$ is the class of $K \in M[I, R[\Omega, \mathbb{R}_+]]$ whose sample Lebesgue integral is bounded with probability 1 (w.p.1).

Consider the first-order IVP of an RDE

$$x' = f(t, x, \omega)$$

$$x(t_0, \omega) = x_0(\omega) \tag{5.18}$$

where $f \in M[\mathbb{R}_+ \times B(z \cdot \rho), R[\Omega, \mathbb{R}]]$.

The results in this section are from Ref. [30, 31].

Definition 5.5.1

A random process $x(t)$ is said to be a sample solution process or a sample solution of the IVP (Eq. 5.18) on an interval $J = [t_0, t_0 + a]$ if it satisfies the following conditions:

(i) $x(t_0) = x_0$,
(ii) $x(t)$ is sample continuous,
(iii) $x(t)$ is product measurable,
(iv) $x'(t, \omega) = f(t, x(t, \omega), \omega)$ w.p.1 for almost every $t \in J$.

The next result gives the basic existence theorem of Carathéodory-type for IVP (Eq. 5.18).

Theorem 5.5.2

Assume that

(i) $f \in M[J \times \bar{B}(z, \rho), R[\Omega, \mathbb{R}]]$ and $f(t, x, \omega)$ is sample continuous for x for each $t \in J$;
(ii) $K \in IB[J, R[\Omega, \mathbb{R}_+]]$ and satisfies

$$\|f(t, x, \omega)\| \le K(t, \omega) \text{ for } (t, x) \in J \times \bar{B}(z, \rho);$$

(iii) $x_0 \in \bar{B}(z, \rho/2)$, ρ being a positive number. Then, the IVP (Eq. 5.18) has at least one solution

$$r(t) = r(t, t_0, r_0) \text{ on } [t_0, t_0 + h] \text{ for some } h > 0.$$

Next, we give a fundamental result concerning random differential inequalities. This is followed by a result pertaining to the basic random comparison theorem, which is useful in estimating sample solutions.

Theorem 5.5.3

Assume that

(H_1) $g \in M\big[[t_0, t_0 + a) \times D, R[\Omega, \mathbb{R}]\big]$, $g(t, u, \omega)$ is almost surely (a.s.) monotone nondecreasing in u for each t, D is an open set in $\mathbb{R}$, and $a > 0$;

(H_2) $u, v \in C\big[[t_0, t_0 + a), R[\Omega, \mathbb{R}]\big]$ for $(t, u(t, \omega)), (t, v(t, \omega)) \in [t_0, t_0 + a) \times D$,

$$D_- v(t, \omega) \le g\big(t, v(t, \omega), \omega\big),$$

and

$$D_- u(t, \omega) > g\big(t, u(t, \omega), \omega\big),$$

(H_3) $v(t_0, \omega) < u(t_0, \omega)$.

Then,

$$v(t, \omega) < u(t, \omega) \quad \text{for } t \in [t_0, t_0 + a).$$

Theorem 5.5.4

Assume that

(H_1) $g \in M[E, R[\Omega, \mathbb{R}]]$ and $g(t, u, \omega)$ is sample continuous in u for fixed t, where $E = [t_0, t_0 + a) \times D$ and D is an open set in $\mathbb{R}$;

(H_2) $g(t, u, \omega)$ is a.s. monotone nondecreasing in u for each fixed t;

(H_3) $r(t, \omega)$ is the sample maximal solution of the random differential system

$$u'(t, \omega) = g\big(t, u(t, \omega), \omega\big), \quad u(t_0, \omega) = u_0(\omega),$$

existing on $[t_0, t_0 + a)$;

(H_4) $m \in C\big[[t_0, t_0 + a), R[\Omega, \mathbb{R}]\big]$, $(t, m(t)) \in E$ w.p.1.

$$m(t_0, \omega) \le u_0(\omega) \le g\big(t, m(t, \omega), \omega\big) \text{ a.e. in } \quad t \in [t_0, t_0 + a).$$

Then,

$$m(t, \omega) \le r(t, \omega) \quad \text{for } t \in [t_0, t_0 + a).$$

We present below a uniqueness result.

Theorem 5.5.5

Assume that

(i) $f \in M\big[J \times \bar{B}(z, \rho), R[\Omega, \mathbb{R}]\big]$ and $f(t, x, \omega)$ is sample continuous in x for each $t \in J$;

(ii) $g \in M[J \times [0, 2\alpha]], R[\Omega, \mathbb{R}]$, $g(t, u, \omega)$ is sample continuous in u for each $t \in J$, and $g(t, u(t), \omega)$ is sample Lebesgue-integrable whenever $u(t)$ is sample absolutely continuous, where $\alpha = \rho$;

(iii) $u(t) \equiv 0$ is the unique solution w.p.1 of the scalar RDE

$$u'(t, \omega) = g(t, u(t, \omega), \omega), \quad u(t_0, \omega) = 0,$$

existing on $[t_0, t_0 + b) \subset J$;

(iv) for $(t, x), (t, y) \in J \times \bar{B}(z, \rho)$,

$$\|f(t, x, \omega) - f(t, y, \omega)\| \leq g(t, \|x - y\|, \omega);$$

(v) $x_0 \in \bar{B}\left(z, \dfrac{1}{2}\rho\right)$;

(vi) $K \in IB[J, R[\Omega, \mathbb{R}_+]]$ satisfies

$$\|f(t, x, \omega)\| \leq K(t, \omega) \quad \text{for } (t, x) \in J \times \bar{B}(z, \rho).$$

Then, the IVP (Eq. 5.18) has a unique solution on $[t_0, t_0 + b]$.

The last set of results in this section deal with stability of the solution of Eq. (5.18).

Let $x(t, \omega) = x(t, t_0, x_0, \omega)$ be a sample solution process of IVP (Eq. 5.18). Without loss of generality, we assume that $x(t, \omega) \equiv 0$ is the unique solution process of IVP (Eq. 5.18) through $(t_0, 0)$. Depending on the mode of convergence in probabilistic analysis, several stability notions, such as stability in probability, stability w.p.1, and stability in the pth mean, relative to the given solution of IVP (Eq. 5.18), can be formulated.

Definition 5.5.6

The trivial solution of IVP (Eq. 5.18) is said to be a "stable in probability" if "for each $\varepsilon > 0$, $\eta > 0$, $t_0 \in \mathbb{R}_+$, there exists a positive function $\delta = \delta(t_0, \varepsilon, \eta)$ that is continuous in t_0 for each ε and η such that the inequality

$$P[\omega : \|x_0(\omega)\| > \delta] < \eta$$

implies

$$P[\omega : \|x(t, \omega)\| \geq \varepsilon] < \eta, \quad t \geq t_0";$$

To study the stability properties of IVP (Eq. 5.18) directly using suitable functions, we need the corresponding random auxiliary comparison differential equation

$$u' = g(t, u, \omega), \quad u(t_0) = u_0(\omega), \tag{5.19}$$

where $g \in M[\mathbb{R}_+ \times \mathbb{R}, K[\Omega, \mathbb{R}]]$ is such that $g(t, u, \omega)$ satisfies the Carathéodory conditions in (t, u) w.p.1 and $g(t, u, \omega)$ is monotone nondecreasing in u for fixed t w.p.1. Suppose that $u(t) \equiv 0$ is the solution of DE (Eq. 5.19) through $(t_0, 0)$ w.p.1.

Corresponding to the above stability definition, we give the stability in probability concerning the stability of the equilibrium solution $u(t) \equiv 0$ of DE (Eq. 5.19) below.

Definition 5.5.7

The trivial solution $u(t) \equiv 0$ of DE (Eq. 5.19) is said to be "stable in probability" if "given $\varepsilon > 0, \eta > 0, t_0 \in \mathbb{R}_+$, there exists a positive function $\delta = \delta(t_0, \varepsilon, \eta)$ such that

$$P[\omega : u_0(\omega) > \delta] < \eta,$$

implies

$$P[\omega : u(t, \omega) \geq \varepsilon] < \eta, \quad t \geq t_0."$$

The next theorem uses the second method of Lyapunov to derive stability results for IVP (Eq. 5.18).

Theorem 5.5.8

Assume that

(i) $g \in M[\mathbb{R}_+ \times \mathbb{R}, R[\Omega, \mathbb{R}]]$ and $g(t, u, \omega)$ is sample continuous and nondecreasing in u for fixed $t \in \mathbb{R}_+$,
(ii) $V \in C[R_+ \times B(\rho), R[\Omega, \mathbb{R}]]$ satisfies a local Lipschitz condition in x w.p.1, and for $(t, x) \in \mathbb{R}_+ \times B(\rho)$,

$$D^+V(t, x, \omega) \leq g(t, V(t, x, \omega), \omega);$$

$$D^+V(t, x, \omega) \equiv \limsup_{h \to 0^+}(1/h)\left[V(t + h, x + hf(t, x, \omega), \omega) - V(t, x, \omega)\right].$$

(iii) for $(t, x) \in \mathbb{R}_+ \times B(\rho)$,

$$b(\|x\|) \le V(t, x, \omega) \le a(t, \|x\|, \omega),$$

where $b \in \mathcal{K}$, $a(t, \cdot, \omega) \in \mathcal{K}$ and $C\left[\mathbb{R}_+ \times \mathbb{R}_+, R[\Omega, \mathbb{R}_+]\right]$. Then, the stability in probability of the zero solution of DE (Eq. 5.19) implies the stability in probability of the zero solution of IVP (Eq. 5.18).

There are many physical processes that can be modeled using stochastic differential equation and RDEs. For example, in electrical engineering they can be used to model communication signals (voice, noise, measurements, etc.), to derive optimal input signal estimators [32], to model the stability of power systems with wind-generated power [33], in ocean engineering to model the rolling motion of a ship in a disturbed sea [34], in civil engineering to model earthquakes [35], in bioengineering to model human metabolic disorders [36], to name just a few applications.

5.6 Impulsive Differential Equations

Some physical processes undergo disturbances whose duration is short compared with the duration of entire process. These disturbances can be considered as impulses. IDEs are a natural description of such phenomena.

There are several types of IDE. We will look at two types: IDEs with impulses at fixed moments and IDEs with impulses at variable times.

An IDE with fixed moments is

$$\begin{aligned} x' &= p(t, x), \quad t \ne t_k, k = 1, 2, 3 \ldots \\ \Delta x &= I_k(x), \quad t = t_k, \end{aligned} \tag{5.20}$$

where the sequence $\{t_k\}$ satisfies $t_k < t_{k+1}$, $k = 1, 2, \ldots$, and $t_k \to \infty$ as $k \to \infty$, $p : \mathbb{R}_+ \times \mathbb{R}^n \to \mathbb{R}^n$ is continuous on $(t_k, t_{k+1}] \times \mathbb{R}^n$, $I_k : \mathbb{R}^n \to \mathbb{R}^n$, $\Delta x(t_k) = x(t_k^+)$ $-x(t_k)$ and $\lim_{h \to 0^+} x(t_k + h) = x(t_k^+)$.

Any solution $x(t)$ of IDE (Eq. 5.20) satisfies $x' = p(t, x(t))$, $t \in (t_k, t_{k+1}]$ and $\Delta x(t_k) = I_k(x(t_k))$, $t = t_k, k = 1, 2, 3, \ldots$

Let $\{S_k\}$ be a sequence of surfaces given by $S_k : t = \tau_k(x)$, $k = 1, 2, \ldots$. Then, an IDE with impulses at variable times is given by

$$\begin{aligned} x' &= p(t, x), \quad t \ne \tau_k(x), \\ \Delta x &= I_k(x), \quad t = \tau_k(x), k = 1, 2, \ldots \end{aligned} \tag{5.21}$$

Systems with impulses at fixed times are comparatively easy to study than systems with variable moments or impulses. This can be known from observing the behavior of a solution at $t = \tau_k(x(t))$ in IDE (Eq. 5.21). A solution may hit a surface S_k once, more than once, or not hit at all. Solutions starting at different times may hit the surfaces at different times, thus complicating the study of these systems.

We give below a simple differential inequality result for IDE with fixed moments of impulse [37].

Theorem 5.6.1

Assume

(A$_1$) the sequence $\{t_k\}$ satisfies $0 \le t_0 < t_1 < t_2 < \cdots$, with $\lim_{k \to \infty} t_k = \infty$;

(A$_2$) $m \in PC^1[\mathbb{R}_+, \mathbb{R}]$ where PC denotes the class of piecewise continuous functions from $\mathbb{R}_+$ to $\mathbb{R}$ with discontinuities of the first kind at t_k, and $m(t)$ is left-continuous at t_k, $k = 1, 2, \ldots$;

(A$_3$) for $k = 1, 2, \ldots, t \ge t_0$,

$$m'(t) \le p(t)m(t) + q(t), \quad t \ne t_k$$
$$m(t_k^+) \le d_k m(t_k) + b_k$$

where $q, p \in C[\mathbb{R}_+, \mathbb{R}]$, $d_k \ge 0$ and b_k are constants.

Then,

$$m(t) \le m(t_0) \prod_{t_0 < t_k < t} d_k \exp\left(\int_{t_0}^{t} p(s)ds\right)$$

$$+ \sum_{t_0 < t_k < t} \left(\prod_{t_k < t_j < t} d_j \exp\left(\int_{t_k}^{t} p(s)ds\right)\right) b_k$$

$$+ \int_{t_0}^{t} \prod_{s < t_k < t} d_k \exp\left(\int_{s}^{t} p(\sigma)d\sigma\right) q(s)ds \quad t \ge t_0.$$

In studying the qualitative behavior of IVPs, it is usually assumed that the initial time remains unchanged. However, this may not be true in some cases. For example, unperturbed differential equations have solutions starting at a time different from those of a perturbed differential equation. This problem led to the concept of differential equations with initial time differences. When such a change in the starting time of each solution is considered, one is faced with the problem of comparing solutions that differ in starting time.

The following result deals with a variation of parameter formula for IDE with fixed moments of impulse involving initial time difference [38].

Consider the unperturbed IDEs

$$x' = p(t, x), \quad t \neq t_k,$$

$$x(t_0^+) = x_0, \tag{5.22}$$

$$x(t_k^+) = x(t_k) + I_{t_k}(x(t_k));$$

$$x' = p(t, x), \quad t \neq t_k,$$

$$x(\tau_0^+) = y_0, \tag{5.23}$$

$$x(t_k^+) = x(t_k) + I_{t_k}(x(t_k)), \quad \text{whenever } t_k \geq \tau_0,$$

together with the perturbed Eq. (5.23).

$$y' = P(t, y), \quad t \neq t_k,$$

$$y(\tau_0^+) = y_0, \tag{5.24}$$

$$y(t_k^+) = y(t_k) + I_{t_k}(y(t_k)), \quad \text{whenever } t_k \geq \tau_0,$$

where

(i) $0 \leq t_0 < t_1 < t_2 < \cdots < t_k < \cdots$, and $\lim_{k \to \infty} t_k = \infty$, $k = 1, 2 \ldots$;
(ii) $\tau_0 > t_0$, $\eta = \tau_0 - t_0$, $\eta \geq$;
(iii) $\bar{t}_k = t_k^+ 0$;
(iv) $S_1 = \{t_k\}$, $S_2 = \{t_k^+\}$, $S = S_1 \cup S_2$;
(v) $t \in \mathbb{R}^+$, $x \in \Omega$, where Ω is an open set in $\mathbb{R}^n$;
(vi) $p, P : \mathbb{R}^+ \times \Omega \to \mathbb{R}^n$;
(vii) $I_{t_k} : \Omega \to \mathbb{R}^n$
(viii) $p(t_0) = 0$, $I_{t_k}(0) = 0$, for all t_k.

Theorem 5.6.2

Assume that
(A_1) the function $p : \mathbb{R} \times \Omega \to \mathbb{R}^n$ is continuous in $[t_{k-1}, t_k] \times \Omega$, $k = 1, 2, \ldots$, and $p_x(t, x)$ is continuous in $[t_{k-1}, t_k] \times \Omega$, $k = 1, 2, \ldots$;
(A_2) for every $x_0 \in \Omega$, $k = 1, 2, \ldots$, there exist finite limits of functions p and p_x as $(t, r) \to (t_k, x_0)$, $t > t_k$;

$$= \det\left(I + \frac{\partial I_k}{\partial x}(x)\right) \neq 0, \quad k = 1, 2, \ldots$$

(A_3) for $k-1,2,\dots$ the mapping $\psi_k : \Omega \to \Omega, x \to z, z = \psi_k(x) \equiv x + I_k(x)$ is a diffeomorphism and for $x \in \Omega$,

$$\det\left(I + \frac{\partial I_k}{\partial x}(x) \right) \neq 0, \quad k = 1,2,\dots$$

Let $x(t,t_0,x_0)$ be a solution of Eq. (5.22). Then, for any solution $y(t) = y(t,\tau_0,y_0)$ of the system (Eq. 5.24), the following variation of parameters formula is valid:

$$y(t+\eta,\tau_0,y_0)$$

$$= x(t,t_0,x_0) + \int_0^1 \Phi(t,t_0,\sigma(s))(y_0 - x_0)ds$$

$$+ \int_{t_0}^t \tilde{p}(s,\tilde{y}(x),\eta)ds$$

$$+ \sum_{t_0 < \bar{t}_k < t} \int_0^1 \Phi\left(t,\bar{t}_k,\tilde{y}(\bar{t}_k) + sI_{\bar{t}_k+\eta}\left(\tilde{y}(\bar{t}_k)\right)\right)ds.$$

$$I_{\bar{t}_k+\eta}\left(\tilde{y}(\bar{t}_k)\right)ds.$$

The last result in this section pertains to Lipschitz stability of the IVP (Eq. 5.24) with

$$P(t+\eta,y) = p(t,y) + R(t+\eta,y).$$

Definition 5.6.3

The solution $y(t+\eta,\tau_0,y_0)$ of the IVP (Eq. 5.24) is said to be initial-time-difference Lipschitz stable (ITDLS) with respect to the solution $x(t,t_0,x_0)$, $t \geq t_0$, where $x(t,t_0,x_0)$ is any solution of the IVP (Eq. 5.22), if and only if there exists an $M = M(\tau_0)$ such that

$$\left\| y(t+\eta,\tau_0,y_0) - x(t,t_0,x_0) \right\| \leq M\left(\|y_0 - x_0\| + \tau_0 - t_0 \right)$$

Theorem 5.6.4

Assume that

(A_1) The assumptions of Theorem 5.6.2 hold with $P(t+\eta,y) = P(t,y) + R(t+\eta,y)$

(A_2) The zero solution of IDS (Eq. 5.23) is Lipschitz stable;

(A_3) $\left\| \phi(t,s,\bar{y}(s))R(s+\eta,\bar{y}(s)) \right\| \leq \gamma(s)\|\bar{y}(s)\|$ for $t_0 < s \leq t$;

(A$_4$) $\left\|\phi\left(t,t_0,\sigma(s)\right)\right\| \le M_1\left(\left\|y_0-x_0\right\|+\eta\right)/\left\|y_0-x_0\right\|$ and M_1 is a constant;

(A$_5$) $\left\|I_{\bar{t}_k+\eta}\left(\tilde{y}\left(\bar{t}_k\right)\right)\right\| \le \beta_k\left\|\left(\tilde{y}\left(\bar{t}_k\right)\right)\right\|$ and $\beta_k \ge 0$ are constants;

(A$_6$) $\left\|\phi\left(t,\bar{t}_k,\tilde{y}\left(\bar{t}_k\right)\right)\right\| + sI_{\bar{t}_{k+\eta}}\left(\tilde{y}\left(\bar{t}_k\right)\right) \ge 0$;

(A$_7$) $\displaystyle\int_{t_0}^{\infty}\gamma(s)ds \in C\left[\mathbb{R}^+,\mathbb{R}^+\right]$ and $\displaystyle\prod_{t_0<\bar{t}_k<t}\left(1+\alpha_k\beta_k\right)<\infty$. Then, the solution $y\left(t+\eta,\tau_0,y_0\right)$ of the IVP (Eq. 5.24) is ITDLS with respect to the solution $x\left(t,t_0,x_0\right)$ of IDS (Eq. 5.22).

Some of the applications of IDE are optimal control models in economics and engineering, feedback systems, and many biological phenomena involving thresholds.

5.7 Fuzzy Differential Equations

Generally when differential equations are used to model physical problems, the parameters, variables, and initial conditions are assumed to be defined exactly, whereas in reality, they may be imprecise. The development of new fields such as robotics, artificial intelligence, and language theory requires new concepts to deal with such imprecise parameters. In 1965, Zadeh initiated the development of a modified set theory in which he defined the concept of a fuzzy set, a tool that makes possible the description of vague or imprecise notions. A fuzzy set is a membership function that describes the gradual transition from membership to nonmembership. Spaces of such fuzzy sets are function spaces with special properties.

Let $K_c\left(\mathbb{R}^n\right)$ denote the family of all nonempty compact convex subsets of $\mathbb{R}^n$. If $\alpha,\beta \in \mathbb{R}$ and $A,B \in K_c\left(\mathbb{R}^n\right)$, then

$$\alpha(A+B)=\alpha A+\alpha B, \quad a(\beta A)=\alpha(\beta A), \quad 1A=A$$

and if $\alpha,\beta \ge 0$, then $(\alpha+\beta)A=\alpha A+\beta A$. Let $I=\left[t_0,t_0+a\right]$, $t_0 \ge 0$ and $a>0$ and denote by $E^n=\left\{u/u:\mathbb{R}^n \to [0,1]\right\}$ such that u satisfies (*i*) to (*iv*) mentioned below:

(i) u is normal, that is, there exists an $x_0 \in \mathbb{R}^n$ such that $u(x_0)=1$;

(ii) u is fuzzy convex, that is, for $x,y \in \mathbb{R}^n$ and $0 \le \lambda \le 1$,

$$u\left(\lambda x+(1-\lambda)y\right) \ge \min\left\{u(x),u(y)\right\};$$

(iii) u is upper semicontinuous;

(iv) $[u]^0=\overline{\left[x\in\mathbb{R}^n:u(x)>0\right]}$ is compact.

For $0 < a \le 1$, we denote $[u]^a = [x \in \mathbb{R}^n : u(x) \ge a]$. Then, from (*i*) to (*iv*), it follows that the α-level sets $[u]^\alpha \in K_c(\mathbb{R}^n)$ for $0 \le \alpha \le 1$. For later purposes, we define $\hat{0} \in E^n$ as $\hat{0}(x) = 1$ if $x = 0$ and $\hat{0}(x) = 0$ if $x \ne 0$.

Let $d_H(A,B)$ be the Hausdorff distance between the sets $A, B \in K_c(\mathbb{R}^n)$. Then, we define

$$d[u,v] = \sup_{0 \le \alpha \le 1} d_H \left[[u]^\alpha, [v]^\alpha \right],$$

which defines a metric in E^n and (E^n, d) is a complete metric space.

Consider the fuzzy differential system

$$u' = p(t,u), \quad u(t_0) = u_0, \tag{5.25}$$

where $p \in C[I \times E^n, E^n]$ and $I = [t_0, t_0 + a]$, $t_0 \ge 0$, $a > 0$. Note that a mapping $u : I \to E^n$ is a solution of the IVP (Eq. 5.25) if and only if it is continuous and satisfies the integral equation

$$u(t) = u_0 + \int_{t_0}^t f(s, u(s)) ds, \quad \text{for } t \in I.$$

An application of the contraction mapping principle yields the following existence and uniqueness result [39].

Theorem 5.7.1

Assume that $p \in C[I \times E^n, E^n]$ and satisfies

$$d[p(t,u), p(t,v)] \le Ld[u,v], \quad L > 0,$$

for $(t,u), (t,v) \in I \times E^n$. Then, the IVP (Eq. 5.25) has a unique solution $u(t) = u(t, t_0, u_0)$ on I.

We note that in order to apply the comparison principle in the proof of this theorem, we need to use the weighted metric

$$H(u,v) = \sup_I d[u(t), v(t)] e^{\lambda t}$$

for $u, v \in C[I, E^n]$ and $\lambda > 0$. Since (E^n, d) is a complete metric space $(C[I, E^n], H)$ is also complete.

Next, we give a comparison result that uses the theory of differential inequalities.

Theorem 5.7.2

Assume that $p \in C[I \times E^n, E^n]$ and for $t \in I, u, v \in E^n$,

$$d[p(t,u), p(t,v)] \leq g(t, d[u,v]),$$

where $g \in C[I \times \mathbb{R}_+, \mathbb{R}_+]$. Suppose further that the maximal solution $r(t, t_0, w)$ of the scalar differential equation

$$w' = g(t,w), \quad w(t_0) = w_0 \geq 0,$$

exists on I. Then, if $u(t), v(t)$ are any two solutions of FDE (Eq. 5.25) through $(t_0, u_0), (t_0, v_0)$, respectively, on I, we have

$$d[u(t), v(t)] \leq r(t, t_0, w_0), \quad t \in I$$

provided $d[u_0, v_0] \leq w_0$.

The next set of results in this section deals with the stability and boundedness of the solution of FDE (Eq. 5.25) using Lyapunov functions. The following is a comparison result using a Lyapunov function. The Lyapunov function serves as a vehicle to transform the FDE into a scalar comparison ODE, and it is enough to consider the stability properties of the simpler comparison equation.

Consider the FDE

$$u' = p(t,u), \quad u(t_0) = u_0, \tag{5.26}$$

where $p \in C[\mathbb{R}_+ \times S(\rho), E^n]$ and $S(\rho) = \left[u \in E^n : d[u, \hat{0}] < \rho\right]$. We assume that $p(t, \hat{0}) = \hat{0}$ so that we have the trivial solution of FDE (Eq. 5.26).

Theorem 5.7.3

Assume that

(i) $V \in C[\mathbb{R}_+ \times S(\rho), \mathbb{R}_+], |V(t, u_1) - V(t, u_2)| \leq L\, d[u_1, u_2], L > 0$ and

(ii) $D^+ V(t,u) \equiv \lim\sup_{h \to 0^+} \frac{1}{h}\left[V(t+h, u+hf(t,u)) - V(t,u)\right] \leq g(t, V(t,u)), where$
$g \in C[\mathbb{R}_+{}^2, \mathbb{R}].$

Then, if $u(t)$ is any solution of FDE (Eq. 5.26) existing on $[t, \infty)$, such that $V(t_0, u_0) \leq w_0$, we have

$$V(t, u(t)) \leq r(t, t_0, w_0), \quad t \geq t_0,$$

where $r(t,t_0,w_0)$ is the maximal solution of the scalar differential equation

$$w' = g(t,w), \quad w(t_0) = w_0 \geq 0,$$

existing on $[t_0,\infty)$.

We provide below the definition of stability and boundedness.

Definition 5.7.4

The trivial solution $u = \hat{0}$ of FDE (Eq. 5.26) is said to be equistable if, for each $\epsilon > 0$ and $t_0 \in \mathbb{R}_+$, there exists a positive function $\delta = \delta(t_0,\epsilon)$ that is continuous in t_0 for each ϵ, such that

$$d\left[u_0,\hat{0}\right] < \delta \text{ implies } d\left[u(t),\hat{0}\right] < \epsilon, \quad t \geq t_0,$$

where $u(t) = u(t,t_0,u_0)$ is the solution of FDE (Eq. 5.26).

Definition 5.7.5

The solution of FDE (Eq. 5.26) is said to be equibounded, if for any $\alpha > 0$ and $t_0 \in \mathbb{R}_+$, there exists $\beta = \beta(t_0,\alpha) > 0$, such that

$$d\left[u_0,\hat{0}\right] < \alpha \text{ implies } d\left[u(t),\hat{0}\right] < \beta, \quad t \geq t_0.$$

The aforementioned comparison theorem can be used to prove the following stability and boundedness results.

Theorem 5.7.6

Assume that the following holds:

(i) $V \in C\left[\mathbb{R}_+ \times S(\rho), \mathbb{R}_+\right], \left|V(t,u_1) - V(t,u_2)\right| \leq L\, d\left[u_1,u_2\right], L > 0$ and for $(t,u) \in \mathbb{R}_+ \times S(\rho)$, where $S(\rho) = \left\{u \in E^n : d\left[u,\hat{0}\right] < \rho\right\}$.

$$D^+V(t,u) \equiv \limsup_{h \to 0^+} \frac{1}{h}\left[V\left(t+h, u + hf(t,u)\right) - V(t,u)\right] \leq 0;$$

(ii) $b\left(d\left[u,\hat{0}\right]\right) \leq V(t,u) \leq a\left(t, d\left[u,\hat{0}\right]\right)$ for $(t,u) \in \mathbb{R}_+ \times S(\rho)$ where $b, a(t,\cdot) \in \mathcal{K}$.

Then, the solution of FDE (Eq. 5.26) is equibounded.

Next, we state nonuniform boundedness property of solutions utilizing perturbing Lyapunov functions.

Theorem 5.7.7

Assume that

(i) For $\rho > 0$, $V_1 \in C[\mathbb{R}_+ \times S(\rho), \mathbb{R}_+]$, V_1 is bounded for $(t, u) \in \mathbb{R}_+ \times \partial S(\rho)$, and

$$|V_1(t, u_1) - V_1(t, u_2)| \le L_1 \, d[u_1, u_2], \quad L_1 > 0,$$

$$D^+V_1(t, u) = \limsup_{h \to 0^+} \frac{1}{h}\left[V_1(t + h, u + hf(t, u)) - V_1(t, u)\right]$$

$$\le g_1(t, V_1(t, u)), (t, u) \in \mathbb{R}^+ \times S^c(\rho)$$

where $g_1 \in C[\mathbb{R}_+{}^2, \mathbb{R}]$.

(ii) $V_2 \in C[\mathbb{R}_+ \times S^c(\rho), \mathbb{R}_+]$,

$$b\left(d\left[u, \hat{0}\right]\right) \le V_2(t, u) \le a\left(d\left[u, \hat{0}\right]\right), \quad a, b \in \mathcal{K}$$

and

$$D^+V_1 + D^+V_2 \le f_2(t, V_1(t, u) + V_2(t, u)), \quad g_2 \in C[\mathbb{R}_+{}^2, \mathbb{R}]$$

(iii) the scalar differential equations

$$w_1' = g_1(t, w_1), \quad w_1(t_0) = w_{10} \ge 0 \tag{5.27}$$

and

$$w_2' = g_2(t, w_2), \quad w_2(t_0) = w_{20} \ge 0 \tag{5.28}$$

are equibounded and uniformly bounded, respectively.

Then, the system (Eq. 5.26) is equibounded.

The role of uncertainties may be found in many subject areas such as physics, biology, and economics, to name a few and FDEs can be used to model such processes. For example, the uncertain rate of burning trees can be modeled as an FDE [40]. Also, many electrical appliances use fuzzy controls.

5.8 Set Differential Equations

In many physical situations, it is not possible to single out an element, and hence we are forced to consider a set of elements. For example, if one were to study the effects of radiation on cancer cells, it is quite difficult to isolate a single cell and one has to consider a set. Another example is in place of considering an element on the real line, an interval in $\mathbb{R}$ would be more reasonable in some situations. The situation where intervals are elements gave rise to interval mathematics and interval differential equations. It can be noted that interval differential equations are a special case of SDEs. While taking sets, it is natural to consider a well-defined set, such as a compact convex set. A systematic introduction of the topic is given in Refs. [41] and [42,43].

Let $K_c\left(\mathbb{R}^n\right)$ be the set of all compact convex subsets of $\mathbb{R}^n$.

The Minkowski addition and scalar multiplication are defined as follows: For $C, D \in K_c\left(\mathbb{R}^n\right)$,

$$C + D = \{c + d \mid c \in C \text{ and } d \in D\}$$

$$\text{and } \lambda C = \{\lambda c \mid c \in C\}.$$

The difference between sets is defined as the Hukuhara difference between two sets and is set as follows:

Given $C, D \in K_c\left(\mathbb{R}^n\right)$, if there exists $E \in K_c\left(\mathbb{R}^n\right)$ such that $C = D + E$, then E is said to be the Hukuhara difference of C and D, and is written as $E = C - D$.

Now $\left(K_c\left(\mathbb{R}^n\right), +, \cdot\right)$ is a semilinear space, and a metric called Hausdorff metric defined using any standard norm $\|\cdot\|$ in $\mathbb{R}^n$, distance from a point to a set, and the notion of Hausdorff separation is given as follows:

For any $x \in \mathbb{R}^n, C \subseteq \mathbb{R}^n$,

$$d(x, C) = \inf\{\|x - c\|, \quad c \in C\}$$

and for $B, C \subseteq \mathbb{R}^n$, the Hausdorff separation of B from C is defined as

$$d_H(B, C) = \sup\{d(b, C), \quad b \in B\}$$

The Hausdorff distance between $B, C \subseteq \mathbb{R}^n$ is defined as

$$D(C, B) = \max\{d_H(C, B), d_H(B, C)\}.$$

The semilinear space $\left(K_c\left(\mathbb{R}^n\right), +, \cdot\right)$ endowed with the Hausdorff metric, $\left(K_c\left(\mathbb{R}^n\right), D\right)$, is a complete separable metric space. Let

$K = \left\{ U \in K_c\left(\mathbb{R}^n\right) \middle/ \text{ for } u \in U, u = (u_1, u_2, \ldots, u_n) \in U, u_i \geq 0, i = 1, 2, \ldots n \right\}$ for be a cone in $K_c\left(\mathbb{R}^n\right)$ and K^0 is the nonempty interior of K. Then, a partial order on $K_c\left(\mathbb{R}^n\right)$ is as follows.

Definition 5.8.1

For any $C, D \in K_c\left(\mathbb{R}^n\right)$, if there exists $U \in K\left(K^0\right)$ and

$$C = D + U.$$

Then, we say $C \geq D (C > D)$. Similarly, one can define $C \leq D (C < D)$.

The derivative of a function defined on an interval J with values in $K_c\left(\mathbb{R}^n\right)$ is called Hukuhara derivative and is defined as follows.

Definition 5.8.2

Let $X : J \to K_c\left(\mathbb{R}^n\right)$ be a multifunction defined on an interval J. For $t_0 \in J$ $\delta t > 0$ sufficiently small, suppose the Hukuhara differences

$$X\left(t_0 + \delta t\right) - X\left(t_0\right) \text{ and } X\left(t_0\right) - X\left(t_0 - \delta t\right)$$

exist.
    We say that X is Hukuhara differentiable at a point $t_0 \in I$, if the limits

$$\lim_{\delta t \to 0+} \frac{X\left(t_0 + \delta t\right) - X\left(t_0\right)}{\delta t}$$

and

$$\lim_{\delta t \to 0+} \frac{X\left(t_0\right) - X\left(t_0 - \delta t\right)}{\delta t}$$

both exist and are equal and the value is equal to $D_H X\left(t_0\right) \in K_c\left(\mathbb{R}^n\right)$. $D_H X\left(t_0\right)$ is called the Hukuhara derivative of X at t_0.
    Now, we can consider the differential equation formed by the Hukuhara derivative called an SDE. The following results are from Ref. [43].
    The IVP for the SDE is given by

$$D_H X = P(t, X), \quad X\left(t_0\right) = X_0 \in K_c\left(\mathbb{R}^n\right), t_0 \geq 0, \tag{5.29}$$

where $P \in C\left[\mathbb{R}_+ \times K_c\left(\mathbb{R}^n\right), K_c\left(\mathbb{R}^n\right)\right]$.

Definition 5.8.3

By a solution of SDE (Eq. 5.29), we mean a function $X \in C^1\left[J, K_c\left(\mathbb{R}^n\right)\right]$, $J = [t_0, t_0 + a)$, $a > 0$, that satisfies SDE (Eq. 5.29) on J.

We present a comparison principle established using differential inequalities.

Theorem 5.8.4

Assume that $P \in C\left[J \times K_c\left(\mathbb{R}^n\right), K_c\left(\mathbb{R}^n\right)\right]$ for $t \in J$ and for $X, Y \in K_c\left(\mathbb{R}^n\right)$,

$$D[P(t,X), P(t,Y)] \le f(t, D[X,Y]), \tag{5.30}$$

where $f \in C[J \times \mathbb{R}_+, \mathbb{R}_+]$. Suppose further that the maximal solution $r(t, t_0, w_0)$ of the scalar differential equation

$$w' = f(t,w), \quad w(t_0) = w_0 \ge 0,$$

exists on J. Then, if $X(t)$, $Y(t)$ are any two solutions through (t_0, X_0), (t_0, Y_0), respectively, on J, it follows that

$$D(X(t), Y(t)) \le r(t, t_0, w_0), \quad t \in J,$$

provided $D[X_0, Y_0] \le w_0$.

Observe that in the above theorem, the scalar function f used to obtain an estimate on P has no additional criteria to satisfy. This is the additional advantage obtained when differential inequalities are used.

Next, we present a generalization of the MIT, which is a constructive approach to obtain existence of solutions of nonlinear SDE from Ref. [43]. The idea is to consider a difference of two monotone—functions—one nondecreasing and the other nonincreasing. This decomposition led to various notions of lower and upper solutions, which are defined below relative to the IVPs of SDEs given by

$$D_H X = P(t, X) + Q(t, X), \quad X(0) = X_0 \in K_c\left(\mathbb{R}^n\right), \tag{5.31}$$

where $P, Q \in C\left[J \times K_c\left(\mathbb{R}^n\right), K_c\left(\mathbb{R}^n\right)\right]$ and $J = [0, T]$.

Definition 5.8.5

Let $X, Y \in C^1\left[J, K_c\left(\mathbb{R}^n\right)\right]$. Then X, Y are said to be

(i) natural lower and upper solutions of SDE (Eq. 5.31) if

$$D_H X \le P(t,X) + Q(t,X), \quad D_H Y \ge P(t,Y) + Q(t,Y), \quad t \in J; \qquad (5.32)$$

(ii) coupled lower and upper solutions of type I of SDE (Eq. 5.31) if

$$D_H X \le P(t,X) + Q(t,Y), \quad D_H Y \ge P(t,Y) + Q(t,X), \quad t \in J; \qquad (5.33)$$

(iii) coupled lower and upper solutions of type II of SDE (Eq. 5.31) if

$$D_H X \le P(t,Y) + Q(t,X), \quad D_H Y \ge P(t,X) + Q(t,Y), \quad t \in J; \qquad (5.34)$$

(iv) coupled lower and upper solutions of type III of SDE (Eq. 5.31) if

$$D_H X \le P(t,Y) + Q(t,Y), \quad D_H Y \ge P(t,X) + Q(t,X), \quad t \in J; \qquad (5.35)$$

Note that whenever $X(t) \le Y(t)$, $t \in J$, if $P(t,X)$ is nondecreasing in X for each $t \in J$ and $Q(t,Y)$ is nonincreasing in Y for each $t \in J$, the lower and upper solutions defined by Eqs. (5.32) and (5.35) reduce to Eq. (5.34). Therefore, it is enough to study the cases Eqs. (5.33) and (5.34).

The following theorem deals with generalized MIT.

Theorem 5.8.6

Assume that

1. $X, Y \in C^1 \left[J, K_c \left(\mathbb{R}^n \right) \right]$ are coupled lower and upper solutions of type I relative to SDE (Eq. 5.31) with $X(t) \le Y(t)$, $t \in J$;
2. $P, Q \in C \left[J \times K_c \left(\mathbb{R}^n \right), K_c \left(\mathbb{R}^n \right) \right] P(t,X)$ is nondecreasing in X and $Q(t,Y)$ is nonincreasing in Y for each $t \in J$:
3. P and Q map bounded sets into bounded sets in $K_c \left(\mathbb{R}^n \right)$.

Then, there exist monotone sequences $\{X_n(t)\}, \{Y_n(t)\}$ in $K_c \left(\mathbb{R}^n \right)$ such that $X_n(t) \to \rho(t)$, $Y_n(t) \to R(t)$ in $K_c \left(\mathbb{R}^n \right)$, and (ρ, R) are the coupled minimal and maximal solutions of Eq. (5.31), respectively, that is, they satisfy

$$D_H \rho = F(t,\rho) + G(t,R), \quad \rho(0) = X_0,$$

$$D_H R = F(t,R) + G(t,\rho), \quad R(0) = X_0, \quad \text{on } J.$$

If the lower and upper solutions of type II are considered, then the sequence of iterates that converge to the solutions of SDE (Eq. 5.31) has some interesting properties.

See Theorem 2.5.3 in Ref. [43]. The remarks following the theorems of MIT in Ref. [43] give insight to the various possible setups.

Sometimes, it is not possible to obtain solutions of the given problem. Then the idea of approximate solutions and Euler solutions may be useful. We now introduce the concept of an approximate solution in this setup and state a result involving it from Ref. [43].

Definition 5.8.7

A function $Y(t) = Y(t, t_0, Y_0, \epsilon)$, $\epsilon > 0$, is said to be an ϵ-approximate solution of IVP of SDE (Eq. 5.29) if $Y \in C^1\left[\mathbb{R}_+, K_c\left(\mathbb{R}^n\right)\right]$, $Y(t_0, t_0, Y_0, \epsilon) = Y_0$ and $D[D_H Y(t)], \quad P(t, Y(t))] \le \epsilon, \quad t \ge t_0$. If $\epsilon = 0$; $Y(t)$ is a solution of SDE (Eq. 5.29).

Theorem 5.8.8

Assume that $P \in C\left[\mathbb{R}_+ \times K_c\left(\mathbb{R}^n\right), K_c\left(\mathbb{R}^n\right)\right]$ and for $t \ge t_0, X, Y, \in K_c\left(\mathbb{R}^n\right)$,

$$D[P(t, X), P(t, Y)] \le f(t, D[X, Y]),$$

where $f \in C\left[\mathbb{R}_+^2, \mathbb{R}_+\right]$.

Let $r(t) = r(t, t_0, w_0, \epsilon)$ be the maximal solution of

$$w' = f(t, w) + \epsilon, \quad w(t_0) = w_0 \ge 0,$$

existing for $t \ge t_0$. Let $X(t, t_0, X_0)$ be any solution of SDE (Eq. 5.29), existing for $t \ge t_0$ and $Y(t, t_0, Y_0, \epsilon)$ be an ϵ-approximate solution of IVP of SDE (Eq. 5.29), existing for $t \ge t_0$. Then, $D[X_0, Y_0] \le w_0$ implies $D[X(t), Y(t)] \le r(t, t_0, w_0, \epsilon), \quad t \ge t_0$.

Once the well-posedness of a problem is established, the equilibrium or stability of solutions can be considered. Of the many types of stability results proved for SDEs, we present the nonuniform stability criteria here. In the following theorem, the method of perturbing Lyapunov functions is used, and the conditions are assumed on $\mathbb{R}_+ \times S(\rho) \cap S^c(\eta)$ for $0 < \eta < \rho$, which is smaller region than the usual $\mathbb{R}_+ \times S(\rho)$.

Theorem 5.8.9

Assume that

(i) $V_1 \in C\left[\mathbb{R}_+ \times S(\rho), \mathbb{R}_+\right], |V_1(t, X_1) - V_1(t, X_2)| \le L_1 D[X_1, X_2], L_1 > 0,$
$V_1(t, X) \le a_0(t, D(X, \theta))$, where $a_0 \in \left[\mathbb{R}_+ \times [0, \rho), \mathbb{R}_+\right]$ and $a_0(t, 0) \in \mathcal{K}$
for each $t \in \mathbb{R}_+$.

(ii) $D^+ V_1(t, X) \le f_1(t, V_1(t, X)), (t, X) \in \mathbb{R}_+ \times S(\rho)$, where $f_1 \in C\left[\mathbb{R}_+^2, \mathbb{R}\right]$ and
$f_1(t, 0) \equiv 0.$

(iii) for every $\eta > 0$, there exists a $V_\eta \in C\left[\mathbb{R}_+ \times S(\rho) \cap S^c(\eta), \mathbb{R}_+\right]$,

$$\left|V_\eta(t, X_1) - V_\eta(t, X_2)\right| \le L_\eta \, D[X_1, X_2]$$

$$b(D(X, \theta)) \le V_\eta(t, X) \le a(D(X, \theta)), \quad a, b \in \mathcal{K};$$

and

$$D^+V_1(t, X) + D^+V_\eta(t, X) \le f_2\left(t, V_1(t, X) + V_\eta(t, X)\right)$$

for $(t, X) \in \mathbb{R}_+ \times S(\rho) \cap S^c(\eta)$, where $f_2 \in C\left[\mathbb{R}_+{}^2, \mathbb{R}\right]$ and $f_2(t, 0) \equiv 0$;

(iv) The trivial solution, $w_1 \equiv 0$, of

$$w_1' = f_1(t, w_1), \quad w_1(t_0) = w_{10} \ge 0,$$

is equistable.

(v) the trivial solution $w_2 \equiv 0$ of

$$w_2' = f_2(t, w_2), \quad w_2(t, 0) = w_{20} \ge 0,$$

is uniformly stable.

Then, the zero solution of SDE (Eq. 5.29) is equistable.

As stated earlier, parallel to the definition of stability concepts, the boundedness notions have been defined. We conclude this subsection with definition of boundedness and a result pertaining to boundedness in the setup of SDEs.

Definition 5.8.10

The solution of SDE (Eq. 5.29) is "equibounded," if, "for any $\gamma > 0$, $t_0 \in \mathbb{R}_+$, there exists $\eta = \eta(t_0, \gamma) > 0$ such that $D(X_0, \theta) < \gamma$ implies $D(X(t), \theta) < \eta, \quad t \ge t_0$."

Theorem 5.8.11

Assume that

(a) $V \in C\left[\mathbb{R}_+ \times K_c(\mathbb{R}^n), \mathbb{R}_+\right], \left|V(t, X_1) - V(t, X_2)\right| \le L\, D(X_1, X_2), L > 0$ and for $(t, X) \in \mathbb{R}_+ \times K_c(\mathbb{R}^n), D^+V(t, X) \le 0$;

(b) $b(D(X, 0)) \le V(t, X) \le u(t, D(X, \theta))$, for $(t, x) \in \mathbb{R}_+ \times K_c(\mathbb{R}^n)$ where $b, a(t, \cdot) \in \mathcal{K}$.

Then, the solution of SDE (Eq. 5.29) is equibounded.

5.9 Fractional Differential Equations

The notion of a fractional derivative (FD) came into existence with the notation of nth order derivative of a function of t, $x(t)$, given by Leibnitz as $\dfrac{d^n x}{dt^n}$.

L'Hopital's famous query, what if $n=1/2$, gave birth to fractional calculus. From then on, Riemann, Fourier, Riesz, Laplace, Grunwald, Letnikov, Liouville, and many other mathematicians contributed to the growth of this field. For a long time, the Riemann–Liouville (RL) FD was popular, but it has the deficiency that for a constant the RL derivative is different from zero unlike Caputo FD (CFD). This was corrected by Caputo in his definition, and then, it became more popular.

There are many types of derivatives of fractional order for a function, and many new definitions are being given. This is being done so as to find an FD that satisfies all the properties satisfied by an ordinary derivative. Recently, two new definitions were introduced [44,45] that were developed from fundamentals using the concept of a limit, like the ordinary derivative.

There are many good books on fractional calculus and FrDEs [46–49]. The book of Oldham and Spanier [50] introduced this concept to engineers and other scientists, and now fractional calculus and FrDEs represent a very active area of both theoretical and applied research.

Fractional calculus and FrDEs have been systematically studied in terms of both RL and Caputo derivatives. We begin by presenting some links between these two derivatives.

(i)   The Caputo FD and RLFD differ by a derivative of a constant [47].

(ii) Any result that holds for an FrDE with RL derivative holds also for an FrDE having Caputo derivative. The converse is not true [51].

(iii) The IVP of RL FrDE has a singularity at the initial value while that of a Caputo FrDE has an initial value parallel to that of ODEs [46].

We present some results involving a Caputo FrDE or an RL FrDE to introduce this very active area of research. Definition of the set, "$C_p\big[[t_0,T],\mathbb{R}\big]$" and RLFD are given below.

Definition 5.9.1

$m \in C_p\big[[t_0,T],\mathbb{R}\big]$ means that "$m \in C\big[(t_0,T],\mathbb{R}\big]$ and $(t-t_0)^p m(t) \in C\big[[t_0,T],\mathbb{R}\big]$ with $p+q=1$."

Definition 5.9.2

For $m \in C_p\big[[t_0,T],\mathbb{R}\big]$, the definition of RLFD of $m(t)$ is

$$D^q m(t) = \frac{1}{\Gamma(p)} \frac{d}{dt} \int_{t_0}^{t} (t-s)^{p-1} m(s) \, ds.$$

The following lemma is vital for proving inequality results and is from Ref. [52].

Lemma 5.9.3

Let $m \in C_p \left[[t_0, T], \mathbb{R} \right]$. Suppose that for any $t_1 \in (t_0, T]$, we have $m(t_1) = 0$ and $m(t) < 0$, for $t_0 \le t < t_1$. Then, $D^q m(t_1) \ge 0$.

The basic differential inequality theorem is as follows.

Theorem 5.9.4

Let $v, w \in C_p \left[[t_0, T], \mathbb{R} \right], P \in C \left[[t_0, T] \times \mathbb{R}, \mathbb{R} \right]$ and

(i) $D^q v(t) \le P(t, v(t))$
    and
(ii) $D^q w(t) \ge P(t, w(t))$, $t_0 < t \le T$.
    Assume P satisfies the one-sided Lipschitz condition

$$P(t, x) - P(t, y) \le L(x-y), \quad x \ge y, L > 0.$$

Then, $v^0 \le w^0$, where $v^0 = v(t)(t-t_0)^{1-q} \big|_{t=t_0}$ and $w^0 = w(t)(t-t_0)^{1-q} \big|_{t=t_0}$ implies

$v(t) \le w(t)$, $t \in [t_0, T]$.

Of the many existence results available, we present a result dealing with Euler solutions for the IVP of RL FrDE given by

$$D^q x = P(t, x), \quad x(t)(t-t_0)^{1-q} \big|_{t=t_0} = x^0, \tag{5.36}$$

where P is from $[t_0, T] \times \mathbb{R}^n \to \mathbb{R}^n$. (Observe that P need not be a continuous function).

Let $\pi = \{t_0, t_1, \ldots, t_n = T\}$ be a partition of $[t_0, T]$. Then, corresponding to the partition π, an Euler curved arc $x = x(t)$ is defined on $[0, T]$ as follows:
On $[t_0, t_1]$ put the value x^0 in the function P in FrDE (Eq. 5.36) to get

$$D^q x = P(t_0, x^0), \quad x(t)(t-t_0)^{1-q} \big|_{t=t_0} = x^0. \tag{5.37}$$

Then, the FrDE (Eq. 5.37) has a constant on the right-hand side and the IVP of FrDE (Eq. 5.37) has a unique solution given by

$$x(t) = \frac{x^0}{\sqrt{(q)}}(t-t_0)^{q-1} + P(t_0, x^0) \frac{(t-t_0)^q}{\sqrt{(1+q)}}, \quad t \in [t_0, t_1].$$

Set the node $x_1 = x(t_1)$ and consider, in $[t_1, t_2]$, the IVP

$$D^q x = P(t_1, x_1), \quad x(t)(t - t_1)^{1-q}\Big|_{t=t_1} = x_1^0.$$

This again has a unique solution in $[t_1, t_2]$. Next, proceeding as above, set $x_2 = x(t_2)$ and continue over the whole partition π defined on $[t_0, T]$. Thus, the Euler arc is constructed on the interval $[t_0, T]$ over a partition π.

Using the Euler arc, an Euler solution is defined below.

Definition 5.9.5

An Euler solution is any curved arc $x = x(t)$, which is the uniform limit of Euler curved arcs x_{π_j}, corresponding to some sequence of partitions π_j such that $\pi_j \to 0$. This means that the diameter $\mu_{\pi_j} \to 0$ as $j \to \infty$, where $\mu_{\pi_j} = \max\{t_i - t_{i-1} : 1 \le i \le N\}$.

The following result deals with Euler solution for FrDE (Eq. 5.36) and is from Ref. [48].

Theorem 5.9.6

Assume that

(i) $|P(t,x)| \le f(t,|x|), \quad (t,x) \in [t_0, T] \times \mathbb{R}^n, \quad$ where $f \in C([t_0, T] \times \mathbb{R}_+, \mathbb{R}_+)$, $f(t,u)$ is nondecreasing in (t,u);

(ii) The maximal solution $r(t) = r(t, t_0, u_0)$ of the scalar FrDE

$$D^q u = f(t,u), \quad u(t)(t - t_0)^{1-q}\Big|_{t=t_0} = u^0 \ge 0$$

exists on $[t_0, T]$.

Then,

(a) there exists at least one Euler solution $x(t) = x(t, t_0, x^0)$ for the IVP of FrDE (Eq. 5.36) which satisfies a Hölder condition:

(b) any Euler solution $x(t)$ of FrDE (Eq. 5.36) satisfies the relation

$$|x(t) - x^0(t)| \le r(t, t_0, u^0) - u^0, \quad t \in [t_0, T]$$

where $u^0 = |x^0|$ and $x^0(t) = \dfrac{x^0 (t - t_0)^{q-1}}{\Gamma(q)}$.

We next proceed to give a result involving perturbed functions.

In perturbation theory, the estimation of the perturbed system is done using the variation of parameters formula or the Lyapunov function.

In either case, sometimes, there is a loss of useful information contained in the perturbed function, as a norm is used [see pg. 285 in Ref. [12]]. In order to overcome this disadvantage, the technique of variational Lyapunov method has been given in Ref. [12].

We present this technique, established in Ref. [52], for FrDEs. Consider

$$^cD^qy = P(t,y), \quad y(t_0) = y_0 \tag{5.38}$$

$$^cD^qx = Q(t,x), \quad x(t_0) = x_0 \tag{5.39}$$

where $P, Q \in C\left[\mathbb{R}_+ \times S(\rho), \mathbb{R}^n\right]$.

(A) Assume that the solutions of FrDE (Eq. 5.38) exist uniquely and depend continuously on initial values and $\|y(t, t_0, x_0)\|$ is locally Lipschitzian in x_0.

The Dini CFD of a Lyapunov function is defined as follows:

Let $\|x_0\| < \rho$ and $\|y(t, t_0, x_0)\| < \rho$ for $t \in [t_0, T]$.

Definition 5.9.7

For any $V \in C\left[\mathbb{R}_+ \times S(\rho), \mathbb{R}_+\right]$, the Dini CFD of $V(s, y(t, s, x))$ is given by

$$^cD_+^qV\big(s, y(t,s,x)\big) = \limsup_{h \to 0_+} \frac{1}{h^q}\Big\{ V\big(s, y(t,s,x)\big)$$

$$- V\Big(s-h, y\big(t, s-h, x - h^qP(s,x)\big)\Big)\Big\},$$

where $s, t \in [t_0, T]$ and $x \in \mathbb{R}^n$,

$$V\Big(s-h, y\big(t, s-h, x - h^qP(s,x)\big)\Big)$$

$$= \sum_{r=1}^{n}(-1)^{r+1}q_{c_r}\, V\Big(s-rh, y\big(t, s-rh, x - h^qP(s,x)\big)\Big).$$

The following comparison theorem is necessary to prove stability results of FrDE (Eq. 5.39) relative to FrDE (Eq. 5.38).

Theorem 5.9.8

Let the hypothesis (A) be satisfied. Further, let

(i) $V \in C\left[\mathbb{R}_+ \times S(\rho), \mathbb{R}\right]$, $V(t, x)$ be locally Lipschitzian in x and for $t_0 \le s \le t$, $x \in S(\rho)$,

$$^cD_+^qV\big(s, y(t,s,x)\big) \le g\big(s, V(s, y(t,s,x))\big); \tag{5.40}$$

(ii) $f \in C\left[\mathbb{R}_+^2, \mathbb{R}\right]$ and the maximal solution $r(t, t_0, u_0)$ of

$$^cD^q u = f(t, u), \quad u(t_0) = u_0 \geq 0, \tag{5.41}$$

exists for $t_0 \leq t \leq T$.

Then, for any solution, $x(t) = x(t, t_0, x_0)$ of FrDE (Eq. 5.41), $V(t, x(t, t_0, x_0)) \leq r(t, t_0, u_0)$, $t_0 \leq t \leq T$, holds, provided $V(t_0, y(t_0, t_0, x_0)) \leq u_0$.

Following theorem establishes the criteria for the stability of the system, FrDE (Eq. 5.39) relative to FrDE (Eq. 5.38) using the above comparison theorem as a tool.

Theorem 5.9.9

Assume that (A) holds and condition (i) of Theorem 5.9.8 is satisfied. Suppose that $f \in C\left[\mathbb{R}^2, \mathbb{R}\right]$, $f(t, 0) \equiv 0$, $P(t, 0) \equiv 0$, $Q(t, 0) \equiv 0$ and for $(t, x) \in \mathbb{R}_+ \times S(\rho)$, $b(\|x\|) \leq V(t, x) \leq a(\|x\|)$, $a, b \in K$. Further, suppose that the trivial solution of FrDE (Eq. 5.41) is uniformly stable and $u \equiv 0$ of FrDE (Eq. 5.38) is uniformly asymptotically stable. Then, the trivial solution of FrDE (Eq. 5.39) is uniformly asymptotically stable.

The Lyapunov function has been successfully used extensively to study stability properties of a variety of problems. But the question as to whether it is advantageous to have more than one Lyapunov function arises. In other words, can a vector Lyapunov function be used to study the stability properties of a given system. The following theorem illustrates the use of this concept to study the stability properties of the trivial solution of the Caputo fractional differential system given by

$$^cD^v x = P(t, x), \quad x(t_0) = x_0, \tag{5.42}$$

where $P \in C\left[\mathbb{R}_+ \times \mathbb{R}^n, \mathbb{R}^n\right]$.

Theorem 5.9.10

Assume that

(i) $f \in C\left[\mathbb{R}_+ \times \mathbb{R}^n, \mathbb{R}^n\right]$, $f(t, 0) \equiv 0$ and $f(t, u)$ is quasimonotone nondecreasing in u for each $t \in R_+$;

(ii) $V \in C\left[\mathbb{R}_+ \times S(\rho), \mathbb{R}_+^n\right]$, $V(t, x)$ is locally Lipschitzian in x and the function

$$V_0(t, x) = \sum_{i=1}^{n} V_i(t, x),$$

is positive definite and decrescent;

(iii) $P \in C\left[\mathbb{R}_+ \times S(\rho), \mathbb{R}^n\right]$, $P(t,0) \equiv 0$ and

$$^c D^q V(t,x) = f\left(t, V(t,x)\right), \quad (t,x) \in R_+ \times S(\rho).$$

Then, the stability properties of the trivial solution of

$$^c D_+^q u = f(t,u), \quad u(t_0) = u_0 \geq 0,$$

imply the corresponding stability properties of a trivial solution of Eq. (5.42).

Applications of this type of mathematical model for the study of various physical phenomena are increasing day by day. This is reflected in the ever-growing number of publications in which the authors opt to model using FDs. The publications cover a wide range of models beginning from as simple a model as the fractional harmonic oscillator to as complex as the fractional Schrodinger equation.

5.10 Differential Equations with Retardation and Anticipation

DEs with R&A are used to model systems whose behavior at a given time are influenced by the memory of their past states as well as by their anticipated future potential states. An example from economics of such models is the Kaldor–Kalecki model of business cycle, which was extended by Dubois [53] to include anticipatory capabilities. In addition to the time delay between a decision of investment and its actual implementation, Dubois inserted into the model an anticipated capital stock, which is connected to the expected value of the current capital stock. Other applications of such models can be found in the study of chaotic systems, and notably in synchronous, coupled chaotic systems, where a feedback loop injects a delayed signal into the system. The coupling and the delayed feedback together result in a phenomenon called anticipated synchronization. Voss [54] first reported this type of phenomenon in 2000. A number of systems with anticipated synchronization were studied subsequently [55–59].

In this section, we begin with a formulation of functional differential systems with anticipation and retardation, introduced in Refs. [60,61]. Then, we present some representative results from the qualitative theory of such systems, namely an existence result and the iterative technique of GOL. We also present a stability result that extends the concept of anticipatory synchronization to variable time delay.

As mentioned above, anticipation refers to future states of the system, while retardation takes into account past states. Such systems are also known as systems with PPF dependence, as they depend on past (P), present (P), and desired future (F) states.

A general IVP with retardation and anticipation is given by

$$x'(t) = p(t, x_t, x^t), \quad t \in [t_0, T], t \geq 0, x_{t_0} = \psi_0, \quad x^T = \eta_0 \qquad (5.43)$$

where $\psi_0 \in C_1$ and $\eta_0 \in C_2$, $C_1 = C[[-\tau_1, 0], \mathbb{R}^n]$, $C_2 = C[[0, \tau_2], \mathbb{R}^n]$, $\tau_1, \tau_2 \geq 0$, $p \in C[[t_0, T] \times C_1 \times C_2, \mathbb{R}^n]$ and $x_t(s) = x(t + s)$, $-\tau_1 \leq s \leq 0$, $x^T(\mu) = x(T + \mu)$, $0 \leq \mu \leq \tau_2$.

The function $\psi_0(s)$, defined on $t_0 - \tau_1 \leq s \leq t_0$, represents the retardation or past information, whereas η_0 represents the desired potential future state, and the function $\psi(t_0)$ the present state. In order to achieve the desired future state, appropriate decisions have to be made. This is represented in Refs. [61,62] by a decision function $z \in C[[t_0, T], \mathbb{R}^n]$ such that $z(t_0) = \psi_0(t_0)$, $z(T) = \eta_0(T)$, and $z^t = a(t + \sigma)$, $0 \leq \sigma \leq \tau_2$ for $t \in [t_0, T]$, with $z^T = \eta_0$, at the tail end.

Under such a representation or model, IVP with R&A (Eq. 5.43) reduces to

$$x'(t) = p(t, x_t, z^t) = P(t, x_t), \quad x_{t_0} = \psi_0, \qquad (5.44)$$

which is a functional differential equation with delay. Note that this equation uses future information from $[t_0, T]$. The following theorem from Ref. [61] gives conditions for the existence of a unique solution.

Theorem 5.10.1

Suppose the functional p in Eq. (5.44) satisfies

$$\left| p(t, \psi_1, z^t) - p(t, \psi_2, z^t) \right| \leq L(\|\psi_1 - \psi_2\|)$$

where $\|\psi_1 - \psi_2\|_0 = \max_{-\tau \leq s \leq 0} |\psi_1(s) - \psi_2(s)|$

for $\psi_1, \psi_2 \in C_1$ and $t \in [t_0, T]$. If $0 < \alpha < \dfrac{1}{2L}$, then for every choice of $z(t)$ there exists a unique solution $x(t_0, \psi_0)(t)$ of Eq. (5.44) on $t_0 \leq t \leq t_0 + \alpha$.

The authors also proved local and global existence theorems in this setup and showed that, for a chosen $z(t)$, the solution must satisfy the condition

$$x(t_0, \psi_0)(T) = \eta_0(T) = z(T), \qquad (5.45)$$

The following theorem gives the existence of $z(t)$ which was proved in Refs. [61,62].

Theorem 5.10.2

Assume that the solutions of DEs with R and A (Eq. 5.44) exist and are unique on $[t_0, T]$. Suppose further that p satisfies the following condition

$$2e^{\alpha t}\left\langle \psi_1(0) - \psi_2(0), p\left(t, \psi_1, z_1^t\right) - p\left(t, \psi_2, z_2^t\right)\right\rangle$$

$$+ e^{\alpha t}\left|\psi_1(0) - \psi_2(0)\right|^2 \leq C_1 e^{\alpha t}\left|z_1 - z_2\right|_0^2$$

where $C_1 > 0, \alpha > 0, |z_1 - z_2|_0 = \max_{t_0 \leq t \leq T + \tau_2}|z_1(t) - z_2(t)|, \psi_1, \psi_2 \in \Omega = \{\psi_1, \psi_2 \in C_1:$

$\max_{-\tau_1 \leq s \leq 0}|\psi_1(s) = \psi_2(s)|e^{\alpha(t+s)} = |\psi_1(0) - \psi_2(0)|e^{\alpha t}\}$. If $\left(\dfrac{C_1}{\alpha}\right)^{\frac{1}{2}} < 1$, then there exists a function $z \in C\left[[t_0, T], \mathbb{R}^n\right)$ such that (Eq. 5.45) holds.

Next, we present a uniqueness result developed in Ref. [63], obtained using the method of QL. This method is an iterative technique that produces monotone sequences of iterates that converge to a unique solution. The sequences of iterates are solutions of a sequence of linear differential equations with retardation and anticipation. The existence and uniqueness of the solutions of the linear differential equation was proved using the basic MIT. The advantage of the QL technique lies in the quadratic convergence of the sequences of iterates. This is achieved by imposing a convexity condition or a convexity-like condition on some of the arguments of p in Eq. (5.44).

We first state a lemma that is needed to prove the existence and uniqueness theorem based on QL.

Lemma 5.10.3

(i) Let $p \in C\left[[t_0 - h_1, T + h_2], \mathbb{R}^n\right]$ be continuously differentiable on $I = [t_0, T]$ and

$$p'(t) \leq -Mp(t) - N\int_{-h_1}^{0} p_t(s)ds \text{ on } I,$$

(ii) $p_{t_0}(s) \leq 0, \quad s \in [-h_1, 0], \quad p \in C^1[t_0 - h_1, t_0, R]$, and $p'(t) \leq \dfrac{\lambda}{T + h_1}$, where $t \in [t_0 - h_1, t_0], \quad \min_{[t_0 - h_1, t_0]} p(s) = -\lambda, \lambda \geq 0$, and $[M + Nh_1](T + h_1) \leq 1$.

Then, $p(t) \leq 0$ on I.

Theorem 5.10.4

(H_1)

    (i) All the second-order Frechet derivatives of $p(t,x,\psi,\eta)$ exist and are bounded;

    (ii) $p(t,x,\psi,\eta)$ is convex in x,ψ and is concave in η,

    (iii) $p_x(t,x,\psi,\eta)$ is increasing in ψ for each (t,x,η) and is independent of η for each (t,x,ψ)

    (iv) $p_\psi(t,x,\psi,\eta)$ is increasing in x for each (t,ψ,η) and is independent of η for each (t,x,ψ)

    (v) $p_\eta(t,x,\psi,\eta)$ is increasing in x for each (t,ψ,η) and is independent of ψ for each (t,x,η);

(H_2)

    (i) $-M \le p_x(t,x,\psi,\eta) \le -N_1, \quad 0 < N_1 < M;$

    (ii) $-L\displaystyle\int_{-\tau_1}^{0} \eta(s)ds \le f_\psi(t,x,\psi,\eta)\eta \le N_2\int_{-\tau_1}^{0} \eta(s)ds, 0 < N_2 < L,$ where $\eta \in C_1;$

    (iii) $0 \le p_\eta(t,x,\psi,\eta)\xi \le N_3\displaystyle\int_{0}^{\tau_2} \xi(s)ds,$ where $\xi \in C_2, N_2 > 0;$

    (iv) $N_2\tau_1 + N_3\tau_2 < 1;$

(H_3)

    α_0, β_0 are lower and upper solutions of Eq. (5.44) that is, $\alpha_0, \beta_0 \in C^1[I,R],$ where $I = [0,T]$

$$\alpha_0'(t) \le p\left(t,\alpha_0,\alpha_{0t},\alpha_0^t\right) \quad \alpha_{00} = \phi_1, \quad \alpha_0^T = \eta_1$$

$$\beta_0'(t) \le p\left(t,\beta_0,\beta_{0t},\beta_0^t\right) \quad \beta_{00} = \phi_1, \quad \beta_0^T = \eta_2$$

where $\alpha_0(t) \le \beta_0(t),$ and $\psi_1,\psi_2 \in C_1, \eta_1,\eta_2 \in C_2$ such that $\psi_1 \le \psi_0 \le \psi_2$ and $\eta_1 \le \eta_0 \le \eta_2$

(H_4)

    $a_{00} - \psi_0, \psi_0 - \beta_{00}$ satisfy the assumptions of Lemma 5.10.3. Then, there exist monotone sequences $\{\alpha_n(t)\}, \{\beta_n(t)\}$, which converge uniformly on I to the unique solution of DEs with R&A (Eq. 5.44) and the convergence is quadratic.

    The concept of anticipation was extended to differential equations with piecewise-constant arguments. An application of IDEs with piecewise-constant arguments to impulsive cellular networks (ICNNs) can be found in Ref. [64]. The authors describe a mathematical model for ICNNs in terms of an impulsive differential system that is of alternately advanced and retarded type, for which they obtain sufficient criteria for a unique equilibrium to exist. The result is important as it is based on easily verifiable conditions.

5.11 Matrix and Graph Differential Equations

It is well known that any system, natural or man-made, that involves interconnections between its members can be represented by a network. This network can be represented by a graph. Graphs are useful when modeling organizations in social sciences [65]. But a static graph cannot represent an organization that is dynamic or is changing with time. Thus, there is a need for a graph that changes with time. This need was addressed by Siljak in Ref. [66] when he introduced a dynamic graph, that is a graph function that varies with time. In Ref. [66], a graph linear space is defined, and a stability result for the predator–prey model is proved in that setup.

In Refs. [66, 67], the building blocks of structures, addition and scalar multiplication, were introduced to define the graph linear space, and a stability result was proved for a prey–predator problem. Observing that the notion of a simple directed graph gives little scope for developing the theory, the idea of a pseudo simple graph was introduced in Ref. [68] as follows.

Definition 5.11.1

A simple graph G having loops is called a pseudo simple graph.

By considering $D_N = \{ G \, / \, G \text{ is (i) weighted (ii) directed (iii) pseudo simple graph} \}$ defined on a set of N vertices $[v_1, v_2, \ldots v_N]$ and using the definitions given for sum and scalar multiplication of pseudo simple graphs, the linear space $(D_N, +, \cdot)$ was introduced.

Corresponding to D_N, the set of adjacency matrices E_N was considered. Then, $(E_N, +, \cdot)$ is a linear space with standard definitions of "+ and ·" defined on matrices.

Considering an IVP of an MDE, the existence results, uniqueness results, and comparison theorems have been obtained in Ref. [69]. In order to study a linear GDE, the notion of a product of two graphs was introduced in Ref. [68] as follows.

Definition 5.11.2

Product graphs: Let G_1 and G_2 be two graphs with edges $\left(e_{ij} \right)_{N \times N}$ and $\left(d_{ij} \right)_{N \times N}$, respectively. Then, the product of the two graphs G_1 and G_2 is the graph G in which the weight g_{ij} of the edge from v_j to v_i is the dot product of the vectors, one having the weights of the edges inwards to v_i and the other having weights of the edges outwards from v_j.

A result concerning a linear GDE is now discussed from Ref. [68].

A linear GDE is given by

$$X' = CX$$
$$X(t_0) = X_0,$$

(5.46)

where C is a graph called a coefficient graph and X_0 is the initial graph.
  Let

$$Y' = AY$$
$$Y(t_0) = Y_0$$

(5.47)

be the corresponding IVP of MDE, where Y_0 is the adjacency matrix corresponding to the initial graph X_0. The following result relating to the solutions of MDE and hence to that of GDEs is from Ref. [68].

Theorem 5.11.3

Suppose $Y(t)$ is a solution of IVP of the linear MDE (Eq. 5.47). Let P be an invertible matrix such that $P^{-1}AP = H$ is a diagonal matrix. Then, the solution $Y(t)$ has the same nature as that of Y_0. In other words, the solution $X(t)$ of the IVP of GDE (Eq. 5.46) has the same nature as that of the initial graph X_0.

  The basic theory, like existence and uniqueness results, comparison theorems, solutions depending continuously on initial values, iterative methods like MIT and QL have been established for GDE and its associated MDE in Refs. [69–71].
  We now proceed to discuss a stability result in a very general setup of two measures given in Ref. [72].
  Consider IVP for MDE

$$X' = P(t, X), \quad X(t_0) = X_0, \quad t \geq t_0,$$

(5.48)

where $P \in [R_+ \times \mathbb{R}^{N \times N}, \mathbb{R}^{N \times N}]$. We assume that P is such that the solution $X(t) = X(t, t_0, X_0)$ of the IVP, MDE (Eq. 5.48) exists uniquely and depends continuously on initial values. Before proceeding further, we introduce the following concepts.

Definition 5.11.4

The differential system (Eq. 5.48) is
  $(h_0 - h)$-equistable if, "for each $\epsilon > 0$, $t_0 \in \mathbb{R}_+$, there exists a positive function $\delta = \delta(t_0, \epsilon)$ that is continuous in t_0 for each ϵ such that $h_0(t_0, X_0) < \delta$ implies $h(t, X(t)) < \epsilon$, $t \geq t_0$ for any solution $X(t) = X(t, t_0, X_0)$ of the system (Eq. 5.48)."

  Next, the following definitions from Ref. [6] are necessary to state a stability result.

Definition 5.11.5

Let $h_0, h \in \Gamma$. Then, "h_0 is finer than h" if "there exist a $\rho > 0$ and a function $\phi \in CK$ such that $h_0(t, X) < \rho$ implies $h(t, X) \leq \phi(t, h_0(t, X))$."

Definition 5.11.6

Let $V \in C\left[\mathbb{R}_+ \times \mathbb{R}^{N \times N}, \mathbb{R}_+\right]$ then V is said to be

(i) "h-positive definite" if "there exist $\rho > 0$ and a function $b \in K$ such that $b\big(h(t, X) \leq V(t, X)\big)$ whenever $h(t, X) \leq \rho$";

(ii) "h-weakly decrescent" if "there exist $\rho > 0$ and a function $a \in CK$ such that $V_0(t, X) \leq a\big(t, h(t, X)\big)$ whenever $h(t, X) < \rho$."

Next, given $V \in C\left[\mathbb{R}_+ \times \mathbb{R}^{N \times N}, \mathbb{R}_+\right]$, Dini derivative of V is given by

$$D^+V(t, X) = \limsup_{\delta \to 0^+} \frac{1}{\delta}\left[V\big(t + \delta, X + \delta P(t, X)\big) - V(t, X)\right] \tag{5.49}$$

for $(t, X) \in \mathbb{R}_+ \times \mathbb{R}^{N \times N}$.

The Lyapunov theorem using two measures is as follows.

Theorem 5.11.7

Assume

(H_1) $V \in C\left[\mathbb{R}_+ \times \mathbb{R}^{N \times N}, \mathbb{R}_+\right], h \in \Gamma, V(t, X)$ is locally Lipschitzian in X *and h-*
positive definite;

(H_2) $D^+V(t, X) \leq 0, (t, X) \in S(h, \rho) = \left\{[t, X) \in \mathbb{R}_+ \times \mathbb{R}^{N \times N}, h(t, X) \langle \rho, \rho \rangle\right\}$;

(H_3) $h_0 \in \Gamma, h_0$ is finer than h and $V(t, X)$ is h_0 weakly decrescent. Then, the system (Eq. 5.48) is (h_0, h)-equistable.

Using the above theorem, we obtain a similar result for the IVP of GDE given by

$$Y' = G(t, Y), \quad Y(t_0) = Y_0, \tag{5.50}$$

where $G \in C\left[\mathbb{R}_+ \times D_N, D_N\right]$. In the following theorem, we investigate the stability of system (Eq. 4.50). The fact that graphs and matrices are isomorphic is used in the proof of the following theorem.

Theorem 5.11.8

Let $F(t, X)$ be a function that is isomorphic to $G(t, Y)$ in GDE (Eq. 4.50), where $F \in C\left[\mathbb{R}_+ \times \mathbb{R}^{N \times N}, \mathbb{R}_+\right]$. Further, let $V \in C\left[\mathbb{R}_+ \times \mathbb{R}^{N \times N}, \mathbb{R}_+\right]$ be a function satisfying the conditions of Theorem 5.11.7. Then, system (Eq. 4.50) is equistable.

In Ref. [72], few examples are given to illustrate the stability concepts. Further, using the comparison theorem, various stability results were proved for GDEs using the associated MDEs in Ref. [72].

The fact that a graph is used to study networks showcases the importance of this budding area of research. MDEs and GDEs are yet to be used as mathematical models. In Refs. [66,68], the prey–predator model was studied using GDE.

5.12 Conclusion

In conclusion, it is clear that the concepts of differential inequalities, comparison principle, and iterative techniques using upper and lower solutions are powerful tools for studying a wide range of differential equations. The results presented in this chapter represent only a small sample of a vast amount of research being done in this area. Also, more complex phenomena were modeled by combining the types of differential equations mentioned in this chapter. Examples of such models are DDE with impulses, SDE with delay and impulses, impulsive FrDEs, and impulsive FDEs, to name only a few.

References

1. Lakshmikantham, V., Leela, S. G., *Differential and Integral Inequalities*, Vol I. Academic Press, New York, 1969.
2. Agarwal, R. P., Lakshmikantham, V., *Uniqueness and Nonuniqueness Criteria for Ordinary Differential Equations*. Series in Real Analysis, Vol 6. World Scientific Publishing Co., Inc., River Edge, NJ, 1993.
3. Lakshmikantham, V., Leela, S., *Nonlinear Differential Equations in Abstract Spaces*. Pergamon Press, Oxford, 1981.
4. Lyapunov, A. M., Problmé general de stabilité du movement, *Annales de la Faculté des sciences de Toulouse*, 9 (1907), 203–474.
5. Lakshmikantham, V., Leela, S., Martynyuk, A. A., *Practical Stability of Nonlinear Systems*. World Scientific Publishing Co., Inc., Teaneck, NJ, 1990.
6. Lakshmikantham, V., Liu, X. Z., *Stability Analysis in Terms of Two Measures*. World Scientific Publishing Co., Inc., River Edge, NJ, 1993.
7. Drici, Z., Vasundhara Devi, J., McRae, F. A., Differential inequalities and the comparison principle: The core of Professor V. Lakshmikantham's research, *Nonlinear World, An International Journal*, 1 (2018), 5–9.
8. Lakshmikantham, V., Leela, S., *Differential and Integral Inequalities*, Vol II. Academic Press, Cambridge, MA, 1969.

9. Ladde, G. S., Lakshmikantham, V., Vatsala, A. S., [Monotone iterative techniques for nonlinear differential equations. Monographs, Advanced Texts and Surveys in Pure and Applied Mathematics, 27]. Pitman (Advanced Publishing Program), Boston, MA; distributed by John Wiley & Sons, Inc., New York, 1985.

10. Bellman, R., Kalaba, R., *Quasilinearization and Nonlinear Boundary Value Problems.* American Elsevier, New York, 1965.

11. Lakshmikantham, V., Matrosov, V. M., Sivasundaram, S., *Vector Lyapunov Functions and Stability Analysis of Nonlinear Systems.* Kluwer Publishers, Dordrecht, the Netherlands, 1991.

12. Lakshmikantham, V., Leela, S., Martynuk, A. A., *Stability Analysis of Nonlinear System.* Marcel Dekker, New York, 1989.

13. Lakshmikantham, V., Vatsala, A. S., Differential inequalities with initial time difference and applications, *Journal of Inequalities and Applications*, 3 (1999), no. 3, 233–244.

14. Lakshmikantham, V., Leela, S., Vasundhara Devi, J., Another approach to the theory of differential inequalities relative to changes in the initial times, *Journal of Inequalities and Applications*, 4 (1999), no. 2, 163–174.

15. McRae, F. A. , Perturbing Lyapunov functions and stability criteria for initial time difference. (English summary), *Applied Mathematics And Computation*, 117 (2001), no. 2–3, 313–320.

16. Pandit, S. G., Dezern, D. H., Adeyeye, J. O., Periodic boundary value problems for nonlinear integro differential equations, *Proceedings of Neural; Parallel; and Scientific Computations*, 4 (2010), 316–320.

17. West, I. H., Vatsala, A. S., Generalized monotone iterative method for initial value problems, *Applied Mathematics Letters*, 17 (2004), 1231–1237, 2444

18. Wang, W. L., Tian, J. F., Generalized monotone iterative method for non-linear boundary value problems with causal operators, *Boundary Value Problems*, 2014 (2014) no. 1, 192.

19. Vasundhara Devi, J., Bharat Iragavarapu, S. N. R. G., Srinivasa Rao, S., Quasilinearization technique for boundary value problem of graph differential equations and its associated matrix differential equations, *Dynamics of Continuous, Discrete and Impulsive Systems, Series B: Applications and Algorithms*, 23 (2016), 287–300. Watam Press.

20. M. Rama Mohana Rao, Ordinary differential equations: Theory and applications. Affiliated East–West Press, New Delhi (1980).

21. Lakshmikantham, V., Vatsala, A. S., *Generalized Quasilinearization for Nonlinear Problems. Mathematics and Its Applications*, Vol 440. Kluwer Academic Publishers, Dordrecht, 1998.

22. Lakshmikantham, V., Leela, S., McRae, F. A., Improved generalized quasilinearization (GQL) method, *Nonlinear Analysis: Theory, Methods & Applications*, 24 (1995) no. 11, 1627–1637.

23. LaSalle, J., Lefchetz, S., *Stability by Liapunov's Direct Method with Applications.* Academic Press, New York/London, 1961.

24. Mira-Cristiana Anisiu. Lotka, Volterra and their Model Didactica Mathematica, Vol. 32(2014), pp. 9–17.

25. Lakshmikantham, V., Rama, M. R. M., *Theory of Integro-Differential Equations. Stability and Control: Theory, Methods and Applications*, Vol 1. Gordon and Breach Science Publishers, Lausanne, 1995.

26. Tang, P., Fan, A., Li, J., Jiang, J., Wang, K., Effect of delay on selection dynamics in long-term sphere culture of cancer stem cells, *Discrete Dynamics in Nature and Society*, 2013 (2013), Article ID 606250, 5, doi: 10.1155/2013/606250.

27. Rihan, F. A., Tunc, C., Saker, S. H., Lakshmanan, S., Rakkiyappan, R., Applications of delay differential equations in biological systems, *Complexity*, Special Issue, 2018 (2018), Article ID 4584389, 3. doi: 10.1155/2018/4584389.

28. Keller, A. A., Time-delay systems with application to mechanical engineering process dynamics and control, *International Journal Of mathematics and Computers in Simulation*, 12 (2018), 64–73.

29. Kazmerchuk, Y. I., The pricing of options for securities markets with delayed response, *Mathematics and Computers in Simulation*, 75 (2007) no. 3–4, 69–79.

30. Ladde, G. S., Lakshmikantham, V., *Random Differential Inequalities. Mathematics in Science and Engineering*, Vol 150. Academic Press, Inc. [Harcourt Brace Jovanovich, Publishers], New York/London, 1980.

31. McRae, F. A., Generalized quasilinearization of stochastic initial value problem, *Stochastic Analysis and Applications*, 13 (1995) no. 2, 205–210.

32. Poor, H. V., *An Introduction to Signal Detection and Estimation*, Springer, New York, 1994, 2nd ed.

33. Wang, X, Chiang, H. D., Wang, J., Liu, H., Wang, T., Long-term stability analysis of power systems with wind power based on stochastic differential equations: Model development and foundations, *IEEE Transactions on Sustainable Energy*, 6 (2015) no. 4, 1534–1542.

34. Üçer, E., Söylemez, M., Stochastic rolling motion of ships in following seas, *Ocean Engineering*, 38 (2011) no. 8, 1001–1006.

35. Zhang, C., Sato, T., Lu, L., A phase model of earthquake motions based on stochastic differential equation, *KSCE Journal of Civil Engineering*, 15 (2011) no. 1, 161–166.

36. Duun-Henriksen, A. K., Schmidt, S., Røge, R. M., Møller, J. B., Nørgaard, K., Jørgensen, J. B., Madsen, H., Model identification using stochastic differential equation grey-box models in diabetes, *Journal of Diabetes Science and Technology*, 7 (2013) no. 2, 431–440.

37. Lakshmikantham, V., Bainov, D. D., Simeonov, P. S., *Theorey of Impulsive Differential Equations*. World Scientific, Singapore, 1989.

38. Wang, P., Liu, X., Nonlinear variation of parameters formula for impulsive differential equations with initial time difference and application, *Abstract and Applied Analysis*, 2014 (2014), Article ID 725832, 6.

39. Lakshmikantham, V., Mohapatra, R. N., *Theory of Fuzzy Differential Equations and Inclusions. Series in Mathematical Analysis and Applications*, Vol 6. Taylor & Francis, Ltd., London, 2003.

40. Chakraverty, S., Tapaswini, S., Behera, D., *Fuzzy Differential Equations and Applications for Engineers and Scientists*. CRC Press, Boca Raton, FL, 2017.

41. Brandao Lopes Pinto, A. J., De Blasi, F. S., Iervolino, F., Uniqueness and existence theorems for differential equations with compact convex-valued solutions, *Bollettino dell'Unione Matematica Italiana* 3 (1970), 47–54.

42. Vasundhara Devi, J., Basic results in set differential equations, *Non-Linear Studies*, 10 (2003) no. 3, 259–272.

43. Lakshmikantham, V., Gnana Bhasker, T., Vasundhara Devi, J., *Theory of Set Differential Equations in Metric Spaces*. Cambridge Scientific Publishers, Cottenham, UK, 2006.

44. Katugampola, U. N., A new approach to generalized fractional derivatives, *Bulletin of Mathematical Analysis and Applications*, 6 (2014), no. 4, 1–15.
45. Khalil, R., Horani, M. A., Yousef, A., Sababheh, M., A new definition of fractional derivative, *Journal of Computational and Applied Mathematics*, 264 (2014), 65–70.
46. Podlubny, I., *Fractional Differential Equations*. Academic Press, San Diego, CA, 1999.
47. Kilbas, A. A., Srivatsava, H. M., Ttujillo, J. J., *Theory and Applications of Fractional Differential Equations*. Elsevier, Amsterdam, 2006.
48. Lakshmikantham, V., Leela, S., Vasundhara Devi, J., *Theory of Fractional Dynamic Systems*. Cambridge Scientific Publishers, Cotterham, 2009.
49. Diethelm, K., *The Analysis of Fractional Differential Equations, An Application Oriented Exposition Using Differential Operators of Caputo Type*. Springer, New York, 1993.
50. Oldham, K. B., Spanier, J., *The Fractional Calculus*. Academic Press, New York/London, 1974.
51. Drici, Z., McRae, F. A., Vasundhara Devi, J., On existence and stability of hybrid Caputo fractional differential equations, *Dynamics of Continuous, Discrete and Impulsive Systems Series A Mathematical Analysis*, 19 (2012) no. 4, 501–512.
52. Vasundhara Devi, J., McRae, F. A., Drici, Z., Variational Lyapunov method for fractional differential equations, *Computers and Mathematics with Applications*, 64 (2012) no. 10, 2982–2989.
53. Dubois, D. M., Extension of the Kalder-Kalecki model of business cycle with computational anticipated capital stock, Journal of Organisational Transformation & Social Change, 1 (2004) 63–80.
54. Voss, H. U., Anticipating chaotic synchronization, *Physical Review*, E 61 (2000), 5115–5119.
55. Leydesdorff, L., Daniel, M. D., Anticipation in social sytems: The incursion and communication of meaning, *International Journal of Computing Anticipatory Systems*, 15 (2004) 203–216.
56. Pyragas, K., Pyragiene, T., Extending anticipation horizon of chaos synchronization schemes with timedelay coupling, *Philosophical Transactions of the Royal Society A*, 368 (2010), 305–317.
57. Ciszak, M., Toral, R., Mirassio, C., Coupling and feedback effects in excitable systems: anticipated synchronization, *Modern Physics Letters B*, 18 (2004) no. 23, 1135–1155.
58. Dubois, D. M., Mathematical foundations of discrete and functional systems with strong and weak anticipations, in *Anticipatory Behavior in Adaptive Learning Systems, State-of-the- Art Survey*, Edited by Martin Butz et al, Lecture Notes in Artificial Intelligence. Springer, Berlin, Heidelberg, LNAI 2684, pp. 110–132, 2003
59. Ambika, G., Amritkar, R. E., Anticipatory synchronization with variable time delay and reset, arXiv:0810.5613v1 [nlin.CD], 31 Oct 2008.
60. Lakshmikantham, V., Vasundhara Devi, J., Functional differential equations with anticipation, *Nonlinear Studies*, 14 (2007) no. 3, 235–240.
61. Gnana Bhaskar, T., Lakshmikantham, V., Functional differential systems with retardation and anticipation, *Nonlinear Analysis; Real World Applications*, 8 (2007), 865–871.
62. Vasundhara Devi, J., Drici, Z., Mc Rae, F. A., Differential equations-retardation, anticipation and synchronized anticipation-A survey, *Mathematics in Engineering, Science And Aerospace (MESA)*, 7 (2016) no. 2, 349–374.

63. Vasundhara Devi, J., Drici, Z., McRae, F. A., Quasilinearization for functional differential equations with retardation and anticipation, *Nonlinear Analysis*, 70 (2009), 1763–1775.
64. Chiu, K. S., Existence and global exponential stability of equilibrium for impulsive cellular neural network models with piecewise alternately advanced and retarded argument, *Abstract and Applied Analysis*, 2013 (2013), Article ID 196139, 13, doi: 10.1155/2013/196139.
65. Radclie-Brown, A. R., On social structure, *Journal of the Royal Anthropological Institute of Great Britain and Ireland*, 70 (1940) 1–12.
66. Siljak, D. D., Dynamic graphs, *Nonlinear Analysis, Hybrid Systems*, 2 (2008), 544–547.
67. Vasundhara Devi, J., Ravi Kumar, R. V. G., Giri Babu, N., On graph Differential equations and its associated matrix differential equations, *Malaya Journal of Mathematik*, 1 (2012) no. 1, 82–90.
68. Vasundhara Devi, J., Ravi Kumar, R. V. G., Modeling the prey predator problem by a graph differential equation, *European Journal of Pure and Applied Mathematics*, 7 (2014) no. 1, 37–44.
69. Ravi Kumar, R. V. G., Srinivasa Rao, S., Existence result for graph differential equations and its associated matrix differential equations, *Global Journal of Mathematical Sciences*, 2 (2015) no. 2, 107–110. IFNA Publishers.
70. Vasundhara Devi, J., Basic theory for graph differential equations and associated matrix differential equations, *Proceedings of Dynamic Systems and Applications*, 7 (2016), 1–7.
71. Ravi Kumar, R. V. G., Bharat, I. S. N. R. G., Generalized quasilinearization for graph differential equations and its associated matrix differential equations, *Global Journal of Mathematical Sciences*, 2 (2015) no. 2, 71–75. IFNA Publishers.
72. Vasundhara Devi, J., Stability in terms of two measures for matrix differential equations and graph differential equations, *Nonlinear Dynamics and Systems Theory*, 16 (2016) no. 2, 179–191.

6

Modeling of Thermal Radiation and Magnetic Effects on Cu–Water Nanofluid Flow Embedded in Porous Medium Nearby a Stagnation Point Past a Stretching/Shrinking Plate with Suction/Blowing and Heat Source/Sink Using Keller-Box Method

Santosh Chaudhary and KM Kanika

Malaviya National Institute of Technology Jaipur

CONTENTS

6.1 Introduction

Without error analyses of the practical and industrial applications in the field of combustion and thermal engineering, effects of thermal radiation cannot be ignored, because thermal radiation strongly impacts the complete heat transfer system. In the advanced energy conversion systems design at high temperature, radiation acts as a relevant character. When surface and ambient temperature difference is large, then the radiation heat transfer becomes more important. Radiation is vital in view of their applications, particularly solar power technology, electrical power generation, gas turbines, missiles, astrophysical ground, reliable equipment design, satellites, and different propulsion devices for aircraft. A study on the influence of kinetic, radiation, and thermal diffusion is pioneered by Elperin and Krasovitov [1]. Further, Yang [2] explored the impacts of thermal radiation on real-based applications. Ever since, several investigations are conducted on thermal radiation problems, such as by Molla and Hossain [3], Jat and Chaudhary [4], Mahapatra et al. [5], Sinha and Shit [6], and Rashad [7].

Flow analysis of electrically conducting fluids such as salted water, electrolytes, plasmas, and liquid metals in the presence of magnetic field is known as magnetohydrodynamic (MHD), which depends on the magnetic induction strength. For the stronger magnetic force, the hall impact cannot be neglected if influence yields over the hall currents. To arrange the nature of boundary layer, MHD principal technique is used, which affect the area of flow in the desired direction. Some novel characteristics of MHD flow have potential attention in different engineering processes like control of boundary layer in aerodynamics, extraction of geothermal energy, MHD power generators, plasma studies, crystal growth, and magnetic drug targeting. Maiellaro and Labianca [8] discovered the stability of MHD flow along the Couette–Poiseuille flow applications. Moreover, Neslituk and Tezer-Sezgin [9] discussed the MHD flow at greater value of Hartmann number by using the finite element analysis. Consequently, Jat and Chaudhary [10], Chaudhary and Kumar [11], Mabood and Khan [12], and Das et al. [13] inspected the numerical analysis of MHD boundary layer flow in various situations.

Water and organic fluids, particularly ethanol, oil, and ethylene glycol, have minor thermal conductivity. So to enhance the thermal performance and improve the heat transfer system, the solid nanoparticles with the diameter of 1–100 nm were added in the ordinary fluid, and this special type of mixture is considered as nanofluid. Nanosolid particles are taken usually, which are made by several materials, specifically copper (Cu), silver (Ag), gold (Au), aluminum oxide (Al_2O_3), copper oxide (CuO), and titanium dioxide (TiO_2). It has been observed that the conventional fluid's thermal conductivity rises with the range of 15%–40% in the presence of nanoparticles. Nanofluids are largely substantial in natural phenomenon via the area of physics, chemistry, biology, and chemical engineering. Analysis of nanofluids is apparent in cancer therapy, vehicle computer, nuclear reactors, advanced cooling system, and transformer cooling applications. Choi [14] developed the term nanofluid and showed that nanofluid is basically an advance fluid of class. Although this investigation is extended by Keblinski et al. [15] for the heat flow mechanism, latterly, Akbarinia and Laur [16], Sheikholeslami and Ganji [17], Safikhani and Abbasi [18], Wang and Su [19], and Sheikholeslami and Rokni [20] have established a broad range literature on nanofluid flow problems.

A matrix has pores identified as porous medium. A matrix structure is generally taken as solid and pores employed by fluid, with pores and solid matrix network are in continuous. Sometimes, medium characteristics such as thermal conductivity, permeability, and heat capacitance are dependent on the pore structure, solid matrix, and porosity of media. Problems of porous medium flow have importance in science, engineering, and geophysical applications such as system of geothermal energy, electrochemistry, regenerative heat exchangers, food processing, thermal and granular insulation, and fibers. Additionally, convection heat transfer can be developed using a porous medium. Initially, Pal and Talukdar [21] presented a numerical study of convection flow over a porous medium in the presence of MHD flow, while Kumar and Prasad [22] instigated same type of problem over the vertical slot. In recent years, numerous studies by Chaudhary and Kumar [23], Ramesh [24], and Xu and Cui [25] created the behavior of porous medium on the fluid flows over different configurations.

The motion of fluid adjacent to the stagnation domain is known as stagnation point flow, and it may exist on all solid bodies that can move in a fluid. At the stagnation point, local velocity vanishes, i.e., fluid velocity is zero, because at that point, the whole kinetic energy is transformed into internal energy. Stagnation region creates the highest pressure, the highest rate of heat transfer, and mass deposition. Fluid flow over the stagnation domain has received considerable attention recently because of its several applications in both industry and boundary layer processes. These include submarines and flow over the aircraft tips. First study in the area of stagnation region is represented by Hiemenz [26]. After that, an extension of this analysis is explored by Homann [27] for the consideration of the axisymmetric stagnation point flow. Later, boundary layer flow problems nearby a stagnation point under some conditions are studied by Jat and Chaudhary [28], Weidman [29], and Hayat et al. [30].

In the recent decades, fluid flow analysis over the stretching surface has attracted the research community along their boundless applications like rubber and plastic sheet manufacturing, processing of crystal growth, glass blowing, polymer extrusion processes, and food processing in the area of biomedical and engineering processes. Various applications of technological processes can be detected for shrinking surface, some examples of these applications are paper production, glass sheet manufacturing, and in textile industries. Moreover, the problem of both types of sheets as stretching and shrinking have numerous applications particularly, metallic plate cooling, aerodynamic and paper films drawing in the field of metallurgy and chemical engineering. Wang et al. [31] illustrated the concept of stretching/shrinking in plate–metal forming. In addition, fluid flow analysis near the stagnation region over a stretching/shrinking plate was inspected by Bachok et al. [32]. Recently, Turkyilmazoglu [33], Chaudhary and Choudhary [34], and Merkin and Pop [35] analyzed the combined illustrations due to stretching and shrinking sheets in boundary layer flow and heat transfer.

Over the bounding surface, the suction/blowing can accordingly change the area of flow. Mostly, suction leads the enhancement in skin friction, although blowing has an opposite phenomenon. Boundary layer behavior over suction/blowing of fluid is an important type of problem in several practical studies and engineering activities such as wires and polymer fiber coating, film cooling, radial diffusers and thrust-bearing design, and recovery of thermal oil. In chemical systems, suction is used to remove reactants, while blowing is applied to add reactants, diminish the drag, prevent scaling or corrosion, and for surface cooling. First, the exploration for the concept of non-Newtonian fluid flow over a channel field along the convective boundary condition and suction/injection impacts is introduced by Rundora and Makinde [36]. Khan et al. [37] and Chaudhary and Choudhary [38] studied different aspects of the flow problem of suction/blowing and described them either analytically or numerically.

For an ecological procedure, heat source and sink represent origin and destination of process, respectively. For example, landscape patches with greater temperature are characterized as heat sources, while landscape patches along minor temperatures are interpreted as heat sinks, which utilized influences of cooling on the surrounding environment. Physical problems of heat source/sink impact in fluids have massive utilization in industrial processes, including flow, fluids undergoing chemical reaction, and heat and mass transfer due to a sheet. Aftereffect, Chaudhary et al. [39] detected an unsteady hydromagnetic flow over a heated porous stretching surface along the thermal radiation and heat source/sink impact. Besides, problems with the heat source/sink condition and penetrative convection over certain geometries have been solved by Harfash [40] and Nandal and Mahajan [41].

In view of the aforementioned literature, the main aim of present investigation is to extend the research of Pal et al. [42] along the influence of viscous dissipation and Joule heating on an electrically conducting copper–water (Cu–water) nanofluid in the presence of magnetic field. Additionally, an efficient Keller-box method of the finite difference scheme is used to solve the resulting differential equations.

6.2 Problem Formulation

Consider an analysis of steady two-dimensional laminar stagnation point flow of Cu–water nanofluid along the porous medium towards a stretching/shrinking surface with free stream velocity $u_e = ax$ and surface velocity $u_w = \pm cx$, where a and c are positive constants and x is the coordinate measured along with the plate. There is no thermal equilibrium and no slip effect between solid spherical shape nanoparticle Cu and ordinary fluid water. A rectangular coordinate system (x,y) is assumed such that the x- and y-axes are taken parallel and perpendicular to the surface, respectively, while flow is confined in the upper half plane region $y \geq 0$ as shown in Figure 6.1. The wall mass flux v_w with suction for $v_w < 0$ or injection for $v_w > 0$ and the surface temperature $T_w = T_\infty + b\left(\dfrac{x}{l}\right)^2$ are surmised, where T_∞ is the constant ambient fluid temperature, b is a positive constant, and l is the reference length.

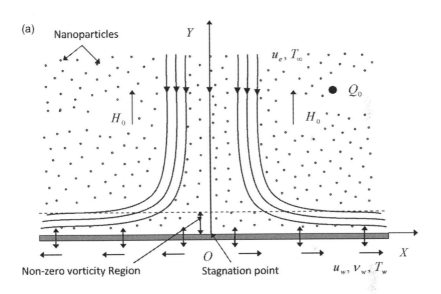

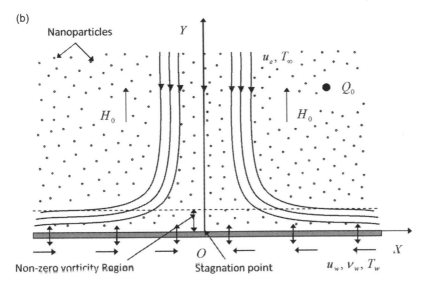

FIGURE 6.1
Flow configuration and coordinate system. (a): Stretching sheet. (b): Shrinking sheet.

TABLE 6.1
Thermophysical Properties of Base Fluid and Solid Nanoparticles

Material	κ(W/m K)	$\rho\left(\text{kg/m}^3\right)$	$C_p\left(\text{J/kg K}\right)$	σ_e(S/m)
Water	0.613	997.1	4,179	0.05
Cu	400	8,933	385	5.96×10^7

A uniform transverse magnetic field with constant intensity H_0 is utilized in the opposite direction of flow. A small value of the magnetic Reynolds number leads that the effect of induced magnetic field may be negligible. Consequently, thermophysical properties of base fluid water and spherical shape nanosolid particle Cu are given in Table 6.1 (Su and Zheng [43]). By applying the earlier considerations, the basic boundary layer equations are

$$\frac{\partial u}{\partial x} + \frac{\partial v}{\partial y} = 0 \tag{6.1}$$

$$\rho_{nf}\left(u\frac{\partial u}{\partial x} + v\frac{\partial u}{\partial y} - u_e\frac{du_e}{dx}\right) = \mu_{nf}\frac{\partial^2 u}{\partial y^2} - \frac{\mu_{nf}}{K}(u - u_e) - (\sigma_e)_{nf}\,\mu_e^{\;2}H_0^{\;2}(u - u_e) \tag{6.2}$$

$$\left(\rho C_p\right)_{nf}\left(u\frac{\partial T}{\partial x} + v\frac{\partial T}{\partial y}\right) = \kappa_{nf}\frac{\partial^2 T}{\partial y^2} + Q_0\left(T - T_\infty\right)$$

$$+ \frac{4\sigma^*}{3k^*}\frac{\partial^2 T^4}{\partial y^2} + \mu_{nf}\left(\frac{\partial u}{\partial y}\right)^2 + (\sigma_e)_{nf}\,\mu_e^{\;2}H_0^{\;2}(u - u_e)^2 \tag{6.3}$$

subjected to the associated boundary conditions

$$u = u_w,\ v = v_w,\ T = T_w \quad \text{at } y = 0$$

$$u \to u_e,\ v \to 0,\ T \to T_\infty \quad \text{as } y \to \infty \tag{6.4}$$

where subscript *nf* denotes the characteristics of nanofluid, u and v are the velocity factors in the direction of x- and y-axes, respectively, ρ is the density, μ is the coefficient of viscosity, K is the permeability of porous medium, σ_e is the electrical conductivity, μ_e is the magnetic permeability, C_p is the specific heat at constant pressure, T is the temperature of nanofluid, κ is the thermal conductivity, Q_0 is the heat generation coefficient when $Q_0 > 0$ or heat absorption coefficient when $Q_0 < 0$, σ^* is the Stefan–Boltzmann constant, and k^* is the mean absorption coefficient. In energy equation (6.3), the third, fourth, and fifth terms represented the impact of thermal radiation, viscous dissipation, and Joule heating, respectively.

Moreover, thermophysical properties of nanofluid, namely coefficient of viscosity, density, electrical conductivity, thermal conductivity, and heat capacitance, followed by Su and Zheng [43] are given as

$$\mu_{nf} = \frac{\mu_f}{(1-\phi)^{5/2}} \tag{6.5}$$

$$\rho_{nf} = (1-\phi)\rho_f + \phi\rho_s \tag{6.6}$$

$$(\sigma_e)_{nf} = \frac{(\sigma_e)_s + 2(\sigma_e)_f + 2\phi\left[(\sigma_e)_s - (\sigma_e)_f\right]}{(\sigma_e)_s + 2(\sigma_e)_f - \phi\left[(\sigma_e)_s - (\sigma_e)_f\right]}(\sigma_e)_f \tag{6.7}$$

$$\kappa_{nf} = \frac{\kappa_s + 2\kappa_f + 2\phi(\kappa_s - \kappa_f)}{\kappa_s + 2\kappa_f + \phi(\kappa_s - \kappa_f)}\kappa_f \tag{6.8}$$

$$(\rho C_p)_{nf} = (1-\phi)(\rho C_p)_f + \phi(\rho C_p)_s \tag{6.9}$$

where subscripts f and s represent the physical properties of base fluid and nanosolid particles, respectively, and ϕ is the nanoparticle volume fraction.

6.3 Analysis

To solve the governing equations (6.2) and (6.3), the stream function $\psi(x,y)$ that is defined in the usual way as $u = \dfrac{\partial\psi}{\partial y}$ and $v = -\dfrac{\partial\psi}{\partial x}$ symmetrically satisfies the continuity equation (6.1), and similarity variable η and nondimensional temperature T are introduced in the following mode as (Pal et al. [42])

$$\psi(x,y) = \sqrt{c\upsilon_f}\, x\, f(\eta), \quad \eta = \sqrt{\frac{c}{\upsilon_f}}\, y, \quad \theta(\eta) = \frac{T - T_\infty}{T_w - T_\infty} \tag{6.10}$$

where $\upsilon = \dfrac{\mu}{\rho}$ is the kinematic viscosity of fluid, $f(\eta)$ is the nondimensional stream function, and $\theta(\eta)$ is the nondimensional temperature.

By using the dimensionless variable equation (6.10), T^4 may be reduced as

$$T^4 = T_\infty^{\,4}\left[1 + \theta_{w_1}\theta(\eta)\right]^4 \tag{6.11}$$

where $\theta_{w_1} = \theta_w - 1$ with $\theta_m = \dfrac{T_w}{T_\infty}$ is the surface temperature excess ratio parameter.

Therefore, using the earlier equations, the governing boundary layer equations (6.2) and (6.3) can be obtained in a dimensionless form

$$f''' + E_1 E_2 \left(f f'' - f'^2 + \varepsilon^2 \right) - K_p \left(f' - \varepsilon \right) - \frac{(\sigma_e)_{nf}}{(\sigma_e)_f} E_1 M \left(f' - \varepsilon \right) = 0 \quad (6.12)$$

$$\left\{ \left[\frac{\kappa_{nf}}{\kappa_f} + Nr \left(1 + \theta_{w_1} \theta \right)^3 \right] \theta' \right\}' + E_3 Pr \left(f\theta' - 2f'\theta \right) + \lambda Pr\theta$$

$$+ \frac{1}{E_1} Br \left[f''^2 + \frac{(\sigma_e)_{nf}}{(\sigma_e)_f} E_1 M \left(f' - \varepsilon \right)^2 \right] = 0 \qquad (6.13)$$

along the corresponding boundary conditions

$$f = S, \, f' = \pm 1, \, \theta = 1 \quad \text{at } \eta = 0$$
$$f' \rightarrow \varepsilon, \, \theta \rightarrow 0 \quad \text{as } \eta \rightarrow \infty \qquad (6.14)$$

where prime ($'$) indicates the differentiation with respect to η, $E_1 = (1 - \phi)^{5/2}$,

$E_2 = \left(1 - \phi + \phi \dfrac{\rho_s}{\rho_f} \right)$, $E_3 = \left[1 - \phi + \phi \dfrac{(\rho C_p)_s}{(\rho C_p)_f} \right]$, $\varepsilon = \dfrac{a}{c}$ is the velocity ratio parameter,

$K_p = \dfrac{\upsilon_f}{cK}$ is the permeability parameter, $M = \dfrac{(\sigma_e)_f \mu_e^2 H_0^2}{c\rho_f}$ is the magnetic

parameter, $Nr = \dfrac{16\sigma^* T_\infty^{\,3}}{3\kappa_f k^*}$ is the radiation parameter, $Pr = \dfrac{(\mu C_p)_f}{\kappa_f}$ is the

Prandtl number, $\lambda = \dfrac{Q_0}{c(\rho C_p)_f}$ is the heat source/sink parameter, $Br = \dfrac{c^2 l^2}{b} \dfrac{\mu_f}{\kappa_f}$

is the Brinkmann number, and $S = -\dfrac{\upsilon_w}{\sqrt{c\upsilon_f}}$ is the mass flux parameter.

6.4 Quantities of Interest

The physical quantities of interest are the local skin friction coefficient C_f and local Nusselt number Nu, which are defined as

$$C_f = \frac{\mu_{nf} \left(\dfrac{\partial u}{\partial y} \right)_{y=0}}{\rho_f u_w^{\,2}/2} \qquad (6.15)$$

$$Nu = -\frac{x\left[\kappa_{nf}\left(\dfrac{\partial T}{\partial y}\right)_{y=0} + \dfrac{4\sigma^*}{3k^*}\left(\dfrac{\partial T^4}{\partial y}\right)_{y=0}\right]}{\kappa_f\left(T_w - T_\infty\right)} \qquad (6.16)$$

Applying the nondimensional similarity variable equation (6.10), equations (6.15) and (6.16) are converted in the following forms:

$$f''(0) = \frac{1}{2}E_1\sqrt{Re}\,C_f \qquad (6.17)$$

$$\theta'(0) = -\frac{1}{\left[\dfrac{\kappa_{nf}}{\kappa_f} + Nr\left(\theta_{w_1} + 1\right)^3\right]}\frac{1}{\sqrt{Re}}Nu \qquad (6.18)$$

where $Re = \dfrac{u_w x}{v_f}$ is the local Reynolds number.

6.5 Solution Technique

The Keller-box method is used to determine the numerical solutions of the governing ordinary differential equations (6.12) and (6.13) with the associated boundary condition equation (6.14). Also, far field boundary conditions as $\eta \to \infty = 6$ is surmised for computational processing.
   Introducing

$$f' = p \qquad (6.19)$$

$$p' = q \qquad (6.20)$$

$$\theta' = s \qquad (6.21)$$

then the Eqs. (6.12) and (6.13) can be reduced in the following form

$$q' + E_1 E_2\left(fq - p^2 + \varepsilon^2\right) - K_p\left(p - \varepsilon\right) - \frac{(\sigma_e)_{nf}}{(\sigma_e)_f}E_1 M(p - \varepsilon) = 0 \qquad (6.22)$$

$$\left\{\left[\frac{\kappa_{nf}}{\kappa_f} + Nr\left(1 + \theta_{w_1}\theta\right)^3\right]s\right\}' + E_3\,Pr(fs - 2p\theta)$$

$$+ \lambda\,Pr\theta + \frac{1}{E_1}Br\left[q^2 + \frac{(\sigma_e)_{nf}}{(\sigma_e)_f}E_1 M(p - \varepsilon)^2\right] = 0 \qquad (6.23)$$

with the relevant boundary conditions, equation (6.14) become

$$f = S, p = \pm 1, \theta = 1 \quad \text{at } \eta = 0$$

$$p \to \varepsilon, \theta \to 0 \quad \text{as } \eta \to \infty$$

$$(6.24)$$

6.5.1 Scheme of Finite Difference

The rectangular model $X\eta$ – plane is shown by Figure 6.2, whose net points are defined in the following way as

$$x_0 = 0, \ x_i = x_{i-1} + k_i, \quad i = 1, 2, 3, \dots I$$

$$\eta_0 = 0, \ \eta_j = \eta_{j-1} + h_j, \quad j = 1, 2, 3, \dots J$$

$$(6.25)$$

By applying the finite difference derivatives, Eqs. (6.19)–(6.23) are reduced into the finite difference form for the midpoint $\left(x_i, \eta_{j-\frac{1}{2}} \right)$ of the line segment BC as

$$f_j - f_{j-1} - \frac{h_j}{2}\left(p_j + p_{j-1} \right) = 0 \qquad (6.26)$$

$$p_j - p_{j-1} - \frac{h_j}{2}\left(q_j + q_{j-1} \right) = 0 \qquad (6.27)$$

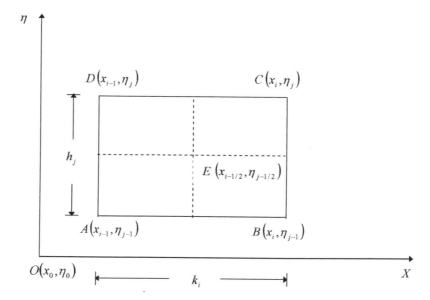

FIGURE 6.2
Net rectangle for difference approximation.

$$\theta_j - \theta_{j-1} - \frac{h_j}{2}\left(s_j + s_{j-1}\right) = 0 \tag{6.28}$$

$$\frac{1}{h_j}\left(q_j - q_{j-1}\right) + \frac{1}{4}E_1E_2\left[\left(f_j + f_{j-1}\right)\left(q_j + q_{j-1}\right) - \left(p_j + p_{j-1}\right)^2 + 4\varepsilon^2\right]$$

$$-\frac{1}{2}K_p\left[\left(p_j + p_{j-1}\right) - 2\varepsilon\right] - \frac{1}{2}\frac{(\sigma_e)_{nf}}{(\sigma_e)_f}E_1M\left[\left(p_j + p_{j-1}\right) - 2\varepsilon\right] = 0 \tag{6.29}$$

$$\left\{\frac{\kappa_{nf}}{\kappa_f} + \frac{1}{8h_j}Nr\left[\theta_{w_1}{}^3\left(\theta_j + \theta_{j-1}\right)^3 + 6\theta_{w_1}{}^2\left(\theta_j + \theta_{j-1}\right)^2 + 12\theta_{w_1}\left(\theta_j + \theta_{j-1}\right) + 8\right]\right\}\left(s_j - s_{j-1}\right)$$

$$+\frac{3}{4h_j}Nr\theta_{w_1}\left[\theta_{w_1}{}^2\left(\theta_j + \theta_{j-1}\right)^2 + 4\theta_{w_1}\left(\theta_j + \theta_{j-1}\right) + 4\right]\left(\theta_j - \theta_{j-1}\right)\left(s_j + s_{j-1}\right)$$

$$+\frac{1}{4}E_3Pr\left[\left(f_j + f_{j-1}\right)\left(s_j + s_{j-1}\right) - 2\left(p_j + p_{j-1}\right)\left(\theta_j + \theta_{j-1}\right)\right] + \frac{1}{2}\lambda Pr\left(\theta_j + \theta_{j-1}\right)$$

$$+\frac{1}{4E_1}Br\left\{\left(q_j + q_{j-1}\right)^2 + \frac{(\sigma_e)_{nf}}{(\sigma_e)_f}E_1M\left[\left(p_j + p_{j-1}\right)^2 - 2\varepsilon\left(p_j + p_{j-1}\right) + \varepsilon^2\right]\right\} = 0$$

$$\tag{6.30}$$

Equations (6.26)–(6.30) are received for $j = 1, 2, 3, \ldots J - 1$, and then from the related boundary condition equation (6.25) at $j = 0$ and $j = J$, it is observed that

$$f_0 = S, \ p_0 = \pm 1, \ \theta_0 = 1, \ p_J \rightarrow \varepsilon, \ \theta_J \rightarrow 0 \tag{6.31}$$

6.5.2 Newton's Method

Nonlinear system is transformed into linear system by using Newton's method. For which some iterations are introduced

$$f_j^{(i+1)} = f_j^{(i)} + \delta f_j^{(i)}, \ p_j^{(i+1)} = p_j^{(i)} + \delta p_j^{(i)}, \ q_j^{(i+1)} = q_j^{(i)} + \delta q_j^{(i)},$$

$$\theta_j^{(i+1)} = \theta_j^{(i)} + \delta\theta_j^{(i)}, \ s_j^{(i+1)} = s_j^{(i)} + \delta s_j^{(i)} \tag{6.32}$$

Substituting the aforementioned expression equation (6.32) into equations (6.26)–(6.30), neglect the quadratic and higher-order terms of $\delta f_j^{(i)}, \delta p_j^{(i)}, \delta q_j^{(i)}, \delta\theta_j^{(i)}$, and $\delta s_j^{(i)}$. Hence, the earlier strategy yields a tridiagonal system as

$$\delta f_j - \delta f_{j-1} - \frac{h_j}{2}\left(\delta p_j + \delta p_{j-1}\right) = \left(r_1\right)_{j-\frac{1}{2}} \tag{6.33}$$

$$\delta p_j - \delta p_{j-1} - \frac{h_j}{2}\left(\delta q_j + \delta q_{j-1}\right) = \left(r_2\right)_{j-\frac{1}{2}} \tag{6.34}$$

$$\delta \theta_j - \delta \theta_{j-1} - \frac{h_j}{2}\left(\delta s_j + \delta s_{j-1}\right) = \left(r_3\right)_{j-\frac{1}{2}} \tag{6.35}$$

$$\left(a_1\right)_{j-\frac{1}{2}}\delta f_j + \left(a_2\right)_{j-\frac{1}{2}}\delta f_{j-1} + \left(a_3\right)_{j-\frac{1}{2}}\delta p_j + \left(a_4\right)_{j-\frac{1}{2}}\delta p_{j-1}$$

$$+ \left(a_5\right)_{j-\frac{1}{2}}\delta q_j + \left(a_6\right)_{j-\frac{1}{2}}\delta q_{j-1} = \left(r_4\right)_{j-\frac{1}{2}} \tag{6.36}$$

$$\left(b_1\right)_{j-\frac{1}{2}}\delta f_j + \left(b_2\right)_{j-\frac{1}{2}}\delta f_{j-1} + \left(b_3\right)_{j-\frac{1}{2}}\delta p_j$$

$$+ \left(b_4\right)_{j-\frac{1}{2}}\delta p_{j-1} + \left(b_5\right)_{j-\frac{1}{2}}\delta q_j + \left(b_6\right)_{j-\frac{1}{2}}\delta q_{j-1} + \left(b_7\right)_{j-\frac{1}{2}}\delta s_j$$

$$+ \left(b_8\right)_{j-\frac{1}{2}}\delta s_{j-1} + \left(b_9\right)_{j-\frac{1}{2}}\delta \theta_j + \left(b_{10}\right)_{j-\frac{1}{2}}\delta \theta_{j-1} = \left(r_5\right)_{j-\frac{1}{2}} \tag{6.37}$$

where

$$\left(a_1\right)_{j-\frac{1}{2}} = \left(a_2\right)_{j-\frac{1}{2}} = \frac{1}{4}E_1E_2\left(q_j + q_{j-1}\right),$$

$$\left(a_3\right)_{j-\frac{1}{2}} = \left(a_4\right)_{j-\frac{1}{2}} = -\frac{1}{4}E_1E_2\left(p_j + p_{j-1}\right) - \frac{1}{2}K_p - \frac{1}{2}\frac{\left(\sigma_e\right)_{nf}}{\left(\sigma_e\right)_f}E_1M,$$

$$\left(a_5\right)_{j-\frac{1}{2}} = \frac{1}{4}E_1E_2\left(f_j + f_{j-1}\right) + \frac{1}{h_j},$$

$$\left(a_6\right)_{j-\frac{1}{2}} = \frac{1}{4}E_1E_2\left(f_j + f_{j-1}\right) - \frac{1}{h_j},$$

$$\left(b_1\right)_{j-\frac{1}{2}} = \left(b_2\right)_{j-\frac{1}{2}} = \frac{1}{4}E_3Pr\left(s_j + s_{j-1}\right),$$

$$\left(b_3\right)_{j-\frac{1}{2}} = \left(b_4\right)_{j-\frac{1}{2}} = -\frac{1}{2}E_3Pr\left(\theta_j + \theta_{j-1}\right) + \frac{1}{4}\frac{\left(\sigma_e\right)_{nf}}{\left(\sigma_e\right)_f}MBr\left[\left(p_j + p_{j-1}\right) - 2\varepsilon\right],$$

$$\left(b_5\right)_{j-\frac{1}{2}} = \left(b_6\right)_{j-\frac{1}{2}} = \frac{1}{4E_1}Br\left(q_j + q_{j-1}\right),$$

$$(b_7)_{j-\frac{1}{2}} = \left\{ \frac{\kappa_{nf}}{\kappa_f} + \frac{1}{8h_j} Nr \left[\theta_{w_1}{}^3 \left(\theta_j + \theta_{j-1} \right)^3 + 6\theta_{w_1}{}^2 \left(\theta_j + \theta_{j-1} \right)^2 + 12\theta_{w_1} \left(\theta_j + \theta_{j-1} \right) + 8 \right] \right\}$$

$$+ \frac{3}{4h_j} Nr\theta_{w_1} \left[\theta_{w_1}{}^2 \left(\theta_j + \theta_{j-1} \right)^2 + 4\theta_{w_1} \left(\theta_j + \theta_{j-1} \right) + 4 \right] \left(\theta_j - \theta_{j-1} \right)$$

$$+ \frac{1}{4} E_3 Pr \left(f_j + f_{j-1} \right),$$

$$(b_8)_{j-\frac{1}{2}} = -\left\{ \frac{\kappa_{nf}}{\kappa_f} + \frac{1}{8h_j} Nr \left[\theta_{w_1}{}^3 \left(\theta_j + \theta_{j-1} \right)^3 + 6\theta_{w_1}{}^2 \left(\theta_j + \theta_{j-1} \right)^2 + 12\theta_{w_1} \left(\theta_j + \theta_{j-1} \right) + 8 \right] \right\}$$

$$+ \frac{3}{4h_j} Nr\theta_{w_1} \left[\theta_{w_1}{}^2 \left(\theta_j + \theta_{j-1} \right)^2 + 4\theta_{w_1} \left(\theta_j + \theta_{j-1} \right) + 4 \right] \left(\theta_j - \theta_{j-1} \right)$$

$$+ \frac{1}{4} E_3 Pr \left(f_j + f_{j-1} \right),$$

$$(b_9)_{j-\frac{1}{2}} = \left\{ \frac{\kappa_{nf}}{\kappa_f} + \frac{1}{8h_j} Nr \left[\theta_{w_1}{}^3 \left(\theta_j + \theta_{j-1} \right)^2 + 6\theta_{w_1}{}^2 \left(\theta_j + \theta_{j-1} \right) + 12\theta_{w_1} \right] \right\} \left(s_j - s_{j-1} \right)$$

$$+ \frac{3}{4h_j} Nr\theta_{w_1} \left[\theta_{w_1}{}^2 \left(\theta_j + \theta_{j-1} \right)^2 + 4\theta_{w_1} \left(\theta_j + \theta_{j-1} \right) + 4 \right] \left(s_j + s_{j-1} \right)$$

$$+ \frac{1}{2} E_3 Pr \left(p_j + p_{j-1} \right) + \frac{1}{2} \lambda Pr,$$

$$(b_{10})_{j-\frac{1}{2}} = \left\{ \frac{\kappa_{nf}}{\kappa_f} + \frac{1}{8h_j} Nr \left[\theta_{w_1}{}^3 \left(\theta_j + \theta_{j-1} \right)^2 + 6\theta_{w_1}{}^2 \left(\theta_j + \theta_{j-1} \right) + 12\theta_{w_1} \right] \right\} \left(s_j - s_{j-1} \right)$$

$$- \frac{3}{4h_j} Nr\theta_{w_1} \left[\theta_{w_1}{}^2 \left(\theta_j + \theta_{j-1} \right)^2 + 4\theta_{w_1} \left(\theta_j + \theta_{j-1} \right) + 4 \right] \left(s_j + s_{j-1} \right)$$

$$+ \frac{1}{2} E_3 Pr \left(p_j + p_{j-1} \right) + \frac{1}{2} \lambda Pr,$$

$$(r_1)_{j-\frac{1}{2}} = -\left(f_j - f_{j-1} \right) + \frac{h_j}{2} \left(p_j + p_{j-1} \right),$$

$$(r_2)_{j-\frac{1}{2}} = -\left(p_j - p_{j-1} \right) + \frac{h_j}{2} \left(q_j + q_{j-1} \right),$$

$$(r_3)_{j-\frac{1}{2}} = -\left(\theta_j - \theta_{j-1}\right) + \frac{h_j}{2}\left(s_j + s_{j-1}\right),$$

$$(r_4)_{j-\frac{1}{2}} = -\frac{1}{h_j}\left(q_j - q_{j-1}\right) - \frac{1}{4}E_1E_2\left[\left(f_j + f_{j-1}\right)\left(q_j + q_{j-1}\right) - \left(p_j + p_{j-1}\right)^2 + 4\varepsilon^2\right]$$

$$+ \frac{1}{2}K_p\left[\left(p_j + p_{j-1}\right) - 2\varepsilon\right] + \frac{1}{2}\frac{(\sigma_e)_{nf}}{(\sigma_e)_f}E_1M\left[\left(p_j + p_{j-1}\right) - 2\varepsilon\right]$$

and

$$(r_5)_{j-\frac{1}{2}} = -\left\{\frac{\kappa_{nf}}{\kappa_f} + \frac{1}{8h_j}Nr\left[\theta_{w_1}{}^3\left(\theta_j + \theta_{j-1}\right)^3 + 6\theta_{w_1}{}^2\left(\theta_j + \theta_{j-1}\right)^2\right.\right.$$

$$\left.+ 12\theta_{w_1}\left(\theta_j + \theta_{j-1}\right) + 8\right]\right\}\left(s_j - s_{j-1}\right)$$

$$- \frac{3}{4h_j}Nr\theta_{w_1}\left[\theta_{w_1}{}^2\left(\theta_j + \theta_{j-1}\right)^2 + 4\theta_{w_1}\left(\theta_j + \theta_{j-1}\right) + 4\right]\left(\theta_j - \theta_{j-1}\right)\left(s_j + s_{j-1}\right)$$

$$- \frac{1}{4}E_3Pr\left[\left(f_j + f_{j-1}\right)\left(s_j + s_{j-1}\right) - 2\left(p_j + p_{j-1}\right)\left(\theta_j + \theta_{j-1}\right)\right] + \frac{1}{2}\lambda Pr\left(\theta_j + \theta_{j-1}\right)$$

$$- \frac{1}{4E_1}Br\left\{\left(q_j + q_{j-1}\right)^2 + \frac{(\sigma_e)_{nf}}{(\sigma_e)_f}E_1M\left[\left(p_j + p_{j-1}\right)^2 - 2\varepsilon\left(p_j + p_{j-1}\right) + \varepsilon^2\right]\right\}.$$

For all iterates, it is assumed that

$$\delta f_0 = 0,\ \delta p_0 = 0,\ \delta\theta_0 = 0,\ \delta p_J = 0,\ \delta\theta_J = 0 \qquad (6.38)$$

6.5.3 Block Elimination Method

The system of equations (6.33)–(6.37) is written in the block tridiagonal matrix form as

$$\begin{bmatrix} [A_1] & [C_1] & & & & \\ [B_2] & [A_2] & [C_2] & & & \\ & & \ddots & \cdots & & \\ & & & [B_{J-1}] & [A_{J-1}] & [C_{J-1}] \\ & & & & [B_J] & [A_J] \end{bmatrix} \begin{bmatrix} [\delta_1] \\ [\delta_2] \\ \vdots \\ [\delta_{J-1}] \\ [\delta_J] \end{bmatrix} = \begin{bmatrix} [r_1] \\ [r_2] \\ \vdots \\ [r_{J-1}] \\ [r_J] \end{bmatrix}$$

i.e.,

$$[A][\delta] = [r] \qquad (6.39)$$

whose elements are

$$[A_1] = \begin{bmatrix} 0 & 0 & 1 & 0 & 0 \\ \dfrac{-h_j}{2} & 0 & 0 & \dfrac{-h_j}{2} & 0 \\ 0 & \dfrac{-h_j}{2} & 0 & 0 & \dfrac{-h_j}{2} \\ a_6 & 0 & a_1 & a_5 & 0 \\ b_6 & b_8 & b_1 & b_5 & b_7 \end{bmatrix},$$

$$[A_j] = \begin{bmatrix} \dfrac{-h_j}{2} & 0 & 1 & 0 & 0 \\ -1 & 0 & 0 & \dfrac{-h_j}{2} & 0 \\ 0 & -1 & 0 & 0 & \dfrac{-h_j}{2} \\ a_4 & 0 & a_1 & a_5 & 0 \\ b_4 & b_{10} & b_1 & b_5 & b_7 \end{bmatrix} \quad \text{for } 2 \le j \le J,$$

$$[B_j] = \begin{bmatrix} 0 & 0 & -1 & 0 & 0 \\ 0 & 0 & 0 & \dfrac{-h_j}{2} & 0 \\ 0 & 0 & 0 & 0 & \dfrac{-h_j}{2} \\ 0 & 0 & a_2 & a_6 & 0 \\ 0 & 0 & b_2 & b_6 & b_8 \end{bmatrix} \quad \text{for } 2 \le j \le J,$$

$$[C_j] = \begin{bmatrix} \dfrac{-h_j}{2} & 0 & 0 & 0 & 0 \\ 1 & 0 & 0 & 0 & 0 \\ 0 & 1 & 0 & 0 & 0 \\ a_3 & 0 & 0 & 0 & 0 \\ b_3 & b_9 & 0 & 0 & 0 \end{bmatrix} \quad \text{for } 1 < j < J-1,$$

$$[\delta_1] = \begin{bmatrix} \delta q_0 \\ \delta s_0 \\ \delta f_1 \\ \delta q_1 \\ \delta s_1 \end{bmatrix}, [\delta_j] = \begin{bmatrix} \delta p_{j-1} \\ \delta \theta_{j-1} \\ \delta f_j \\ \delta q_j \\ \delta s_j \end{bmatrix} \quad \text{for } 2 \le j \le J \text{ and}$$

$$[r_j] = \begin{bmatrix} (r_1)_j \\ (r_2)_j \\ (r_3)_j \\ (r_4)_j \\ (r_5)_j \end{bmatrix} \quad \text{for } 1 \le j \le J.$$

6.5.3.1 Forward Sweep

Matrix A is considered as a nonsingular matrix to solve Eq. (6.39). As usual, nonsingular matrix A can be written as the product of lower triangular and upper triangular matrices, such as

$$[A] = [L][U] \tag{6.40}$$

with

$$[L] = \begin{bmatrix} [\alpha_1] & & & & \\ [B_2] & [\alpha_2] & & & \\ & & \ddots & & \\ & & & [\alpha_{J-1}] & \\ & & & [B_J] & [\alpha_J] \end{bmatrix} \quad \text{and}$$

$$[U] = \begin{bmatrix} I & [\beta_1] & & & \\ & I & [\beta_2] & & \\ & & \ddots & \ddots & \\ & & & I & [\beta_{J-1}] \\ & & & & I \end{bmatrix}$$

where I is the identity matrix of order 5×5, and $\left[\alpha_j \right]$ and $\left[\beta_j \right]$ are the matrices of order 5×5, whose elements are found by the following equations

$$[\alpha_1] = [A_1] \tag{6.41}$$

$$[A_1][\beta_1] = [C_1] \tag{6.42}$$

$$[\alpha_j] = [A_j] - [B_j][\beta_{j-1}], \quad j = 2,3,...J \tag{6.43}$$

6.5.3.2 Backward Sweep

In view of

$$[L][U][\delta_j] = [r_j] \tag{6.44}$$

Along to the assumptions as

$$[U][\delta_j] = [W_j] \tag{6.45}$$

$$[L][W_j] = [r_j] \tag{6.46}$$

where $\left[W_j \right]$ are the column matrices of order 5×1, the elements of $\left[W_j \right]$ can be determined by Eq. (6.46) as

$$[\alpha_1][W]_1 = [r_1] \tag{6.47}$$

$$[\alpha_j][W_j] = [r_j] - [B_j][W_{j-1}], \quad 2 \leq j \leq J \tag{6.48}$$

Equation (6.45) gives the solution with the help of given relations

$$[\delta_j] = [W_j] - [\beta_j][\delta_{j+1}], \quad 1 \leq j \leq J-1 \tag{6.49}$$

$$[\delta_J] = [W_J] \tag{6.50}$$

This iterative procedure is repeated until convergence condition is fulfilled, keeping up an accuracy of 10^{-7}, while process is closed when $\left| \delta q_0^{(i)} \right| \leq \zeta$, where ζ is a small prescribed value.

TABLE 6.2

Comparison for Computational Values of $f''(0)$ for Several Values of Considering Parameters over Stretching/Shrinking Sheet with the Previous Published Data When $M = 0.0$

φ	K_p	S	ε	Stretching Sheet		Shrinking Sheet	
				Pal et al. [42]	Present Results	Pal et al. [42]	Present Results
0.1	0.5	0.1	1.5	1.1630	1.1631	3.9161	3.9183
			2.0	2.5491	2.5492	5.6777	5.6562
0.2			1.5	1.2034	1.2035	4.0382	4.0398
	0.6			1.2139	1.2140	4.1188	4.1202
		0.2		1.2561	1.2562	4.3531	4.3545

TABLE 6.3

Dual Solution Comparison with the Earlier Published Outcomes of $f''(0)$ for Various Values of φ and ε over Shrinking Surface When $K_p = 0.2$, $M = 0.0$, and $S = 3.5$

φ	ε	Upper Solution		Lower Solution	
		Pal et al. [42]	Present Results	Pal et al. [42]	Present Results
0.2	0.1	5.4948	5.4949	−5.6291	−5.2187
	0.3	6.6483	6.6483	−7.8403	−7.7871
0.3	0.1	5.1324	5.1324	−5.0155	−4.4477
	0.3	6.2702	6.2202	−6.8484	−6.7554

6.6 Validation of Proposed Method

In Table 6.2, the present computational values of surface gradient $f''(0)$ along several values of pertinent physical parameter due to the stretching/ shrinking sheet are compared with some already published works of Pal et al. [42] in the absence of magnetic field. Further, for dual solutions of surface gradient $f''(0)$ with various values of the nanoparticle volume fraction φ and the velocity ratio parameter ε along the shrinking surface where the other pertinent parameters are constant, the comparison between earlier results and present results are exhibited in Table 6.3. From these tables, it can be noted that the results are in excellent agreement with previous published data and that the proposed method is valid and effective.

6.7 Discussion of Numerical Analysis

Effects of various values of specified parameters, particularly the nanoparticle volume fraction ϕ, the permeability parameter K_p, the magnetic parameter

M, the mass flux parameter S, the velocity ratio parameter ε, the radiation parameter Nr, the heat source/sink parameter λ, and the Brinkmann number Br on the velocity $f'(\eta)$ as well as temperature $\theta(\eta)$ profiles for Cu–water nanofluid due to stretching/shrinking plate are presented via graphs in this section. Later, numerical values of the surface shear stress $f''(0)$ and the surface heat flux $\theta'(0)$ are enumerated for considering parameters, which are given through tables.

Impacts of the nanoparticle volume fraction φ on the velocity $f'(\eta)$ and the temperature $\theta(\eta)$ distributions are exhibited via Figures 6.3 and 6.4, respectively, while the remaining parameters are considered as constant.

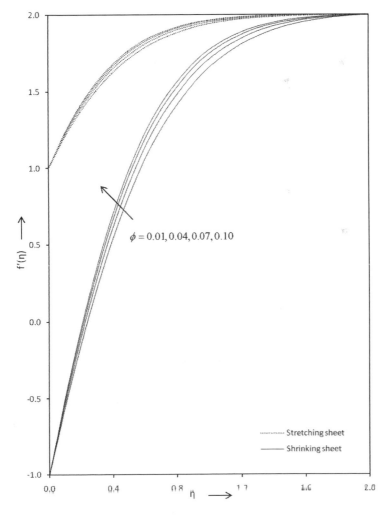

FIGURE 6.3
Behavior of velocity distributions of Cu–water nanofluid for several values of ϕ when $K_p = 0.1$, $M = 0.1$, $S = 0.1$, and $\varepsilon = 2.0$.

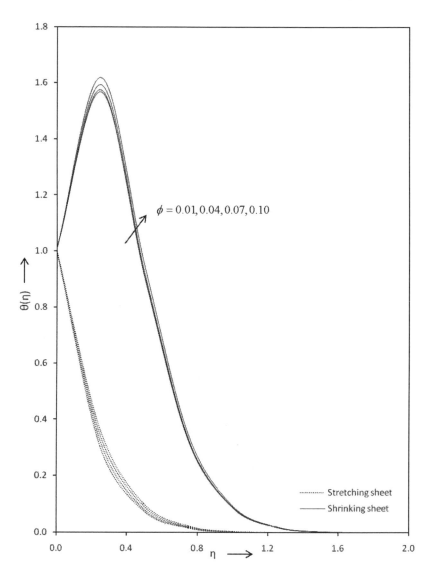

FIGURE 6.4
Behavior of temperature distributions of Cu–water nanofluid for several values of ϕ when $K_p = 0.1$, $M = 0.1$, $S = 0.1$, $\varepsilon = 2.0$, $Nr = 0.5$, $\theta_w = 0.1$, $Pr = 6.8$, $\lambda = 0.1$, and $Br = 1.36$.

Figures 6.3 and 6.4 show that the fluid flow as well as fluid temperature is the increasing function of nanoparticle volume fraction φ along the stretching/shrinking plate. From the physical point of view, due to the raising volume fraction of solids, the nanofluid has greater viscosity, which leads that velocity of nanofluid depreciates, although velocity of nanofluid enhances near the stagnation point. Moreover, an increment in the nanoparticle volume fraction implies the enhancement in the thermal conductivity. So the thermal boundary layer thickness enlarges.

Figures 6.5 and 6.6 sketch for the influence of permeability parameter K_p on the velocity $f'(\eta)$ and temperature $\theta(\eta)$ profiles, respectively, keeping the rest of the controlling parameters fixed. It is observed from these figures that the momentum boundary layer boosts and the thermal boundary

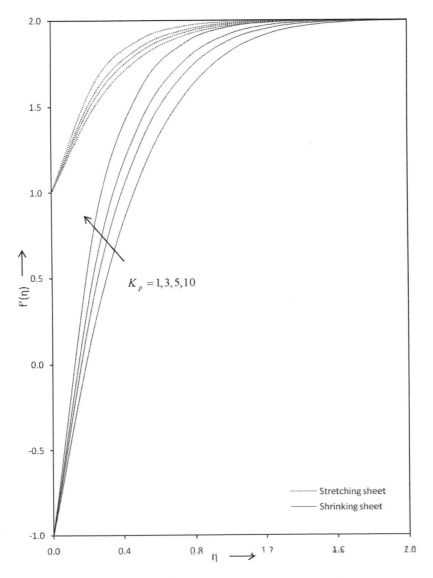

FIGURE 6.5
Behavior of velocity distributions of Cu–water nanofluid for several values of K_p when $\phi = 0.1$, $M = 0.1$, $S = 0.1$, and $\varepsilon = 2.0$.

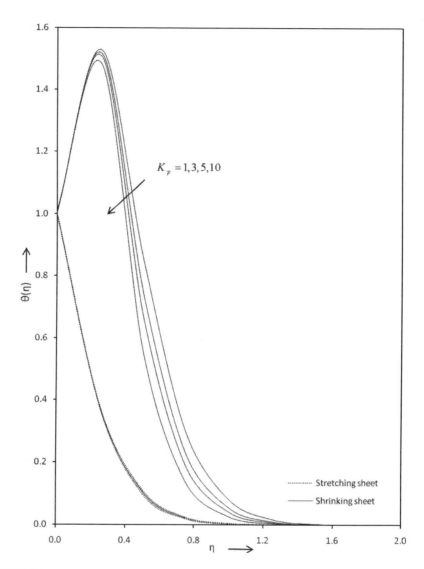

FIGURE 6.6
Behavior of temperature distributions of Cu–water nanofluid for several values of K_p when $\phi = 0.1$, $M = 0.1$, $S = 0.1$, $\varepsilon = 2.0$, $Nr = 0.5$, $\theta_w = 0.1$, $Pr = 6.8$, $\lambda = 0.1$, and $Br = 1.36$.

layer depreciates along the booming values of the permeability parameter K_p towards both types of sheets. Because of higher values of the permeability parameter, the fluid flow is firmly uniform from which fluid velocity is greatly a thin domain. This is appreciated among the inference of complete Darcy flow.

Behaviors of velocity field $f'(\eta)$ and temperature field $\theta(\eta)$ for distinct values of the magnetic parameter M are mentioned in Figures 6.7 and 6.8,

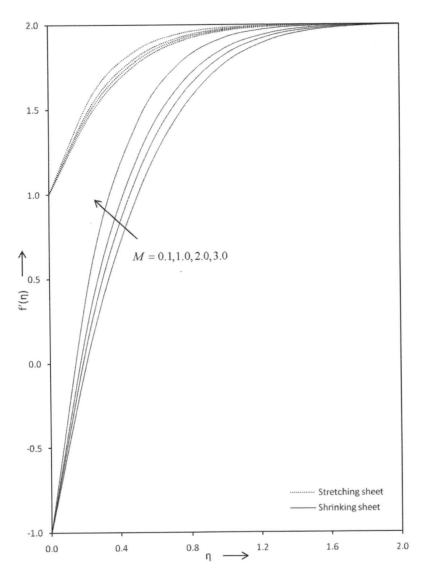

FIGURE 6.7
Behavior of velocity distributions of Cu–water nanofluid for several values of M when $\phi = 0.1$, $K_p = 0.1$, $S = 0.1$, and $\varepsilon = 2.0$.

respectively, besides the constant value of the remaining specified parameters. Figures 6.7 and 6.8 reveal that due to an enlargement in magnetic parameter M, fluid flow and temperature develop subjected to stretching/ shrinking surface, until a reverse can be seen in temperature when $\eta > 0.5$. Physically, an enhancement in the magnetic parameter tends to accelerate the Lorentz force. This type of force leads the resistive development to the

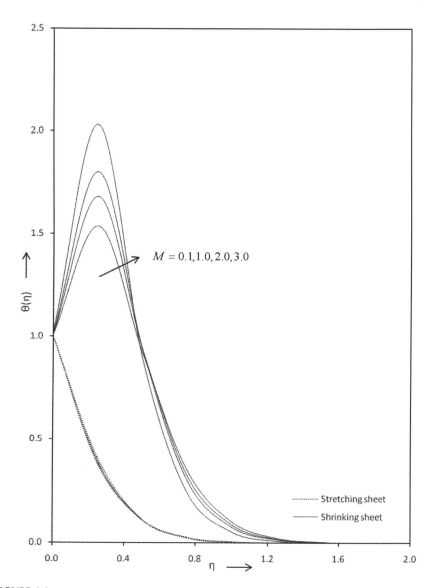

FIGURE 6.8
Behavior of temperature distributions of Cu–water nanofluid for several values of M when $\phi = 0.1$, $K_p = 0.1$, $S = 0.1$, $\varepsilon = 2.0$, $Nr = 0.5$, $\theta_w = 0.1$, $Pr = 6.8$, $\lambda = 0.1$, and $Br = 1.36$.

fluid motion in the boundary layer and in turn creates more heat, resulting in increment of thermal boundary layer.

Figures 6.9–6.12 depict the fluid flow $f'(\eta)$ and the fluid temperature $\theta(\eta)$ for the influence of the mass flux parameter S, respectively, while the remaining relevant parameters are taken as constant. From Figures 6.9 and 6.11, it can be seen that the fluid velocity increases for stretching/shrinking sheet

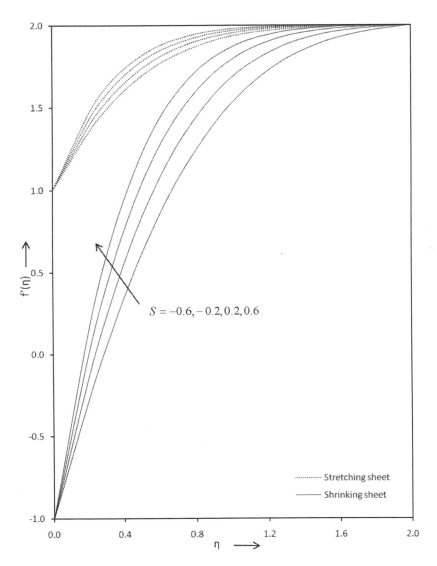

FIGURE 6.9
Behavior of velocity distributions of Cu–water nanofluid for several values of S when $\phi = 0.1$, $K_p = 0.1$, $M = 0.1$, and $\varepsilon = 2.0$.

with the rising values of the mass flux parameter S, but the reverse happens for the temperature field. Consequently, dual solutions of the velocity $f'(\eta)$ and the temperature $\theta(\eta)$ profiles along shrinking surface are demonstrated in Figures 6.10 and 6.12, respectively. It can be observed from these figures that there are two solution branches, specifically upper and lower solution branches, which conclude that an enhancement in the mass flux parameter S leads the increment in the fluid velocity for both solution branches,

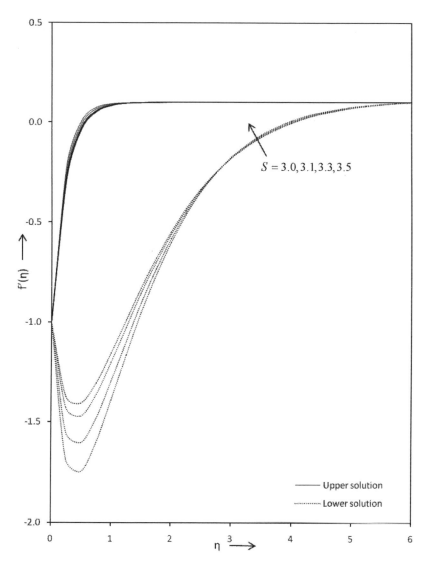

FIGURE 6.10
Behavior of velocity distributions of Cu–water nanofluid for several values of S when $\phi = 0.1$, $K_p = 0.1$, $M = 0.01$, and $\varepsilon = 0.1$ over shrinking sheet.

while for a lower solution branch, an opposite behavior can be found when $\eta < 2.5$. Moreover, the fluid temperature declines with the booming values of mass flux parameter S for upper and lower solution branches, although a reverse phenomenon occurs when $\eta > 1.75$ in lower solution branch. Physically, fluid is thrown out/received adjacent to stagnation region in the defense of suction/blowing, so flow field step-up and due to suction/injection if hot fluid is discarded/retained over the plate leads to temperature falls down.

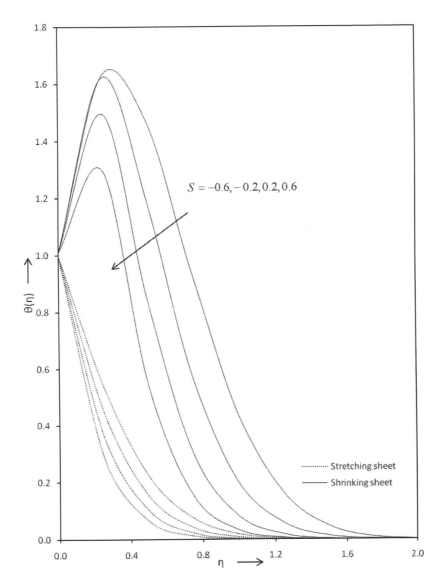

FIGURE 6.11
Behavior of temperature distributions of Cu–water nanofluid for several values of S when $\phi = 0.1$, $K_p = 0.1$, $M = 0.1$, $\varepsilon = 2.0$, $Nr = 0.5$, $\theta_w = 0.1$, $Pr = 6.8$, $\lambda = 0.1$, and $Br = 1.36$.

The effects of the velocity ratio parameter ε on the dimensionless velocity $f'(\eta)$ and the dimensionless temperature $\theta(\eta)$ are portrayed in Figures 6.13–6.16, respectively, keeping additional physical parameters fixed. It is clear from Figures 6.13 and 6.15 that over to stretching/shrinking surface the velocity and temperature develop with the evolving values of the velocity ratio parameter ε. Even though a reverse trend is found in the

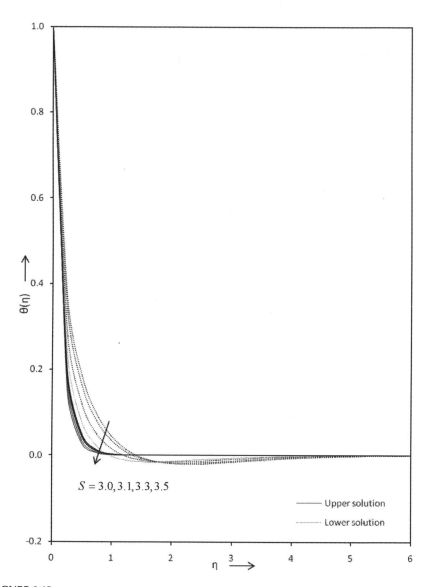

FIGURE 6.12
Behavior of temperature distributions of Cu–water nanofluid for several values of S when $\phi = 0.1$, $K_p = 0.1$, $M = 0.01$, $\varepsilon = 0.1$, $Nr = 1.0$, $\theta_w = 0.5$, $Pr = 6.8$, $\lambda = 0.1$, and $Br = 0.068$ over shrinking sheet.

temperature over stretching plate when $\eta > 0.75$ and over to shrinking plate when $\eta > 0.5$. Subsequently, dual solutions exist for the impact of the velocity ratio parameter ε on the fluid velocity $f'(\eta)$ and the fluid temperature $\theta(\eta)$ due to shrinking surface, which are presented by Figures 6.14 and 6.16, respectively. From these figures, it is noted that an enhancement in the

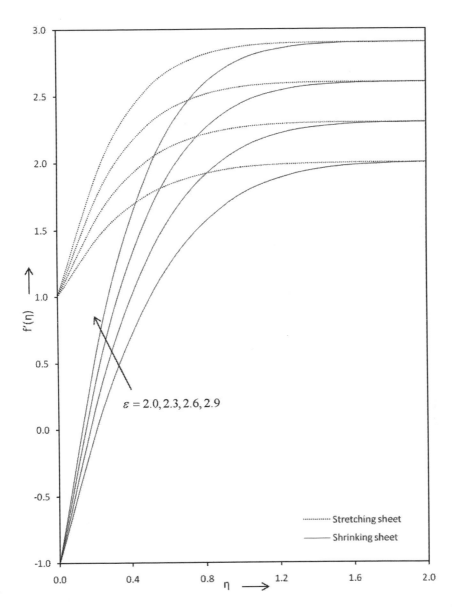

FIGURE 6.13
Behavior of velocity distributions of Cu–water nanofluid for several values of ε when $\phi = 0.1$, $K_p = 0.1$, $M = 0.1$, and $S = 0.1$.

velocity ratio parameter ε implies an increment in velocity for upper and lower solution branches, while an opposite phenomenon can be seen for lower solution branch when $\eta < 1.5$. However, for the rising values of the velocity ratio parameter ε, the temperature field reduces, but a reverse reaction is found for a lower solution branch when $\eta > 2.5$. This is a consequence

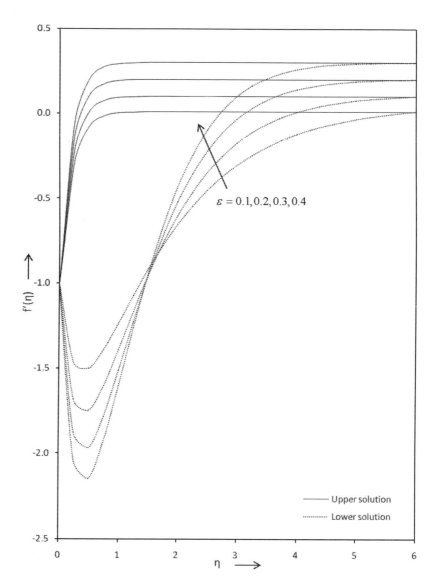

FIGURE 6.14
Behavior of velocity distributions of Cu–water nanofluid for several values of ε when $\phi = 0.1$, $K_p = 0.1$, $M = 0.01$, and $S = 3.5$ over shrinking sheet.

that when the free stream velocity is greater than the stretching velocity the retarding force decays. So the fluid flow grows up and the temperature of fluid steps down.

Figure 6.17 shows the reflection of the radiation parameter Nr on the fluid temperature $\theta(\eta)$ when the values of other pertinent parameters are kept constant. It declares that along the rising values of the radiation parameter Nr,

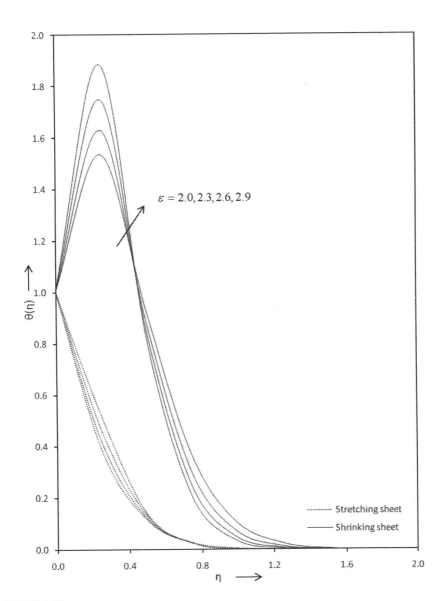

FIGURE 6.15
Behavior of temperature distributions of Cu–water nanofluid for several values of ε when $\phi = 0.1$, $K_p = 0.1$, $M = 0.1$, $S = 0.1$, $Nr = 0.5$, $\theta_w = 0.1$, $Pr = 6.8$, $\lambda = 0.1$, and $Br = 1.36$.

the temperature of fluid raises, while declines over the stretching plate when $\eta < 0.3$ and over to the shrinking plate when $\eta < 0.9$. This occurs because, in the operating fluid, heat is produced in large amount due to the thermal radiation process, which motivates the enlargement in the temperature field.

Nature of the dimensionless temperature $\theta(\eta)$ for various values of heat source/sink parameter λ is plotted in Figure 6.18 while other relevant

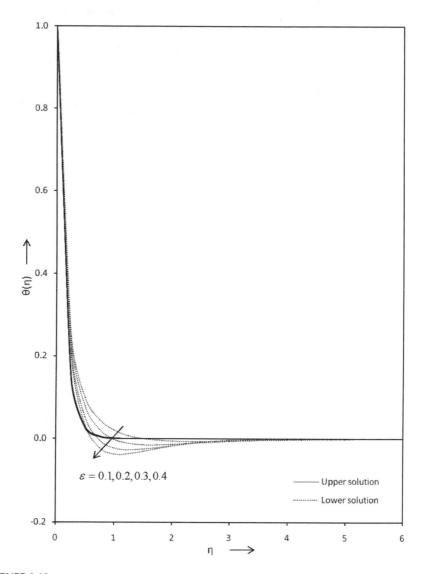

FIGURE 6.16

Behavior of temperature distributions of Cu–water nanofluid for several values of ε when $\phi = 0.1$, $K_p = 0.1$, $M = 0.01$, $S = 3.5$, $Nr = 1.0$, $\theta_w = 0.5$, $Pr = 6.8$, $\lambda = 0.1$, and $Br = 0.068$ over shrinking sheet.

parameters are taken as fixed. This figure illustrated that the increasing behavior of the heat source/sink parameter λ signifies the increment in the thermal boundary layer thickness towards the stretching/shrinking sheet. This concludes that the development in fluid temperature indicates that there is greater induced flow along the surface with thermal buoyancy impact. Hence, thermal boundary layer thickness evolves with the heat source/sink parameter.

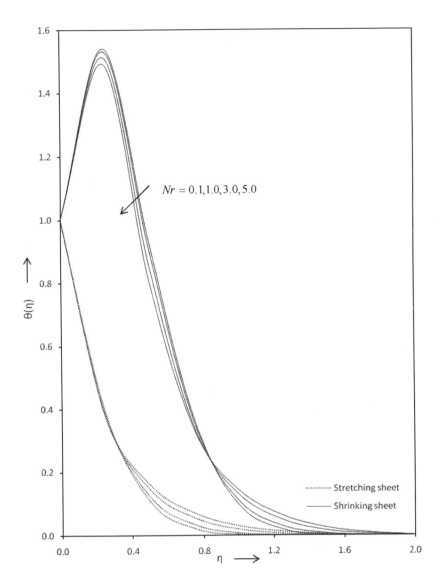

FIGURE 6.17
Behavior of temperature distributions of Cu–water nanofluid for several values of Nr when $\phi = 0.1$, $K_p = 0.1$, $M = 0.1$, $S = 0.1$, $\varepsilon = 2.0$, $\theta_w = 0.1$, $Pr = 6.8$, $\lambda = 0.1$, and $Br = 1.36$.

Figure 6.19 represents the impact of the Brinkmann number Br on the temperature $\theta(\eta)$ profile, keeping additional dimensionless parameters fixed. From this figure, it is detected that the fluid temperature increases with the values of Brinkmann number Br for both types of surfaces as stretching/ shrinking. Physically it appears because, at the dissipation process, the energy is conserved in the fluid domain over viscosity and elastic deformation.

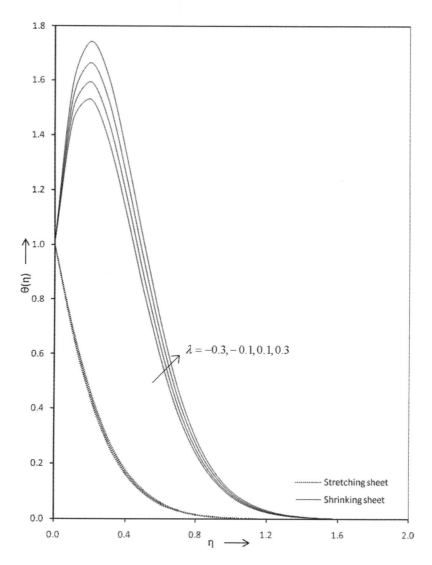

FIGURE 6.18
Behavior of temperature distributions of Cu–water nanofluid for several values of λ when $\phi = 0.1$, $K_p = 0.1$, $M = 0.1$, $S = 0.1$, $\varepsilon = 2.0$, $Nr = 0.5$, $\theta_w = 0.1$, $Pr = 6.8$, and $Br = 1.36$.

The values of the surface gradient $f''(0)$ and heat transfer rate $\theta'(0)$ with Cu–water nanofluid are given in Table 6.4 for several values of nanoparticle volume fraction φ, the permeability parameter K_p, the magnetic parameter M, the mass flux parameter S, the velocity ratio parameter ε, the radiation parameter Nr, the heat source/sink parameter λ, and the Brinkmann number Br, while other considering parameters are taken as constant. It is noteworthy from Eqs. (6.17) and (6.18) that $f''(0)$ and $\theta'(0)$ are proportional to the

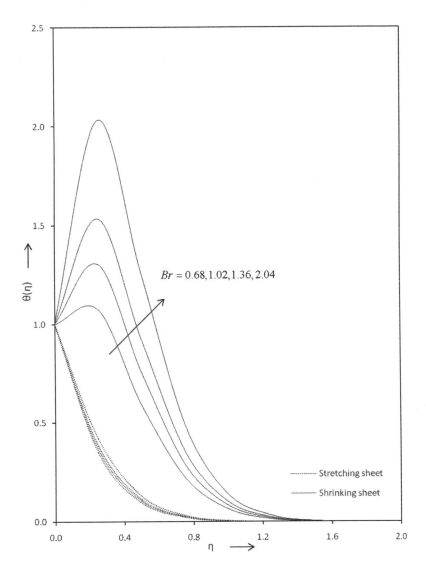

FIGURE 6.19
Behavior of temperature distributions of Cu–water nanofluid for several values of Br when
$\phi = 0.1,, K_p = 0.1, M = 0.1, S = 0.1, \varepsilon = 2.0, Nr = 0.5, \theta_w = 0.1, Pr = 6.8,$ and $\lambda = 0.1$.

local skin friction C_f and the local Nusselt number Nu, respectively. From this table, it can be concluded that the local skin friction C_f and the local Nusselt number Nu increase along with increasing values of the nanoparticle volume fraction ψ, the permeability parameter K_p, the magnetic parameter M, the mass flux parameter S, and the velocity ratio parameter ε toward the stretching/shrinking plate, while the opposite happens in the local Nusselt number Nu for the rising values of mass flux parameter S over the stretching

TABLE 6.4

Results of $f''(0)$ and $\theta'(0)$ for Various Values of Physical Parameters over Stretching/Shrinking Surface with Cu–Water Nanofluid When $\theta_w = 0.1$ and $Pr = 6.8$

								Stretching Sheet		Shrinking Sheet	
φ	K_p	M	S	ε	Nr	λ	Br	$f''(0)$	$-\theta'(0)$	$f''(0)$	$\theta'(0)$
0.01	0.1	0.1	0.1	2.0	0.5	0.1	1.36	2.17649	4.06460	4.73988	7.26772
0.04								2.31290	3.76004	5.02882	7.44760
0.07								2.41518	3.48252	5.24657	7.63638
0.10								2.48937	3.22776	5.40485	7.82123
	1.0							2.66615	3.15577	6.13235	8.63401
	3.0							3.02281	3.00361	7.49310	10.32848
	5.0							3.34169	2.86065	8.63693	11.88310
	10.0							4.03003	2.53429	10.97838	15.32087
	0.1	1.0						2.67034	3.02176	6.14920	10.38765
		2.0						2.85826	2.80689	6.87917	12.89718
		3.0						3.03468	2.60423	7.53691	15.15612
		0.1	−0.6					1.98619	1.91060	3.79137	4.86002
			−0.2					2.26228	2.58697	4.66680	6.60190
			0.2					2.56880	3.46550	5.66505	8.18826
			0.6					2.90378	4.52446	6.76861	9.45333
			0.1	2.3				3.40190	2.51105	6.51939	9.87072
				2.6				4.38110	1.50750	7.68551	12.36994
				2.9				5.42286	0.19530	8.90218	15.29922
				2.0	0.1			2.48936	3.15008	5.40488	7.84630
					1.0				3.32248		7.79027
					3.0				3.68095		7.66891
					5.0				4.01487		7.54933
					0.5	−0.3			3.49187		6.86123
						−0.1			3.36151		7.32277
						0.3			3.09039		8.36302
						0.1	0.68		3.76743		3.48722
							1.02		3.49747		5.65538
							2.04		2.68910		12.15162

plate. Consequently for stretching/shrinking surface, enhancing values of the heat source/sink parameter λ and the Brinkmann number Br imply an enlargement in the surface heat flux $\theta'(0)$, although a reverse phenomenon occurs for the radiation parameter Nr. Physically, positive sign of the wall shear stress implies that fluid adapts a drag force on the wall, even though the negative values of heat flux leads that there is a heat flow to the surface and vice versa.

Dual solutions exist of the surface shear stress $f''(0)$ and heat transfer rate $\theta'(0)$ with the different values of the mass flux parameter S and the velocity ratio parameter ε past a shrinking sheet, which are demonstrated via

TABLE 6.5

Dual Solutions of $f''(0)$ and $\theta'(0)$ for Distinct Values of S and ε over Shrinking Sheet with Cu–Water Nanofluid When $\phi = 0.1$, $K_p = 0.1$, $M = 0.01$, $Nr = 1.0$, $\theta_w = 0.5$, $Pr = 6.8$, $\lambda = 0.1$, and $Br = 0.068$

		Upper Solution		Lower Solution	
S	ε	$f''(0)$	$-\theta'(0)$	$-f''(0)$	$-\theta'(0)$
3.0	0.1	4.27329	12.66198	2.24662	13.52391
3.1		4.43533	13.15079	2.61160	14.34385
3.3		4.75718	14.12056	3.43128	16.00278
3.5		5.07657	15.08185	4.37590	17.73592
	0.2	5.61162	15.08529	5.61557	18.46158
	0.3	6.15576	15.08467	6.65840	19.39226
	0.4	6.70889	15.08008	7.58021	20.75837

Table 6.5, keeping other parameters fixed. This table characterized that the surface shear stress grows up for upper solution branch with the booming values of mass flux parameter S and the velocity ratio parameter ε, although for a lower solution branch, an opposite formation is created. Also, the heat transfer rate $\theta'(0)$ declines with raising values of the mass flux parameter S as well as the velocity ratio parameter ε for both solution branches, whereas the reverse can be seen in upper solution branch when $\varepsilon > 0.2$.

6.8 Conclusions

Investigation of a numerical problem of steady, incompressible MHD stagnation point flow of an electrically conducting Cu–water nanofluid in a porous medium over a stretching/shrinking plate with radiation and heat source/sink impacts is inspected. Also, the influences of suction/blowing, viscous dissipation and Joule heating are examined. The similarity variables to reduce the set of nonlinear partial differential equations into a system of nonlinear ordinary differential equations are applied. Solutions of the resulting equations are carried out by using the Keller-box scheme. Following main findings are compiled from this problem:

i. Due to stretching/shrinking surface, the momentum boundary layer thickness, the fluid temperature, the velocity gradient, and the heat flux develop for the rising values of nanoparticle volume fraction, the permeability parameter, and the magnetic parameter, while a reverse phenomenon occurs in temperature field with the enlargement in the permeability parameter and also with the enhancement in the magnetic parameter when $\eta > 0.5$.

ii. Increasing nature of the mass flux parameter and the velocity ratio parameter for stretching/shrinking sheet leads to the enlargement in the fluid velocity, the local skin friction coefficient, and the local Nusselt number, until an opposite behavior is found in the local Nusselt number for the mass flux parameter over stretching sheet. Further, the thermal boundary layer thickness declines with the enhancing values of the mass flux parameter and the velocity ratio parameter for both types of sheet, even though a reverse happens with the grow up in the velocity ratio parameter, due to the stretching plate when $\eta < 0.75$ and due to shrinking plate when $\eta < 0.5$.

iii. Dual solution branches exist for the influence of the access values of mass flux parameter and velocity ratio parameter past a shrinking surface. The momentum boundary layer thickness and the local skin friction coefficient increase corresponding to the raising values of the mass flux parameter and the velocity ratio parameter for upper solution branch, but reversal aspect creates for lower solution branch in momentum boundary layer thickness with enlargement in the mass flux parameter when $\eta < 2.5$ and the velocity ratio parameter when $\eta < 1.5$. Subsequently, for upper and lower solution branch, an increment in the mass flux parameter and the velocity ratio parameter define the fall down in the fluid temperature as well as the surface heat flux, while the opposite is true in the dimensionless temperature for the mass flux parameter and the velocity ratio parameter when $\eta > 1.75$ and $\eta > 2.5$, respectively, along the lower solution branch. Further, it is also noted that the surface heat flux enhances with the increment in the velocity ratio parameter when $\varepsilon > 0.2$ over upper solution branch.

iv. Reaction of the access values of the radiation parameter, the heat source/sink parameter, and the Brinkmann number is directed towards to cause an increment in the fluid temperature and heat transfer rate for both cases like stretching and shrinking surface, whereas opposite reflection can be observed for the radiation parameter in the thermal boundary layer for stretching sheet when $\eta < 0.3$, for shrinking sheet when $\eta < 0.9$ and also an opposite effect is seen in the surface heat flux due to both considering sheets.

References

1. T. Elperin and B. Krasovitov (1995), Radiation, thermal diffusion and kinetic effects in evaporation and combustion of large and moderate size fuel droplets, *Int J Heat and Mass Transfer*, 38, 409–418.

2. K. T. Yang (2002), Thermal radiation effects and their determinations room fires, *Adv Building Technol*, I, 83–93.
3. M. M. Molla and M. A. Hossain (2007), Radiation effect on mixed convection laminar flow along a vertical wavy surface, *Int Therm Sci*, 46, 926–935.
4. R. N. Jat and S. Chaudhary (2010), Radiation effects on the MHD flow near the stagnation point of a stretching sheet, *Z Angew Math Phys*, 61, 1151–1154.
5. T. R. Mahapatra, D. Pal and S. Mondal (2013), Mixed convection flow in an inclined enclosure under magnetic field with thermal radiation and heat generation, *Int Commun Heat Mass Transfer*, 41, 47–56.
6. A. Sinha and G. C. Shit (2015), Electromagnetohydrodynamic flow of blood and heat transfer in a capillary with thermal radiation, *J Magn Magn Mater*, 378, 143–151.
7. A. M. Rashad (2017), Impact of thermal radiation on MHD slip flow of a ferro-fluid over a non-isothermal wedge, *J Magn Magn Mater*, 422, 25–31.
8. M. Maiellaro and A. Labianca (2002), On the nonlinear stability in anisotropic MHD with application to Couette–Poiseuille flows, *Int J Eng Sci*, 40, 1053–1068.
9. A. I. Neslituork and M. Tezer-Sezgin (2005), The finite element method for MHD flow at high Hartmann numbers, *Comput Method Appl Mech Eng*, 194, 1201–1224.
10. R. N.Jat and S.Chaudhary (2009), Unsteady magnetohydrodynamic boundary layer flow over a stretching surface with viscous dissipation and Joule heating, *IL Nuovo Cimento B*, 124, 53–59.
11. S. Chaudhary and P. Kumar (2015), Magnetohydrodynamic stagnation point flow past a porous stretching surface with heat generation, *Indian J Pure Appl Phys*, 53, 291–297.
12. F. Mabood and W. A. Khan (2016), Analytical study for unsteady nanofluid MHD flow impinging on heated stretching sheet, *J Mol Liq*, 219, 216–223.
13. S. Das, B. Tarafdar and R. N. Jana (2018), Hall effects on unsteady MHD rotating flow past a periodically accelerated porous plate with slippage, *Euro J Mech-B/Fluids*, 72, 135–143.
14. S. U. S. Choi (1995), Enhancing thermal conductivity of fluids with nanoparticles, *Publ Fed*, 231ASME, 99–106.
15. P. Keblinski, S. R. Phillpot, S. U. S. Choi and J. A. Eastman (2002), Mechanisms of heat flow in suspensions of nano-sized particles (nanofluids), *Int J Heat Mass Transfer*, 45, 855–863.
16. A. Akbarinia and R. Laur (2009), Investigating the diameter of solid particles effects on a laminar nanofluid flow in a curved tube using a two phase approach, *Int J Heat Fluid Flow*, 30, 706–714.
17. M. Sheikholeslami and D. D. Ganji (2013), Heat transfer of Cu-water nanofluid flow between parallel plates, *Powder Technol*, 235, 873–879.
18. H. Safikhani and F. Abbasi (2015), Numerical study of nanofluid flow in flat tubes fitted with multiple twisted tapes, *Adv Powder Technol*, 26, 1609–1617.
19. Y. Wang and G. H. Su (2016), Experimental investigation on nanofluid flow boiling heat transfer in a vertical tube under different pressure conditions, *Exp Therm Fluid Sci*, 77, 116–123.
20. M. Sheikholeslami and H. B. Rokni (2017), Effect of melting heat transfer on nanofluid flow in existence of magnetic field considering Buongiorno Model. *Chin J Phys*, 55, 1115–1126.

21. D. Pal and B. Talukdar (2010), Buoyancy and chemical reaction effects on MHD mixed convection heat and mass transfer in a porous medium with thermal radiation and Ohmic heating, *Commun Nonlinear Sci Numer Simul*, 15, 2878–2893.
22. R. Kumar and B. G. Prasad (2012), Magnetohydrodynamics convective flow in a vertical slot through a porous medium, *Theor Appl Mech Lett*, 2, Article 052003.
23. S. Chaudhary and P. Kumar (2014), MHD forced convection boundary layer flow with a flat plate and porous substrate, *Meccanica*, 49, 69–77.
24. K. Ramesh (2016), Effects of slip and convective conditions on the peristaltic flow of couple stress fluid in an asymmetric channel through porous medium, *Comput Methods Programs Biomed*, 135, 1–14.
25. H. Xu and J. Cui (2018), Mixed convection flow in a channel with slip in a porous medium saturated with a nanofluid containing both nanoparticles and microorganisms, *Int J Heat Mass Transfer*, 125, 1043–1053.
26. K. Hiemenz (1911), Die Grenzschicht an einem in den gleichformigen Flussigkeitsstrom eingetauchten geraden Kreiszylinder, *Dingler's Polytech J*, 326, 321–324.
27. F. Homann (1936), Der einfluss grosser zahigkeit bei der stromung um den zylinder und um die kugel, *Z Angew Math Phys*, 16, 153–164.
28. R. N. Jat and S. Chaudhary (2007), MHD stagnation flows with slip, *IL Nuovo Cimento B*, 122, 823–831.
29. P. Weidman (2016), Axisymmetric rotational stagnation point flow impinging on a radially stretching sheet, *Int J Non-Linear Mech*, 82, 1–5.
30. T. Hayat, M. Z. Kiyani, I. Ahmad, M. I. Khan and A. Alsaedi (2018), Stagnation point flow of viscoelastic nanomaterial over a stretched surface, *Results Phys*, 9, 518–526.
31. C. T. Wang, G. Kinzel and T. Altan (1995), Failure and wrinkling criteria and mathematical modeling of shrink and stretch flanging operations in sheet-metal forming, *J Mater Process Tech*, 53, 759–780.
32. N. Bachok, A. Ishak and I. Pop (2010), Melting heat transfer in boundary layer stagnation-point flow towards a stretching/shrinking sheet, *Phys Lett A*, 374, 4075–4079.
33. M. Turkyilmazoglu (2014), Three dimensional MHD flow and heat transfer over a stretching/shrinking surface in a viscoelastic fluid with various physical effects, *Int J Heat Mass Transfer*, 78, 150–155.
34. S. Chaudhary and M. K. Choudhary (2016), Heat and mass transfer by MHD flow near the stagnation point over a stretching or shrinking sheet in a porous medium, *Indian J Pure Appl Phys*, 54, 209–217.
35. J. H. Merkin and I. Pop (2018), Stagnation point flow past a stretching/shrinking sheet driven by Arrhenius kinetics, *Appl Math Comput*, 337, 583–590.
36. L. Rundora and O. D. Makinde (2013), Effects of suction/injection on unsteady reactive variable viscosity non-Newtonian fluid flow in a channel filled with porous medium and convective boundary conditions, *J Pet Sci Eng*, 108, 328–335.
37. M. I. Khan, M. Tamoor, T. Hayat and A. Alsaedi (2017), MHD boundary layer thermal slip flow by nonlinearly stretching cylinder with suction/blowing and radiation, *Result Phys*, 7, 1207–1211.

38. S. Chaudhary and M. K. Choudhary (2018), Partial slip and thermal radiation effects on hydromagnetic flow over an exponentially stretching surface with suction or blowing, *Therm Sci*, 22, 797–808.
39. S. Chaudhary, M. K. Choudhary and R. Sharma (2015) Effects of thermal radiation on hydromagnetic flow over an unsteady stretching sheet embedded in a porous medium in the presence of heat source or sink, *Meccanica*, 50, 1977–1987.
40. A. J. Harfash (2016), Resonant penetrative convection in porous media with an internal heat source/sink effect, *Appl Math Comput*, 281, 323–342.
41. R. Nandal and A. Mahajan (2018), Penetrative convection in couple-stress fluid via internal heat source/sink with the boundary effects, *J Non-Newtonian Fluid Mech*, 260, 133–141.
42. D. Pal, G. Mandal and K. Vajravelu (2014), Flow and heat transfer of nanofluids at a stagnation point flow over a stretching/shrinking surface in a porous medium with thermal radiation, *Appl Math Comput*, 238, 208–224.
43. X. Su and L. Zheng (2013), Hall effect on MHD flow and heat transfer of nanofluids over a stretching wedge in the presence of velocity slip and Joule heating, *Central Euro J Phys*, 11, 1694–1703.

7

Thermal Radiation Effects on the
Fundamental Flows of a Ree–Eyring
Hydromagnetic Fluid through Porous
Medium with Slip Boundary Conditions

Katta Ramesh

Symbiosis Institute of Technology

CONTENTS

7.1 Introduction

The past few decades have witnessed a wide research and study on Newtonian and non-Newtonian fluids. These two types of fluids can be distinguished from one another, as the former follows the Newton's law of viscosity. Due to their large number of applications in engineering, physiology, biomedical industry, etc., an enormous number of researchers have studied the flow of these fluids and tried to understand the behavior of these fluids. As it has not yet been possible to put forward a single constitutive equation that can represent the behavior of all real fluids in general, so that many researchers have been introduced diverse non-Newtonian fluid models. Using this concept, different researchers have been studying the flow of these fluids with different geometries. An important fluid among non-Newtonian fluids that has been receiving the attention of many researchers is the Ree–Eyring fluid. This fluid has a special property: it can be reduced to a viscous fluid model. Examples of such fluids can be seen in common substances such as blood, butter, ketchup, jam soup, etc. In view of these applications, many researchers have been sharing their work on Ree–Eyring fluids, and the research is still ongoing to understand the behavior of various flow situations. Taha [1] investigated the flow of Casson and Ree–Eyring fluids and made a comparison between the analytical and variational solutions. Maryam and Arif [2] have studied the blood flow of Ree–Eyring fluid with the effect of heat and mass transfer in a compliant channel. Bhatti et al. [3] have analyzed the Ree–Eyring fluid flow in a porous medium. Hayat et al. [4] have examined the transfer of heat in Ree–Eyring fluid in a rotating frame and illustrated the variation of physical parameters. Shawky [5] has used Light hill method to discuss the flow of a Ree–Eyring dusty fluid in a channel with heat transfer. Magnetohydrodynamics (MHDs) can be defined as a study of magnetofluids, i.e., fluids with magnetic properties. So MHDs is governed by the principles of both fluid dynamics and electromagnetism. From the records, it has been found that the term was first used by Alfeven [6]. Due to the huge number of applications in the fields of engineering, such as cooling of reactors, MHD pumps, power generators, and petroleum industries, this field has been a hub for researchers all over the world from the last few decades. Moreover, the applications of MHDs can also be seen in agriculture, geophysics, metrology, etc. Hayat et al. [7] have discussed the Jeffrey fluid flow under the influence of MHD and Newtonian heating over a stretching surface. They employed homotopy analysis method to get the solutions. Usman et al. [8] have studied incompressible Newtonian fluid with the effect of magnetic field. Wahab and Salem [9] have made an analytical approach to investigate the blood flow in a narrow vessel with the magnetic field. Faisal et al. [10] have studied the flow of a subclass of non-Newtonian fluids over a suddenly moved flat plate. They obtained new exact solutions of the governing problem by approaching with Fourier sine and Laplace transformations. Shehzad et al. [11] have investigated the MHD flow of Casson fluids. Hamid et al. [12] have numerically and analytically investigated the squeezing flow of a Casson fluid with the influence of MHDs.

A porous medium can be defined as a material having interstices, which usually contain a liquid or gas. Porous medium is a concept of great importance due to its wide variety of applications in science, engineering, biology, geology, etc. Porous medium has played a vital role in fluid dynamics, where it has helped the researchers to study the behavior of different fluids. Devakar et al. [13] have employed a numerical approach to analyze and investigate the flow properties of couple stress and Jeffery fluids in a square duct. Khaled et al. [14] have studied the mass diffusion in tissues and flow convection in biological processes to show the role of porous media and its applications in biology. Santhosh and Radhakrishnamacharya [15] have studied the flow of

Jeffery fluid in the presence of porous medium and magnetic field in a tube with small diameter. Eldesoky [16] examined the blood flow through a porous medium under the influence of magnetic field and slip conditions. El-Shehawey et al. [17] have discussed the pulsatile blood flow through a porous medium. Siddiqui et al. [18] have discussed the unsteady viscous MHD fluid flow between two parallel plates using the homotopy perturbation method. Ismail et al. [19] have studied the unsteady flow of MHD viscous fluid between two parallel plates in a porous medium. Chand and Kumar [20] have investigated the flow of a viscoelastic fluid under the effects of heat and mass transfer with the help of Laplace transformations. Ganesh and Krishnambal [21] have investigated the Stokes flow of viscous fluid between two parallel plates.

Heat transfer can be defined as a mechanism of transfer of heat between two systems or within a system that are at different temperatures. This transfer takes place as per the second law of thermodynamics, i.e., heat travels from an object with high temperature to an object with low temperature. There are three modes of heat transfer, viz. conduction, convection, and radiation. The applications of heat transfer can be seen everywhere, in everyday life, in fields of engineering like mechanical engineering (heat exchangers, boilers), electrical engineering (transformers and generators), chemical engineering, aerospace engineering, and in industrial fields. The concept has played a vital role in the field of fluid dynamics, where it has proven to be very useful in studying, analyzing, and investigating the behavior of fluids. In the last few decades, considerable attention, by many researchers, has been paid in analyzing fluid flows with heat transfer. A recent work in this field can be seen in the Hamilton and Crosser model by Yang and Xu [22], where the model undergoes modifications over time to time. A modified model can be seen in the works of Taylor, who introduced a dispersed model of energy equation to account for random movements in the main flow. However, a modified model of this version was developed by Xuan and Li [23]. In recent works, a modified model from Azari et al. [24] can be seen to have a connection to the experimental data. Shasmsi et al. [25] have numerically studied the laminar flow of a non-Newtonian fluid in a rectangular microchannel. Barnoon and Toghraie [26] have also made a numerical approach to investigate the heat transfer of nanofluid in a porous medium. Yasir and Smarda [27] have used the homotopy method to analyze the flow and heat transfer of a non-Newtonian fluid in the region of stagnation point. Rudraiah et al. [28] have discussed the free and forced convective effects on electrically conducting fluid through a channel. Kang et al. [29] have used homotopy analysis method to acquire the analytical solutions of the heat transfer effects in the flow fluid through a porous medium. To the best of author's cognizance, none of the studies focused on the Ree–Eyring fluid flow situation under the effects of heat transfer, thermal radiation, porous medium, magnetic field, and slip boundary conditions. In view of this, the author has inspired to investigate the influence of radiation and viscous dissipation on a MHD Ree–Eyring fluid in porous structures with radiation and porous medium. This chapter is organized as follows: in Section 7.2, mathematical modeling as well as closed-form solutions are presented for the velocity and temperature distributions. In Section 7.3, the numerical results are discussed for the velocity and temperature profiles through graphs. In Section 7.4, the important conclusions are presented for the studied problems.

7.2 Formulation and Solution of the Problem

The governing equations associated with the Ree–Eyring fluid in the presence of porous medium, MHDs, and radiation can be presented as

$$\nabla \cdot \bar{q} = 0, \tag{7.1}$$

$$\rho\left(\frac{\partial \bar{q}}{\partial t} + (\bar{q}.\nabla)\bar{q}\right) = -\nabla P + \nabla \cdot \mathbf{S} - \sigma B_0{}^2 \bar{q} - \frac{\mu_s}{k_0}\bar{q}, \tag{7.2}$$

$$\rho c_p\left(\frac{\partial \bar{T}}{\partial t} + \bar{q} \cdot \nabla \bar{T}\right) = \nabla\left(k^* \nabla \bar{T}\right) - \frac{\partial q_r}{\partial y} + \frac{\bar{J}^2}{\sigma} + \mu_s \phi, \tag{7.3}$$

where ρ is the density, $\bar{q}$ denotes the velocity vector, t represents the time, P is the pressure, σ is the electrical conductivity, c_p is the specific heat at constant pressure, B_0 represents the uniform magnetic field, k^* is the thermal conductivity, $\bar{T}$ denotes the temperature, q_r defines the radiation, μ is the viscosity coefficient, $\bar{J}$ is the current vector due to Ohm's law, k_0 denotes the permeability parameter, ϕ is the viscous dissipation, and $\mathbf{S}$ is the extra tensor of Ree–Eyring fluid model, and it is given by

$$\mathbf{S} = \mu_s A_1 + \frac{1}{\bar{B}}\sinh^{-1}\left(\frac{1}{\bar{C}}A_1\right), \tag{7.4}$$

where

$$A_1 = \nabla\bar{q} + (\nabla\bar{q})^T, \tag{7.5}$$

we have $\sinh^{-1}(x) \approx x \; \forall \; x \leq 1$.

Using the earlier expression in Eq. (7.4), it becomes

$$\mathbf{S} = \mu_s A_1 + \frac{1}{\bar{B}}\left(\frac{1}{\bar{C}}A_1\right). \tag{7.6}$$

The fluid is optically very thin with a relatively low density and radiative heat flux, and it is given by

$$\frac{\partial q_r}{\partial y} = 4\alpha^2 \left(T_0 - \overline{T}_i\right), \tag{7.7}$$

where α denotes the mean radiation absorption coefficient.

In the present investigation, we have considered the Ree–Eyring fluid flow between two parallel plates (see Figure 7.1) in various situations (such as letting the motion of plates in opposite directions, keeping one plate at rest by letting the other plate move, and letting both the plates move in the same direction). For these problems, we have chosen the velocity field as $\overline{q} = \left(u_i(y), 0, 0\right)$. This velocity field choice automatically satisfies the continuity equation (7.1). In steady case, the governing equations (7.2)–(7.7) become

$$\left(1 + \frac{1}{\mu_s \overline{BC}}\right)\frac{d^2 u_i}{dy^2} - \sigma B^2 u_i - \frac{\mu_s}{k_0} u_i + G = 0, \tag{7.8}$$

$$k^* \frac{d^2 \overline{T}_i}{dy^2} + \sigma B_0{}^2 u_i{}^2 - 4\alpha^2 \left(T_0 - \overline{T}_i\right) + \mu_s \left(1 + \frac{1}{\mu_s \overline{BC}}\right)\left(\frac{du_i}{dy}\right)^2 = 0, \tag{7.9}$$

with the corresponding boundary conditions

$$u_i - L_1 \left(1 + \frac{1}{\mu_s \overline{BC}}\right)\frac{du_i}{dy} = U_i, \quad \overline{T}_i - L_2 \frac{d\overline{T}_i}{dy} = T_0 \quad \text{at } y = -h, \tag{7.10}$$

$$u_i + L_1 \left(1 + \frac{1}{\mu_s \overline{BC}}\right)\frac{du_i}{dy} = -U_i \,/\, 0 \,/\, U_i, \quad \overline{T}_i + L_2 \frac{d\overline{T}_i}{dy} = T_1 \quad \text{at } y = h, \tag{7.11}$$

where L_1 is the velocity slip parameter and L_2 denotes the temperature slip parameter. Now we introduce the nondimensional quantities as follows:

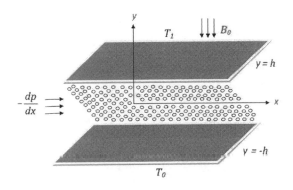

FIGURE 7.1
Physical configuration of the problem.

$$\bar{x} = \frac{x}{h}, \quad \bar{y} = \frac{y}{h}, \quad \bar{u}_i = \frac{u_i}{U_i}, \quad \bar{p} = \frac{hP}{\mu_s U_i}, \quad Ha = \sqrt{\frac{\sigma}{\mu_s}} B_0 h, \quad Da = \frac{k_0}{h^2},$$

$$N = \frac{2\alpha h}{\sqrt{k^*}}, \quad \beta = \frac{1}{\mu_s \bar{B} \bar{C}}, \quad \theta = \frac{\bar{T}_i - T_0}{T_1 - T_0}, \quad Br = \frac{\mu_s U_i^2}{k^*(T_1 - T_0)}, \quad \gamma = \frac{L_1}{h}, \quad \beta_1 = \frac{L_2}{h},$$

where Ha denotes the Hartmann number, Da represents the Darcy number, N is the thermal radiation parameter, β is the Ree–Eyring fluid parameter, θ is the nondimensional temperature, Br is the Brinkman number, β_1 is the dimensionless temperature slip parameter, and γ is the nondimensional velocity slip parameter. Using the Eqs. (7.8)–(7.11) with the dimensionless quantities, the governing equations can be written as

$$(1+\beta)\frac{d^2 u_i}{dy^2} - M^2 u_i + G = 0, \tag{7.12}$$

$$\frac{d^2 \theta_i}{dy^2} + N^2 \theta_i + Br(1+\beta)\left(\frac{du_i}{dy}\right)^2 + Br\, Ha^2\, u_i^2 = 0, \tag{7.13}$$

with corresponding boundary conditions:

$$u_i - \gamma(1+\beta)\frac{du_i}{dy} = 1; \quad \theta_i - \beta_1 \frac{d\theta_i}{dy} = 0 \quad \text{at } y = -1, \tag{7.14}$$

$$u_i + \gamma(1+\beta)\frac{du_i}{dy} = -1/0/1; \quad \theta_i + \beta_1 \frac{d\theta_i}{dy} = 1 \quad \text{at } y = 1, \tag{7.15}$$

where $M = \sqrt{Ha^2 + \dfrac{1}{Da}}$ and $G = \left(-\dfrac{\partial p}{\partial x}\right)$ denotes the nondimensional pressure gradient.

7.2.1 Fluid Flow due to Motion of Plates in Opposite Directions

Let us consider the flow of a Ree–Eyring fluid between two parallel, infinite plates placed horizontally. The plates are moving in opposite directions, where the upper plate represented by $y = h$ is caused to move with a thrust in the positive direction and the lower plate represented by $y = -h$ is moving in the opposite direction of flow. The effects of magnetic field, heat transfer, and porous medium are considered into account. The influence of Joule heating, thermal radiation, and viscous dissipation are taken in an energy equation. It is assumed that the lower plate is maintained at temperature T_0, while the upper plate is at temperature T_1. The uniform magnetic field of strength B_0 is acted in the transverse direction of the flow. In this situation, the velocity profile is considered as $(u_1, 0, 0)$. For this case, the momentum and energy equations can be written as

$$(1+\beta)\frac{d^2u_1}{dy^2} - M^2u_1 + G = 0, \tag{7.16}$$

$$\frac{d^2\theta_1}{dy^2} + N^2\theta_1 + Br(1+\beta)\left(\frac{du_1}{dy}\right)^2 + Br\,Ha^2\,u_1{}^2 = 0, \tag{7.17}$$

with corresponding boundary conditions

$$u_1 - \gamma(1+\beta)\frac{du_1}{dy} = 1, \quad \theta_1 - \beta_1\frac{d\theta_1}{dy} = 0 \quad \text{at } y = -1, \tag{7.18}$$

$$u_1 + \gamma(1+\beta)\frac{du_1}{dy} = -1, \quad \theta_1 + \beta_1\frac{d\theta_1}{dy} = 1 \quad \text{at } y = 1. \tag{7.19}$$

From the Eqs. (7.16) and (7.17), with the help of boundary conditions (7.18) and (7.19), the solutions for the velocity and temperature are given by

$$u_1 = C_1 e^{Ay} + C_2 e^{-Ay} + \frac{G}{A^2(1+\beta)}, \tag{7.20}$$

$$\theta_1 = D_1 \cos Ny + D_2 \sin Ny + S_1 e^{2Ay} + S_2 e^{-2Ay} + S_3 e^{Ay} + S_4 e^{-Ay} + S_5, \tag{7.21}$$

where $C_i's(i = 1,2)$, $D_j's(j = 1,2)$ and $S_k's(k = 1-5)$ are algebraic expressions, which are calculated using the computational software Mathematica.

7.2.2 Flow due to Motion of the Lower Plate

In this case, we consider the flow of a Ree–Eyring fluid between two infinite and parallel plates that are placed horizontally. Here we assumed that the lower plate $y = -h$ is moving while the upper plate $y = h$ is at rest. The lower plate $y = -h$ is caused to move with a constant velocity. In this case, the effects of magnetic field, heat transfer, porous medium, Joule heating, viscous dissipation, and thermal radiation are considered into account. Here, the lower plate is maintained at temperature T_0 and the upper plate is maintained at temperature T_1. The uniform magnetic field of strength B_0 is applied in the transverse direction of the flow. Here we considered the velocity profile as $(u_2, 0, 0)$. Using these, the resulting dimensionless problem can be presented as

$$(1+\beta)\frac{d^2u_2}{dy^2} - M^2u_2 + G = 0, \tag{7.22}$$

$$\frac{d^2\theta_2}{dy^2} + N^2\theta_2 + Br(1+\beta)\left(\frac{du_2}{dy}\right)^2 + Br\ Ha^2\ u_2{}^2 = 0, \qquad (7.23)$$

with corresponding slip boundary conditions

$$u_2 - \gamma(1+\beta)\frac{du_2}{dy} = 1; \quad \theta_2 - \beta_1\frac{d\theta_2}{dy} = 0 \quad \text{at } y = -1, \qquad (7.24)$$

$$u_2 + \gamma(1+\beta)\frac{du_2}{dy} = 0; \quad \theta_2 + \beta_1\frac{d\theta_2}{dy} = 1 \quad \text{at } y = 1, \qquad (7.25)$$

from Eqs. (7.22) and (7.23) with the help of boundary conditions (7.24) and (7.25), the solutions for the velocity and temperature are given by

$$u_2 = C_3 e^{Ay} + C_4 e^{-Ay} + \frac{G}{A^2(1+\beta)}, \qquad (7.26)$$

$$\theta_2 = D_3 \cos Ny + D_4 \sin Ny + S_6 e^{2Ay} + S_7 e^{-2Ay} + S_8 e^{Ay} + S_9 e^{-Ay} + S_{10}, \quad (7.27)$$

where $C_i's(i = 3,4)$, $D_j's(j = 3,4)$, and $S_k's(k = 6-10)$ are algebraic expressions, which are obtained from the computational mathematical software Mathematica.

7.2.3 Fluid Flow due to Motion of Plates in the Same Direction

In this case, like the two earlier cases, we again consider the flow of a Ree–Eyring fluid between two infinite and parallel plates. But in this case, we proceed with the observation that both the plates, i.e., the upper plate $y = h$ and the lower plate $y = -h$, are moving in the same direction. The plates are caused to move with a constant velocity. Similarly, the effects of magnetic field, heat transfer, porous medium, Joule heating, etc. have been taken into account as in the first two cases. Here also, we assumed the lower plate to be at a temperature T_0 and the upper plate at a temperature T_1. The uniform magnetic field of strength B_0 is applied in the transverse direction of flow. Here we considered the velocity profile as $(u_3,0,0)$. From these, the governing dimensionless problem can be written as

$$(1+\beta)\frac{d^2u_3}{dy^2} - M^2u_3 + G = 0, \qquad (7.28)$$

$$\frac{d^2\theta_3}{dy^2} + N^2\theta_3 + Br(1+\beta)\left(\frac{du_3}{dy}\right)^2 + Br\ Ha^2\ u_3{}^2 = 0, \qquad (7.29)$$

with corresponding slip boundary conditions

$$u_3 - \gamma(1+\beta)\frac{du_3}{dy} = 1; \quad \theta_3 - \beta_1 \frac{d\theta_3}{dy} = 0 \quad \text{at } y = -1, \tag{7.30}$$

$$u_3 + \gamma(1+\beta)\frac{du_3}{dy} = 1; \quad \theta_3 + \beta_1 \frac{d\theta_3}{dy} = 1 \quad \text{at } y = 1, \tag{7.31}$$

from Eqs. (7.28) and (7.29), with the help of slip boundary conditions (7.30) and (7.31), the closed-form solutions for the velocity and temperature can be written as

$$u_3 = C_5 e^{Ay} + C_6 e^{-Ay} + \frac{G}{A^2(1+\beta)}, \tag{7.32}$$

$$\theta_3 = D_5 \cos Ny + D_6 \sin Ny + S_{11}e^{2Ay} + S_{12}e^{-2Ay} + S_{13}e^{Ay} + S_{14}e^{-Ay} + S_{15}, \tag{7.33}$$

where $C_i's(i=5,6)$, $D_j's(j=5,6)$, and $S_k's(k=11-15)$ are algebraic expressions, which are calculated by using the computational mathematical software Mathematica.

7.3 Results and Discussion

This section deals with the discussion of various flow parameters on the velocity and temperature distributions. The results have been presented graphically for velocity and temperature distributions in all three cases when the plates are moving in opposite directions, when lower plate is moving and upper plate is at rest, and when both the plates are moving in the same direction. The effect of various parameters, such as radiation parameter N, Ree–Eyring fluid parameter β, temperature slip parameter β_1, velocity slip parameter γ, Hartmann number Ha, Darcy number Da, and pressure gradient G, has been discussed for velocity and temperature profiles. The velocity and temperature distributions in all three cases have been treated as follows: u_1 and θ_1, u_2 and θ_2, and u_3 and θ_3, when both the plates are moving in the opposite direction, when the upper plate is moving and other is at rest, and when both the plates are moving in the same direction, respectively.

7.3.1 Velocity Profile

Figures 7.2–7.5 have been developed to study the effect of Hartmann number Ha, Darcy number Da, Ree–Eyring fluid parameter β, and the velocity slip parameter γ on the velocity of the Ree–Eyring fluid

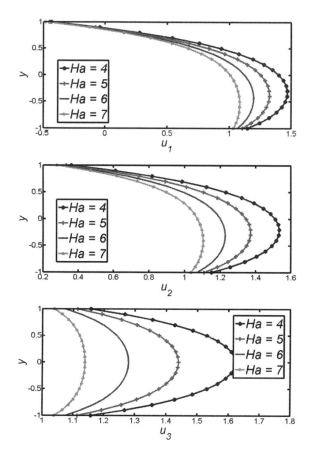

FIGURE 7.2
Impact of Hartmann number on the velocity profiles for fixed values of $\beta = 0.3$, $\gamma = 0.1$, $Da = 1$, $G = 9$.

flows. Figure 7.2 demonstrates the nature of velocity under the effect of Hartmann number in the aforementioned three cases, respectively. It is clear from Figure 7.2, with an increase in Hartmann number Ha, there is a decrease in velocity in all the cases. It is evident that this is due to the resistance of the flow by the effect of Lorentz force. Because an increase in Hartmann number increases the Lorentz force. To understand the reaction of Darcy number on the velocity profile, Figure 7.3 has been provided. The figure demonstrates that for a higher Darcy number higher will be the velocity. This depends on penetrability. For more penetrability in porous medium, we observe more resistance to the fluid flow, and as a result, there is a decrease in the velocity. Figure 7.4 has been developed to display the behavior of Ree–Eyring fluid parameter β on the velocity in distinct

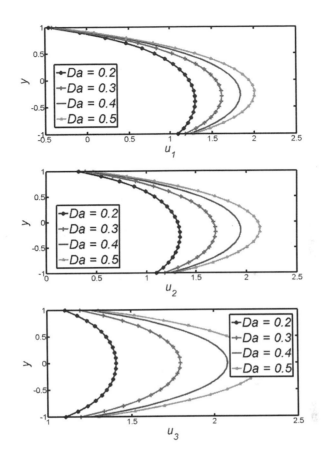

FIGURE 7.3
Effect of Darcy number on the velocity profiles for fixed values of $\beta = 0.3$, $\gamma = 0.1$, $Ha = 1$, $G = 9$.

flow situations. It can be visualized that, near the immobile plate, there is a decrease in velocity, and an increase in velocity can be observed near the moving plate. It can also be observed that, as we approach near to the center of the plates, the velocity decreases. The increasing effect of the slip parameter on the velocity of the fluid can be seen in Figure 7.5. The higher values of the slip parameter show that the fluid velocity component increases. In other words, the more the fluid slip gets stronger, the less effect is its velocity.

7.3.2 Temperature Field

In order to understand the effect of Hartmann number Ha, Ree–Eyring fluid parameter β, radiation parameter N, and the temperature slip

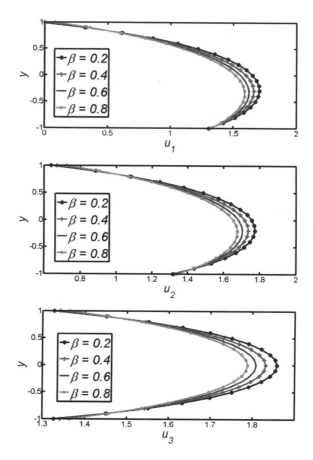

FIGURE 7.4
Effect of Ree–Eyring fluid parameter on the velocity for fixed values of $Ha = 1$, $\gamma = 0.2$, $Da = 0.3, G = 9$.

parameter β_1 on temperature, Figures 7.6–7.9 have been prepared. To examine the behavior of temperature on different values of the Hartmann number Ha, Figure 7.6 has been presented. It can be noticed from Figure 7.6 that, in all the three cases, the temperature is a decreasing function of Hartmann number. This decrease in temperature is due to the fact that the presence of magnetic field produces a Lorentz force. This force increases the friction between the layers of fluid and thus decreases the temperature. Then, Figure 7.7 demonstrates the behavior of the Ree–Eyring fluid parameter β on temperature. The expression of the Ree–Eyring fluid parameter is given as $B = \left(\dfrac{1}{\mu_s \overline{BC}} \right)$. As Ree–Eyring fluid parameter increases, the viscosity decreases, and whenever viscosity

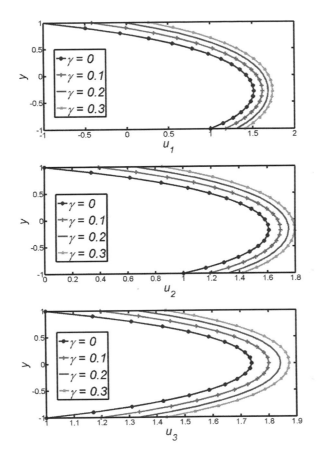

FIGURE 7.5
Effect of slip parameter on the velocity profiles for fixed values of $Ha = 1$, $\beta = 0.3$, $Da = 0.3$, $G = 9$.

decreases in the medium, the flow of heat is generated. Due to this fact, the temperature of fluid decreases with increasing values of Ree–Eyring fluid parameter. The impact of radiation parameter N on temperature can be visualized in Figure 7.8. It can be easily visualized from the figure that an increase in the radiation parameter values leads to a rise in temperature. It is due to the fact that more heat is transferred to the fluid on increasing the values of radiation parameter. The effect of temperature slip parameter on the temperature profile is shown in Figure 7.9. It is clear from the figure that, in all cases, with an increase in temperature slip parameter, the temperature of Ree–Eyring fluid increases. It is also worth mentioning that higher temperatures are observed in the cases of temperature slip parameter when compared with that of no-slip parameters.

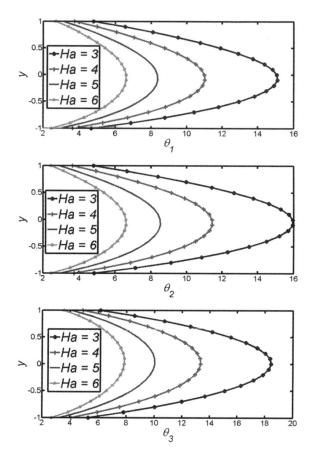

FIGURE 7.6
Effect of Hartmann number on the temperature profiles for fixed values of $\beta = 0.3$, $\gamma = 0.2$, $Da = 03$, $Br = 3$, $G = 9$, $\beta_1 = 0.2$, $N = 0.5$.

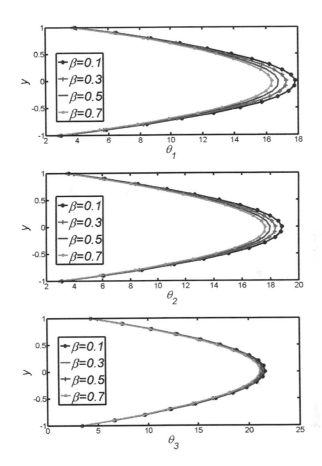

FIGURE 7.7
Impact of Ree–Eyring fluid parameter on the temperature profiles for fixed values of $Ha = 2$, $\gamma = 0.1$, $Br = 3$, $Da = 0.3$, $N = 0.5$, $G = 9$, $\beta_1 = 0.1$.

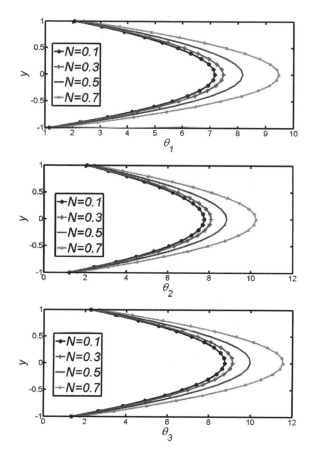

FIGURE 7.8
Effect of radiation parameter on the temperature profiles for fixed values of $Ha = 2$, $\beta = 0.3$, $\gamma = 0.2$, $Br = 1$, $Da = 3$, $G = 9$, $\beta_1 = 0.1$.

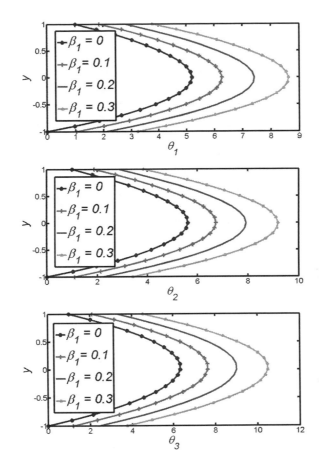

FIGURE 7.9

Effect of temperature slip parameter on the temperature profiles for fixed values of $Ha = 2$, $\beta = 0.3$, $\gamma = 0.2$, $Br = 1$, $Da = 0.3$, $N = 0.5$, $G = 9$.

7.4 Conclusions

The Ree–Eyring fluid flow between two infinite parallel plates (both the plates are moving in opposite direction, one plate is moving while the other is at rest, both the plates are moving in same direction) has been studied. The effects of porous medium, magnetic field, thermal radiation, viscous dissipation, and Joule heating have been considered in all the mentioned cases. In all the three cases, we have considered the uniform magnetic field acting in the transverse direction of flow. The plates are also considered at different temperatures. The governing equations for each of the problem are solved analytically, and closed-form solutions have been obtained in each case. The effects of various parameters on velocity and temperature for problem are presented using the graphs. The main results of this study are summarized later:

- The velocity decreases with an increase in Hartmann number in all the cases.
- The velocity is an increasing function of Darcy number.
- The velocity decreases with an increase of Ree–Eyring fluid parameter.
- Higher velocities are noticed in the cases of velocity slip when compared with no-slip and Newtonian fluid model, respectively.
- The temperature is a decreasing function of Hartmann number and Ree–Eyring fluid parameter.
- The temperature increases with an increase of thermal radiation and temperature slip parameters.

Nomenclature

$\bar{q}$:   Velocity vector
ρ: Density
p: Pressure
t: Time
σ: Electrical conductivity
B_0:   Uniform magnetic field
k^*:   Thermal conductivity
T: Temperature
q_r: Radiation parameter
J: Current vector
μ: Viscosity coefficient
k_0:   Permeability parameter

ϕ: Viscous dissipation
α: Mean radiation absorption coefficient
β: Ree–Eyring fluid parameter
β_1: Temperature slip parameter
N: Radiation parameter
γ: Velocity slip parameter
Ha: Hartmann number
Da: Darcy number
G: Pressure gradient

References

1. Taha, S. Variational approach for the flow Ree-Eyring and Casson fluids in pipes. *International Journal of Modeling, Simulation and Scientific Computing*, 2016, 7: 1650007.
2. Maryam, J., Arif, R. MHD peristaltic blood flow of Ree-Eyring fluid in a compliant channel with influence of heat and mass transfer. In *14th International Bhurban Conference on Applied Sciences and Technology*, Islamabad, Pakistan: Centres of Excellence in Science and Applied Technologies, 2017, doi: 10.1109/IBCAST.2017.7868113.
3. Bhatti, M.M., Abbas, M.A., Rashidi, M.M. Combine effects of MHD and partial slip on peristaltic blood flow of Ree-Eyring fluid with wall properties. *Engineering Sciences and Technology an International Journal*, 2016, 19: 1497–1502.
4. Hayat, T., Zahir, H., Alsaedi, A., Ahmad, B. Heat transfer analysis on peristaltic transport of Ree-Eyring fluid in rotating frame. *Chinese Journal of Physics*, 2017, 55: 1894–1907.
5. Shawky, H.M. Pulsatile flow with heat transfer of dusty magnetohydrodynamic Ree-Eyring fluid through a channel. *Heat and Mass Transfer*, 2009, 45(10): 1261–1269.
6. Alfaven, H. Existence of electromagnetic-hydrodynamic waves. *Nature*, 1942, 150: 405–406.
7. Hayat, T., Awais, M., Alsaedi, A. Newtonian heating and MHD effects in flow of a Jeffery fluid over a radially stretching surface. *International Journal of Physical Sciences*, 2012, 7: 2838–2844.
8. Usman, M., Naheed, Z., Nazir, A., Mohyud-Din, S.T. On MHD flow of an incompressible viscous fluid. *Journal of Egyptian Mathematical Society*, 2014, 22(2): 214–219.
9. Wahab, A.M., Salem, S.I. Magnetohydrodynamic blood flow in a narrow tube. *World Research Journal of Biomaterials*, 2012, 1(1): 1–7.
10. Faisal, S., Aziz, Z.A., Ching, D.L.C. New exact solutions for MHD transient rotating flow of a second grade fluid in a porous medium. *Journal of Applied Mathematics*, 2011, 2011: 823034.
11. Shehzad, S.A., Hayat, T., Qasim, M., Asghar, S. Effects of mass transfer on MHD flow of Casson Fluid with chemical reaction and suction. *Brazilian Journal of Chemical Engineering*, 2013, 30(1): 187–195.

12. Hamid, K., Qayyum, M., Khan, O., Ali, M. Unsteady squeezing flow of Casson fluid with MHD effect and passing through porous medium. *Mathematical Problems in Engineering*, 2016, 2016: 4293721.
13. Devakar, M., Ramesh, K., Chouhan, S., Raje, A. Fully developed flow of non-Newtonian fluids in a straight uniform square duct through porous medium. *Journal of the Association of Arab Universities for Basic and Applied Sciences*, 2017, 23: 66–74.
14. Khaled, A.R.A., Vafai, K. The role of porous media in modeling flow and heat transfer in biological tissues. *International Journal of Heat and Mass Transfer*, 2003, 46: 4989–5003.
15. Santhosh, N., Radhakrishnamacharya, G. Jeffery fluid flow through porous medium in the presence of magnetic field in narrow tubes. *International Journal of Engineering Mathematics*, 2014, 2014: 713831.
16. Eldesoky, I.M. Slip effects on the unsteady MHD pulsatile blood flow through porous medium in an artery under the effect of body acceleration. *International Journal of Mathematics and Mathematical Sciences*, 2012, 2012: 860239.
17. El-Shehawey, E.F., Elbarbary, E.M.E., Afifi, N.A.S., El-Shahed, M. Pulsatile flow of blood through a porous medium under periodic body acceleration. *International Journal of Theoritical Physics*, 2000, 39(1): 183–188.
18. Siddiqui, A.M., Irum, S., Ansari, A.R. Unsteady squeezing flow of a viscous MHD fluid between parallel plates, a solution using the homotopy perturbation method. *Mathematical Modelling and Analysis*, 2008, 13(4): 565–576.
19. Ismail, A.M., Ganesh, S., Kirubhashankar, C.K. Unsteady magnetohydrodynamic flow between two parallel plates through a porous medium. *International Journal on Design and Manufacturing Technologies*, 2013, 7(2): 1–6.
20. Chand, K., Kumar, R. Hall effect on heat and mass transfer in the flow of oscillating viscoelastic fluid through porous medium with wall slip conditions. *Indian Journal of Pure and Applied Physics*, 2012, 50(3): 149–155.
21. Ganesh, S., Krishnambal, S. Unsteady magnetohydrodynamic Stokes flow of viscous fluid between two parallel porous plates. *Journal of Applied Sciences*, 2007, 7(3): 374–379.
22. Yang, L., Xu, X. A renovated Hamilton-Crosser model for the effective thermal conductivity of CNTs nanofluids. *International Communications in Heat and Mass Transfer*, 2017, 81: 42–50.
23. Xuan, Y., Li, Q. Heat transfer enhancement of nanofluids. *International Journal of Heat and Fluid Flow*, 2000, 21(1): 58–64.
24. Azari, A., Kalbasi, M., Moazzeni, A., Rahman, A. A thermal conductivity model for nanofluids heat transfer enhancement. *Petroleum Science and Technology*, 2014, 32(1): 91–99.
25. Shamsi, M.R., Akberi, O.A., Marzban, A., Toghraie, D., Mashayekhi, R. Increasing heat transfer of non-Newtonian nanofluid in rectangular microchannel with triangular ribs. *Physica E: Low dimensional Systems and Nanostructures*, 2017, 93: 167–178.
26. Barnoon, P., Toghraie, D. Numerical investigation of laminar flow and heat transfer of non-Newtonian nanofluid within a porous medium. *Power Technology*, 2018, 325: 78–91.
27. Yasir, K., Smarda, Z. Heat transfer analysis on the Heimenz flow of a non-Newtonian fluid, a homotopy Method solution. *Abstract and Applied Analysis*, 2013, 2013: 342690.

28. Rudraiah, N., Kumudini, V., Unno, W. Theory of non-linear magneto-convection and its applications to solar convection problems. *Astronomical Society of Japan*, 1985, 37(2): 183–233.

29. Kang, J., Xia, T., Liu, Y. Heat transfer and flow of convection in a fluid saturated rotating porous medium. *Mathematical Problems in Engineering*, 2015, 2015: 905458.

8

The Minimum Spanning Tree with Node Index ≤ 2 Is Equivalent to the Minimum Traveling Salesman Tour

Santosh Kumar
University of Melbourne
RMIT University

Elias Munapo
North West University

Caston Sigauke
University of Venda

Masar Al-Rabeeah
RMIT University
Basrah University

CONTENTS

8.1 Introduction

In graph theory, a well-known problem is to find the minimum salesman tour i.e., to find a way of moving from a given origin node and returning back to the same node after visiting each other node only once, where the total distance or cost is minimal. This problem has been classified as a nondeterministic polynomial hard (NP-hard) combinatorial optimization problem; see Nemhauser and Wolsey (1989). One more well-known problem in graph theory is to find the minimum spanning tree (MST) of all nodes, which has been classified as an easy problem for which the greedy approach works. In an "n" node network, the MST can be obtained in $(n–1)$ greedy iterations. Many approaches to find the MST has been suggested, see for example, Anupam (2015), Garg and Kumar (1968), and Kruskal (https://www.youtube.com/watch?v=3rrNH_ AizMA). Recently, a variation of the MST was proposed, and a solution approach was provided by Munapo et al. (2016). They obtained the MST of a given graph under the restriction that the index of each node in the spanning tree has to be ≤2. A spanning tree with index ≤2 turns out to be a path between two nodes, which passes through all other remaining nodes of the network. Motivation to find the index-restricted MST was to establish a relationship between the MST and the minimum traveling salesman tour (MTST). Earlier attempts to establish this relationship between MTST and index-restricted MST have appeared in Kumar et al. (2016, 2018). Note that if any edge from the TST is removed, the remaining TST becomes a spanning tree between the two nodes joining the removed edge. These earlier attempts by Kumar et al. (2016, 2018) solved several index-restricted MST problems to find the required MTST. In this chapter, yet another approach is proposed, which establishes a strong equivalence between TST and index-restricted MST.

The TST has many industrial applications, see Berman and Karpinski (2006), Cowen (2016), Garg et al. (1970), Kumar et al. (2014), Munapo (2013), Razali and Geraghty (2011) and Nemhauser and Wolsey (1989).

This chapter has been organized in the following sections. Section 8.2 presents a brief review of earlier approaches. Essential concepts are discussed in Section 8.3. Section 8.4 deals with the modification and mathematical development required for the proposed approach. Section 8.5 presents a few numerical illustrations. This chapter concludes with Section 8.6.

8.2 A Brief Review of Earlier Approaches

Kumar et al. (2018) considered a connected network, $G(n,m)$, where n denotes the number of nodes and m denotes the number of edges in the network $G(n,m)$. From the given network $G(n,m)$, Kumar et al. (2018)

developed a modified network denoted by $G'\big((n-1),(m-d_p)\big)$, where the number of nodes was reduced by one, i.e., the number of nodes in the modified network is given by $(n–1)$ and the number of edges was reduced from m to $(m-d_p)$, where p is the selected node removed and d_p represents the number of edges joining the node p with other nodes in the given network $G(n,m)$. Since TST passes through each node, the selected node p can be any one node of the given network $G(n,m)$. However, if the number of edges joining the node p with other nodes is given by d_p, the modified graph, after removing the node p and d_p edges, can be denoted by $G'\big((n-1),(m-d_p)\big)$. It is essentially the same network without the node p and without all the edges that join the node p to all other nodes of the network $G(n,m)$. If arbitrarily any two edges (p,q) and (p,k) from the node p are selected, the index-restricted MST of the network joining the nodes q and k in the modified network $G'\big((n-1),(m-d_p)\big)$ along with the edges (p,q) and (p,k) forms a feasible TST and hence gives an upper bound of the required minimum TST. It may be noted that the index of nodes q and k is one, and for all other remaining nodes, the index is 2. Since the total number of edges joining the node p to other nodes is assumed to be equal to d_p, the possible number of such combinations of two at a time will be given by $(d_p)_{C_2}$. In the case of a completely connected network $G(n,m)$, the number of edges from every node will be given by $(n–1)$, and consequently, for any arbitrarily selected node p, the complexity will be given by $(n-1)_{C_2}$. Here complexity means the number of index-restricted MSTs one has to solve before the optimal TST can be concluded.

Since the complexity of this approach depends on the number of edges joining other nodes from the selected node p, Kumar et al. (2018) selected the node p as the one with the minimum number of edges, which will be equal to $(n–1)$ in the case of a completely connected graph.

This complexity dependence on the number d_p motivated the authors, and they reconsidered the problem of finding the TST through the MST and modified the network slightly in a different way. That approach appeared in Kumar et al. (2016). For the network $G(n,m)$, the modified network was considered as $G'\big((n),(m-d_p)\big)$, i.e., no node was removed from the modified graph, and only the edges joining the selected node to other nodes were removed. Let these edges from the node p be denoted by $\{(p,q_1),(p,q_2),....,(p,q_{dp})\}$. If we consider one edge at a time, say the edge (p,q_1), then we find an index-restricted MST with node index as one at the nodes p and q_1, and thus solve d_p number index-restricted MST problems for each possibility of joining the node p to node q_k, where $k-1,?,\ ,d_p$. Thus the complexity was reduced from $(d_p)_{C_2}$ to d_p. For example, if $d_p = 10$, the requirement of solving index-restricted MST problem reduces from 45 to 10, which is a significant reduction.

As mentioned in the introduction, the proposed approach solves only one index-restricted MST to find the TST. It is now independent of d_p, and thus is truly establishing equivalence between index-restricted MST and conventional TST.

8.3 Mathematical Background

Some concepts concerning the MST with index ≤ 2 are necessary for developing the method discussed in Section 8.4. Details of the MST with index restriction are given in Munapo et al. (2016).

- Node index: The MST of a given graph can be obtained by any known method, for example, Anupam (2015), Garg and Kumar (1968), and Kruskal (https://www.youtube.com/watch?v=3rrNH_AizMA). The MST obtained by these methods will have a node index at any typical node i, denoted by n_i, which will be an integer number such that $1 \leq n_i \leq (n-1)$, where n is the total number of nodes in that network.

- High index node classification: In the index-restricted MST, when the index of a node is >2, that node is classified as a high index node. Since the index-restricted spanning tree is a path between two given nodes and passing through all other nodes, the index value at the origin and destination nodes will be one. Hence, these two nodes will be classified as high index nodes if the index value is >1.

- Basic edge: If an edge belongs to MST, it is classified as a basic edge; otherwise, it is nonbasic.

- Total index value: In an "n" node network, the MST will have $(n–1)$ selected edges. The total index value of all these selected edges will be a fixed value given by $2(n–1)$.

- Index balancing theorem: Adding the same constant to all edges emanating from the same node does not change their relative merit; however, it can create alternatives for the selection of an edge for inclusion in the MST. Since in MST, ties are broken arbitrarily, one can replace an edge from a high index node by an edge that increases the index value of a low index node. Therefore, each application of the index balancing theorem reduces the number of imbalances, and repeated applications will give rise to a balanced MST under the index restriction ≤ 2. This MST will be a path joining two given nodes and passing through all other remaining nodes of the given network.

8.4 The Proposed Approach for Complexity Reduction

The proposed approach for reducing the complexity is based on the following concepts and observations.

8.4.1 Network Modification

8.4.1.1 Observation 1

A network $G(n,m)$ can be modified to form a graph $G'\big((n+1),(m+d_p)\big)$. This is easily explained by using an illustration. For example, consider a graph shown in Figure 8.1, which can be represented as $G(4,6)$, i.e., the network has four nodes and six edges. Here the number of edges meeting at each node is equal, i.e., the value of $d_p = 3$, which is independent of the node position. Let the lengths of these six edges be denoted by l_{ij}, which for Figure 8.1 is assumed as $l_{12} = 2$, $l_{13} = 4$, $l_{14} = 5$, $l_{23} = 3$, $l_{24} = 5$, and $l_{34} = 4$.

From Figure 8.1, a new network $G\big((n+1),(m+d_p)\big)$, i.e., $G(5,9)$ is developed as shown in Figure 8.2. We have arbitrarily assumed the starting node "p" as node 1. A new node 1' has been added. Note that the node 1' has no real existence; it has been created for convenience to achieve our objective, i.e., to reduce the complexity and establish the equivalence of MTST and MST. Like the node 1', the edges $\{(1',2), (1',3),$ and $(1',4)\}$ also have a conditional existence. In other words, either $(1,2)$ or $(1',2)$ will exist if this link is a part of MST but not both. In other words, both edges $(1,2)$ and $(1',2)$ cannot be the basis at the same time; however, they both can be nonbasic. Similar will be the case for other two pairs, i.e., $\{(1,3)$ or $(1',3)\}$ and $\{(1,4)$ or $(1',4)\}$; only one link can be basic, but both can remain nonbasic. This is controlled by giving them conditional weights.

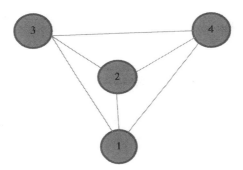

FIGURE 8.1
The network $G(4,6)$.

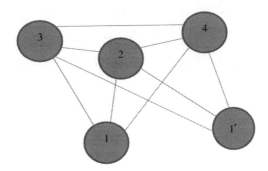

FIGURE 8.2
Modified network from the network in Figure 8.1, with node 1′ as a duplicate node.

First, we develop a new network related to the network in Figure 8.1, which is shown in Figure 8.2. Note that

i. There will be only one index associated with the nodes 1 and 1′ in the index-restricted MST joining the nodes 1 and 1′, i.e., an MST path between the nodes 1 and 1′ will pass through the nodes 2, 3, and 4. Since nodes 1 and 1′ are essentially the same, the index-restricted MST between these two nodes will be the required MTST for the given graph.

ii. If any edge from (1,2) or (1,3) or (1,4) is basic, we make the corresponding edge of infinite length, thus such a pair of edges will not be basic at the same time.

To make our point clearly, let us assign some values to the edges in Figure 8.1, which were assumed as given later:

$$l_{12} = 2, l_{13} = 4, l_{14} = 5, l_{23} = 3, l_{24} = 5, l_{34} = 4$$

The edges in Figure 8.2 will have two kinds of edges, and their lengths will be as follows:

i. Length of the duplicate edges as a function of the original length is given by

$$l_{1',2} = \frac{l_{12}}{\left(1 - s_{12}\right)}, l_{1',3} = \frac{l_{13}}{\left(1 - s_{13}\right)}, \ l_{1',4} = \frac{l_{14}}{\left(1 - s_{14}\right)}. \tag{8.1}$$

The $s_{1j}, j = 2, 3$, and 4 are defined in (iii).

ii. Lengths of original edges, which have been duplicated, are also not independent. If any duplicate edge is basic, the original edge will not exist. Their lengths will be given by

$$l_{12} = \frac{l_{1',2}}{\left(1 - s_{1',2}\right)}, l_{13} = \frac{l_{1',3}}{\left(1 - s_{1',3}\right)}, l_{14} = \frac{l_{1',4}}{\left(1 - s_{1',4}\right)} \tag{8.2}$$

iii. For the remaining edges, which have not been duplicated, lengths will not alter, i.e., $l_{23} = 3, l_{24} = 5, l_{34} = 4$ will remain the same in the altered network.

Here $s_{1,k} = \begin{cases} 1, & \text{if the edge } (1,k) \text{ is basic} \\ 0, & \text{if the edge } (1,k) \text{ is not basic} \end{cases}$

And $s_{1',k} = \begin{cases} 1, & \text{if the edge } (1',k) \text{ is basic} \\ 0, & \text{if the edge } (1',k) \text{ is not basic} \end{cases}$

In other words, an edge, if included in the MST, will either be associated with node 1 or with node 1′ but not with both. Here onwards, we call these edges as type 1 and type 2. Type 1 edges are those which do not change and are shown in (iii) earlier and Type 2 edges are shown in (i) and (ii) earlier, i.e., edges emanating from the node "p" and its duplicate node "p′." As aforementioned, the origin node is represented by node 1.

The condition that both edges in Type 2, i.e., the edge and its duplicate, cannot be basic is easily taken care by imposing a constraint that

$$s_{1,k} + s_{1',k} \le 1, \quad k = 2, 3, 4.$$

For obtaining MST, we arrange these lengths as given in Table 8.1. The first edge to be selected is (1,2) as it is the minimum cost edge. Since the edge (1,2) is now basic, the value of $s_{1,2}$ will become 1 and the edge (1′,2) will become ∞. This value is shown in Table 8.1, see the element (1′,2) or (2,1′). The duplicate links or Type 2 links can change their edge weight. Next minimum is 3, associated with the edge (2,3), which can be included in MST. Next minimum is 4,

TABLE 8.1

Edge Distances for the graph in Figure 8.2

From	1	1′	2	3	4
			To		
1	–	–	2*	4	5
1′	–	–	$\frac{2}{(1-s_{12})} = \infty \text{ or } 2$	$\frac{4}{(1-s_{13})} = \infty \text{ or } 4$	$\frac{5}{(1-s_{14})} = \infty \text{ or } 5$
2	2	$\frac{2}{(1-s_{12})} = \infty \text{ or } 2$	–	3*	5
3	4	$\frac{4}{(1-s_{13})} = \infty \text{ or } 4$	3	–	4*
4	5	$\frac{5}{(1-s_{14})} = \infty \text{ or } 5*$	5	4	–

which corresponds to the edge (3,4), hence included. Next minimum is 5, i.e., edge (4,1′) is selected. This is sown in Figure 8.3. Since nodes 1 and 1′ are in fact only one node, the MST gives the required TST, which is 1->2->3->4->1′. The link (1,4) becomes infinite, but MST is complete, and hence no consequence. The total minimum cost is 14.

The TST for the network in Figure 8.1 with the help of the MST network in Figure 8.3 is shown in Figure 8.4.

8.4.2 MST by Kruskal's Greedy Approach: Justification for a Slight Modification

Normal steps of the Kruskal's (2017) MST algorithm, as taken from https://www.youtube.com/watch?v=3rrNH_AizMA, are

1. *Sort all the edges in nondecreasing order of their weight.*
2. *Pick the smallest edge. Check if it forms a cycle with the spanning tree formed so far. If a cycle is not formed, include this edge. Else, discard it.*
3. *Repeat step #2 until there are (n−1) edges in the spanning tree.*

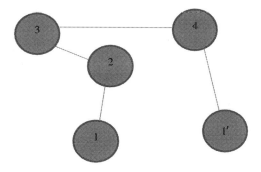

FIGURE 8.3
MST of the network in Figure 8.2 with TST interpretation.

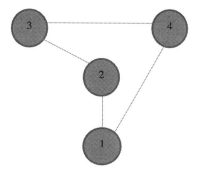

FIGURE 8.4
The TST of the network in Figure 8.1. Total cost being 14.

To illustrate why a simple modification is required in the aforesaid simple steps, we consider a trivial illustration shown in Figure 8.5, where
$l_{12} = 1, l_{13} = 1, l_{14} = 1, l_{23} = 2, l_{24} = 2$, and $l_{34} = 3$.

The distance matrix for the earlier network is given in Table 8.2.

When the network in Figure 8.5a is modified as discussed in Section 8.4.1,
we get a network in Figure 8.6.

Thus, in view of the modification, the greedy approach can lead to an
isolated node. Therefore, a modification to Kruskal's approach is desired
with respect to the inclusion of Type 2 edges in the MST. Note that, in an

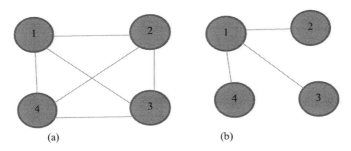

(a)   (b)

FIGURE 8.5
(a) A trivial network G(4,6), (b) The MST of the network in (a).

TABLE 8.2

Distance Matrix

		To		
From	**1**	**2**	**3**	**4**
1	–	1	1	1
2	1	–	2	2
3	1	2	–	3
4	1	2	3	–

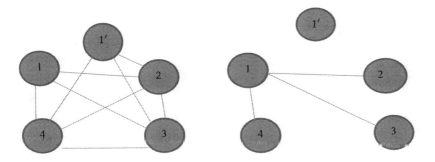

FIGURE 8.6
Modified graph and MST obtained by using the Kruskal method, which is not an MST.
All edges from node 1′ will become of infinite length.

n node network, if a selection of a link is forming a path with less than n links between the node and its duplicate, it will result in a subcycle; hence, it should not be selected. More details are given in the illustrative example.

8.4.3 Proposed Modification to Kruskal's Greedy Approach for the MST

As noted in Figure 8.6, the Kruskal's approach needs a slight modification to overcome the possible difficulty shown in Figure 8.6; Kruskal's steps remain valid with respect to Type 1 edges. Modified steps are required to handle inclusion of Type 2 edges. In view of Type 2 edges, it may be noted that a sub-cycle also comes up when selection of an edge forms a path of less than n links between the nodes p and p'. Such a link can be discarded. These steps are described later.

8.4.3.1 *Modified Steps for Obtaining the MST of* $G'\left((n+1),\left(m+d_p\right)\right)$

Step 1. For the given network $G(n,m)$, construct a modifies network $G'\left((n+1),\left(m+d_p\right)\right)$ and arrange the links in increasing order. Initially, all links are nonbasic, and hence the lengths of duplicated edges will be as given in the network $G(n,m)$. The total number of Type 1 edges will be $\left(m-d_p\right)$ and the number of Type 2 edges will be $\left(2d_p\right)$. Type 2 edges will be such that at least two edges will have equal lengths initially.

Step 2. Set a counter $K = 1$. Select the edge of minimum length and include it in the MST and go to Step 3.

Step 3. If the selected edge was Type 1 and $K < n$, set $K = K+1$ and select the next minimum length. If $K = n$, go to Step 7.

Step 4. If the selected edge was Type 2 and $K < n$, first set its duplicate length equal to ∞ and rearrange lengths in increasing order. Set $K = K + 1$ and select the next minimum length. If $K = n$, go to Step 7.

Step 5: If the selected edge forms a cycle with the spanning tree formed so far, discard it, else include it if it is Type 1 and return to Step 3. If the selected edge is Type 2, go to Step 6.

Step 6. With respect to the selected edge, check

1. That the edge does not lead to an isolated node in the network; if so, discard it.
2. Its alternatives; select in favor of the edge that maintains a balance among index of nodes forming the MST.
3. Since the selected edge is of Type 2, return to Step 4.

Step 7. Stop. MST has been obtained.

In view of the earlier discussion, return to Figure 8.6, where the links (1,2) will be selected, the next will be the link (1,3) or (1,4). If we go for the link (1,3), the modified step will prefer to select the link (1',3) so that node 1 does not become a high index node. The next link will be (1,4) or (1',4), it can be placed either with node 1 or with node 1', and there is no choice but that one of them will become a high index node to be adjusted later by the application of the index balancing theorem.

8.5 Numerical Illustration

Let us reconsider for illustration the network shown in Figure 7 from Kumar et al. (2018), reproduced here as Figure 8.7.

Edge lengths for the graph in Figure 8.7 are given in Table 8.3.

We now redevelop a network based on the network in Figure 8.7, assuming the node 1 as the node to be duplicated. The modified network will be given

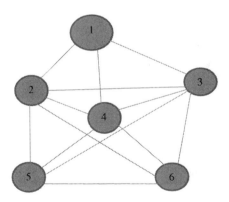

FIGURE 8.7
Network for illustration of the proposed method.

TABLE 8.3

Showing Edge Lengths

From			To			
	1	2	3	4	5	6
1	–	11	9	9	–	–
2	11	–	14	10	10	15
3	9	14	–	6	13	11
4	9	10	6	–	9	10
5	–	10	13	6	–	8
6	–	15	11	10	8	–

by $G((6 + 1),(13 + 3)) = G(7,16)$. This modified network is shown as Figure 8.8. From this modified network, first find its MST using the modified steps as described in Section 8.4.3, convert that MST to index-restricted MST which will be a path joining the nodes 1 and 1' and passing through all nodes of the network in Figure 8.8. Since nodes 1 and 1' are the same, this path will have an interpretation of TST for the network in Figure 8.7.

The edge lengths are given in Table 8.4. Note that, in the initial stage, values of three edges emanating from node 1 and those emanating from node 1' are the original values as all $s_{1,k} = s_{1',k} = 0$. They will change only if any of these edges become basic.

From Table 8.4, the Type 2 edges are: (1,2), (1,3), (1,4), (1',2), (1',3), and (1',4). If arranged in increasing order, they will be of length:

9 for (1,3), (1',3), (1,4) and (1',4) and
11 for (1,2) and (1',2).

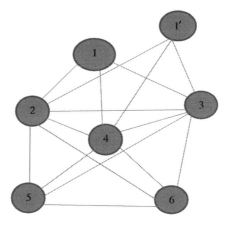

FIGURE 8.8
Modified network of the network in Figure 8.7 for determination of MST.

TABLE 8.4

Edge Lengths for the Modified Network

	To						
From	**1**	**1'**	**2**	**3**	**4**	**5**	**6**
1	–	–	11	9	9	–	–
1'	–	–	11	9	9	–	–
2	11	11	–	14	10	10	15
3	9	9	14	–	6	13	11
4	9	9	10	6	–	9	10
5	–	–	10	13	9	-	8
6	–	–	15	11	10	8	–

Type 1 links will be (2,3), (2,4), (2,5), (2,6), (3,4), (3,5), (3,6), (4,5), (4,6), and (5,6). If arranged in increasing order, they will be of lengths:

6 for the link (3,4),

8 for the link (5,6),

9 for the link (4,5),

10 for the links (2,4), (2,5) and (4,6)

11 for the link (3,6),

14 for the link (2,3) and

15 for the link (2,6).

As aforesaid, the greedy approach will select links and start to include them in the MST. The selected links will be (3,4) and (5,6) with lengths 6 and 8, respectively. It may be noted that these links are independent of $s_{1,k}$ and $s_{1',k}$ values. In other words, they are Type 1 links. The next link to be considered for selection will be of length 9, which provides many choices, i.e., (1,3) or (1′,3), (1,4) or (1′,4), and (4,5). Let us arbitrarily select the link (4,5) first and then (1,3), which makes $s_{1,3} = 1$, and therefore changes the length of the link (1′,3) as shown in Tables 8.5 and 8.6.

TABLE 8.5

Determination of the MST

S. No	Link	Weight	MST Remarks
1	(3,4), Type 1	6	Included
2	(5,6), Type 1	8	Included
3	(4,5), Type 1	9	Included
4	(1,3), Type 2	9	Included[a]

[a] Note the edge length (1′,3) will change to ∞.

TABLE 8.6

Updated Edge Lengths for the Modified Network with Included Links in Bold

From				To			
	1	1′	2	3	4	5	6
1	–	–	11	9	9	–	–
1′	–	–	11	∞	9	–	–
2	11	11	–	14	10	10	15
3	9	..	11		6	13	11
4	9	9	10	6	–	9	10
5	–	–	10	13	9	–	8
6	–	–	15	11	10	8	–

The greedy approach will identify one more link of length 9, i.e., the link (1,4) or (1',4). In the case of a conventional MST, one has no preference in the selection of (1,4) or (1',4), and we would have resolved the tie arbitrarily, but here, preference will be in favor of (1',4) to avoid node 1 being a high index node. However, the links (1',4) will be excluded from MST as it will make a subcycle among the selected nodes, i.e., 1, 3, 4, and 1'. Hence, both (1,4) and (1',4) are excluded. Note that both links can remain as nonbasic.

Next higher value is 10, which corresponds to links (2,4), (2,5), and (4,6). A close examination with respect to these three links indicate that link (4,6) is excluded due to subcycle, the link (2,4) makes index of node 4 heavy; hence, selection is in favor of the link (2,5).

Next higher value is 11, which corresponds to links (1,2), (1',2), and (3,6). Selection of links (1,2) or (3,6) leads to a cycle with already selected edges. Hence, it is discarded and we select the link (1',2). This selection also changes the length of link (1,2) to ∞. This is reflected in Tables 8.7 and 8.8.

The total selection process is summarized in Table 8.9.

This MST is shown in Figure 8.9. The total length is 53, which is the lower bound on TST.

TABLE 8.7

Determination of the MST

S. No	Link	Weight	MST Remarks
1	(3,4), Type 1	6	Included
2	(5,6), Type 1	8	Included
3	(4,5), Type 1	9	Included
4	(1,3), Type 2	9	Included
5	(2,5), Type 1	10	Included
6	(1',2), Type 2	11	Included

TABLE 8.8

Updated Edge Lengths for the Modified Network with Included Links in Bold

	To						
From	1	1'	2	3	4	5	6
1	–	–	∞	9	9	–	–
1'	–	–	11	∞	9	–	–
2	∞	11	–	14	10	**10**	15
3	9	∞	14	–	**6**	13	11
4	9	9	10	**6**	–	**9**	10
5	–	–	**10**	13	**9**	–	**8**
6	–	–	15	11	10	**8**	–

TABLE 8.9

Determination of the MST

S. No	Link	Weight	Remarks	K Value
1	(3,4)	6	Included	1
2	(5,6)	8	Included	2
3	(4,5)	9	Included	3
4	(1,3)	9	Included	4
5	(1,4)	9	Exclude	
6	(1',4)	9	Excluded	
7	(2,5)	10	Included	5
8	(2,4)	10	Excluded	
9	(4.6)	10	Excluded	
10	(1,2)	11	Excluded	
11	(3,6)	11	Excluded	
12	(5,3)	11	Excluded	
13	(1',2)	11	Included	6

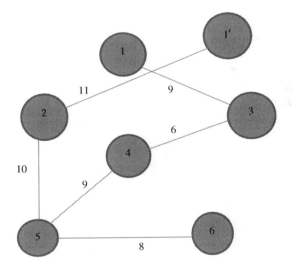

FIGURE 8.9
MST using the modified process discussed in this chapter.

Note that node 5 is a high index node with index value 3 and node 6 is a low index node with index value 1. If we add 1 to all edges emanating from node 5, the edge (4,5) will be of length 10 and it can be replaced by its alternative link (4,6), which has the same length, i.e., 10 units. This is shown in Figure 8.10.

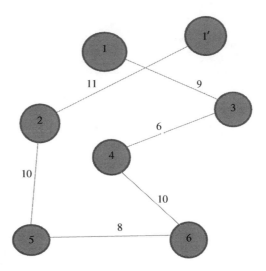

FIGURE 8.10
MST with index restriction ≤2.

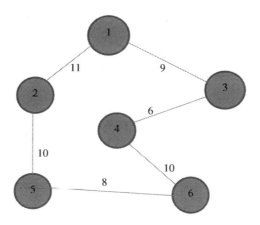

FIGURE 8.11
The TST of the network in Figure 8.7.

Now we rearrange the network in Figure 8.10 as in Figure 8.11 by merging nodes 1 and 1'. This is shown in Figure 8.11.

The TST shown in Figure 8.11 is of length 54, which is the required TST of the network in Figure 8.7 and is of the same value as was obtained earlier by Kumar et al. (2018).

8.6 Concluding Remarks

- As a future study, a general code will be developed to perform some meaningful computational experiments to establish advantages of the approach discussed here.

- In this chapter, a strong relationship between the MST under the index restriction and MTST has been established.

- It may be noted that the MTST has an NP-hard classification, whereas the conventional MST is classified as an easy problem. However, an equivalence between the MST with index restriction has been established equivalent to the minimum traveling salesman problem. The index-restricted MST can be obtained from the MST and the index balancing theorem.

References

Anupam, G., (2015), Deterministic MSTs, Lecture #1, Advanced algorithms, CMU Spring 15–859E, pp. 1–6.

Berman, P. and Karpinski, M., (2006), Approximate algorithms for TSP, In *Proceedings of the Seventeenth Annual ACM-SIAM Symposium on Discrete Algorithms*, Society for Industrial and Applied Mathematics, Miami, FL, pp. 641–648.

Cowen, L., (2016), Metric TSP and knapsack part 1, Lecture 3, Comp 260, Advanced Algorithms, Tufts University, pp. 1–8.

Garg, R.C. and Kumar, S., (1968), Shortest connected graph through dynamic programming, *Mathematics Magazine*, Vol. 41, No. 4, 170–173.

Garg, R.C., Kumar, S., Dass, P., and Sen, P., (1970), Generalized travelling salesman problem through n sets of nodes in a competitive market, *AKOR, Ahlanf und Planningforschung*, Band II, Heft 2, 116–120.

Kumar, S., Munapo, E., Lesaoana, M., and Nyamugure, P., (2014), A minimum spanning tree approximation to the routing problem through K specified nodes, *Journal of Economics*, Vol. 5, No. 3, 307–312.

Kumar, S., Munapo, E., Lesaoana, M., and Nyamugure, P., (2016), Is the travelling salesman actually NP Hard? Chapter 3 in *Engineering and Technology: Recent Innovations and Research*, Editor Ashok Matani, International Research Publication House, Delhi, ISBN: 978-93-86138-06-4, pp. 37–58.

Kumar, S., Munapo, E., Lesaoana, M., and Nyamugure, P., (2018), A minimum spanning tree based heuristic for the travelling salesman tour, *Opsearch, Journal of the Operational Research Society of India*, Vol 55, No. 1, 150–164. doi: 10.1007/s12597-017-0318-5.

Munapo, E., (2013), A return to the travelling salesman model: A network Branch and Bound Approach, *African Journal of Economics and Management Science*, Vol. 16, No. 1, 52–63.

Munapo, E., Kumar, S, Lesaoana, M., and Nyamugure, P., (2016), A minimum spanning tree with index ≤2, *ASOR Bulletin*, Vol 34, No. 1, 1–14.

Nemhauser, G.L. and Wolsey, L.A., (1989), *Integer and Combinatorial Optimization*, Wiley Inter-science Series in Discrete Mathematics and Optimization, North Holland, Amsterdam.

Razali, N.M. and Geraghty, J., (2011), Genetic Algorithm performance with different selection strategies in solving the TSP, In *Proceedings of the World Congress on Engineering*, July 6–8, London.

9

Pattern Formation Dynamics of Predator–Prey System with Hunting Cooperation in Predators

Teekam Singh

Graphic Era (Deemed to be University)

Ramu Dubey

J. C. Bose University of Science & Technology, YMCA

CONTENTS

The objective of this chapter is to study one- and two-dimensional spatial patterns in predator–prey model with hunting cooperation under zero-flux boundary conditions and circular initial conditions. Spatial patterns occur when diffusion-driven instability takes place in predator–prey model with hunting cooperation. We performed analytical and numerical analysis in both one- and two-dimensional spatial domains. The spatial dynamics of prey and predator populace in this model are discussed extensively. The heterogeneous dynamics of the system have both ecological and mathematical connotation. This chapter investigates meticulousness in the spatial dynamics of predator–prey model with hunting cooperation via diffusion-driven instability. The outcomes presented here help us to comprehend dynamics of predator–prey interplays in real-world phenomenon.

9.1 Introduction

The impact of space may be disregarded in a certain extent, particularly when the population of a given species stays fixed in space at any moment of time. Albeit, this assumption is not completely realistic. Individuals of an ecological species do not stay fixed at all times in space, and their dispersion in space changes incessantly by the self-movement of individuals [1–5].

All of us are living in a spatial real world, and spatial patterns are found everywhere in nature, and these spatial patterns transform the nonspatial (temporal) dynamical qualitative properties of densities of the population at a spectrum of spatial scale. Spatial pattern formation is a dissipative procedure giving growth to spatiotemporal behavior ruled by internal characteristics or external restrictions into a model. The spatial pattern formation factor of ecological as well as tumor–immune interplays have been recognized as a vital component, of how ecological communities and tumors are composed, and is one of the pivotal subject of natural sciences [6–9].

The idea of diffusion may be considered as the natural propensity for a cluster of particles at the beginning, concentrated close to a location in space to spread out in time, slowly occupying an ever-sizable area close to the initial point. Here, the word "particles" mention not only to physical portion of the matter but also to biological populations or to any other recognizable elements as well. Moreover, the word "space" does not mention only to general Euclidean n-space but can also be an hypothetical living space (such as ecological space) [1,4,5,10].

Diffusion is a natural phenomenon, where physical material moves from an area of high concentration to an area of low concentration, that is, diffusion is a natural process by which the particle clusters as entire dispersions according to the nonuniform movement of every particle. Basically, diffusion can be defined as an invariant process by which particle clusters, population, etc., diffuse inside a given space according to individual random movement [5].

Reaction-diffusion partial differential equation systems can be used to represent mathematical models, which describe how the individuals of one or more species distributed in space changes under the effect of two procedures, first is local interaction, in which the species interact with each other, and second is the diffusion, which causes the species to spread out over a surface in space. Mathematically, reaction-diffusion systems take the form of semilinear parabolic partial differential equations [1,5,6].

Using mathematical modeling as a viable tool, complex biological processes are studied. Mathematical modeling can be extremely helpful in analyzing factors that may contribute to the complexity intrinsic in insufficiently understood tumor–immune as well as predator–prey interactions. Likewise, the primary objective of the mathematical modeling of tumor–immune and predator–prey models are, briefly, the analysis of the interplay inside and between biological species and their artificial surrounding, and the examination of the temporal transformation of clusters of individuals of different biological species. It is however true that space and time are indivisible "sibling coordinates," and only when population densities (tumor–immune system or predator–prey system) are contemplated in both space and time, actual dynamics can be understood [1,11–14,15–19].

Majority of models in mathematical ecology or tumor–immune interaction deals with nonspatial variants. The rate of change of the number of individuals u in a population may be manifested as the derivative with respect to time "t," du/dt. The model equations of a biological community of interacting individuals and their environment are then founded by equating this derivative to another relation expressing the effect of species interaction on population. Same is the situation with tumor–immune interacting models. This type of straightforward analysis is not practicable when spatial models are considered. Directly connected to species interplay is the net population via an arbitrary infinitesimal piece of space rather than the spatial rate of change of the population itself, and thus a reasonable manifestation is unreachable without knowledge of the mechanism of motion of individuals.

9.1.1 Predator–Prey System

Predator–prey system represents the functional dependence of one species on another, where the first species depends on the second species for food. Predation is a mode of life in which food is primarily obtained by killing and consuming organisms. The prey is part of the predator's habitat, and if the predators do not get any prey for food, then they become extinct. The functional dependence in general depends on many factors, namely the various species densities, the efficiency with which the predator can search out and kill the prey, and the handling time [10,20,21].

9.1.1.1 Hunting Cooperation

Hunting is one of the highly fascinating tactical natural instincts in the animal kingdom. A successful hunt requires a great deal of cooperation and coordination within the group. Hunting cooperation in animal kingdom are very frequent; for example, group hunting enables lionesses to have greater success in capturing prey, and it involves both divisions of work and role specialization. It has been connected to the social system of animal species and the evolution of society and thus provides a unique approach to study cooperative behavior [22–31].

9.2 Mathematical Preliminaries

Some mathematical methods and ideas have been depicted in this segment, which are used to examine the nonlinear dynamics and pattern formation (spatiotemporal models), introduced in this thesis. The details of the ideas are depicted for nonspatial (ordinary differential equations) as well as spatial (reaction-diffusion partial differential equations) systems.

9.2.1 Stability Analysis

Let $X(t) \in R^n$ depicts the states of a system at time "t." The dynamics of the system is ruled by a system of first-order nonlinear ordinary differential equations:

$$\dot{X}(t) = G(X(t), \theta), \ X(0) = X_0, \tag{9.1}$$

where $X = [x_1, x_2, \ldots, x_n]^T$ stands for the n state variables, θ holds the parameter values, and G is a nonlinear function of the state variables and parameter values. If $G(X_*) = 0$, then X_* is an equilibrium solution of the system. Consider X_0 to be its neighboring point. The equilibrium solution X_* is stable if, for all $\epsilon > 0$, there is a $\delta > 0$ such that

$$\|X(t) - X_*\| < \epsilon, \quad \text{whenever } \|X_* - X_0\| < \delta.$$

That is, X_* is stable if the equilibrium solutions go ahead to X_* at a said time stay close to X_* for each future time. X_* is asymptotically stable if neighboring solutions not only stay close but also approach X_* as t goes to infinity, for each future time. That is, X_* is stable and

$$\lim_{t \to \infty} X(t) = X_*,$$

then the solution X_* is asymptotically stable.

Asymptotic Stability $\Rightarrow$ Stability.

Stability of an equilibrium solution is a local property. An equilibrium solution X_* that is not stable is called unstable.

9.2.1.1 Local Stability Analysis

The system of interacting populations

$$\frac{dX_i(t)}{dt} = G_i(X_1, X_2, \ldots, X_n), \tag{9.2}$$

with initial conditions

$$X_i(0) = X_{i0} \geq 0, \quad i = 1, 2, \ldots, n. \tag{9.3}$$

Let us suppose that the function G_i is such that the solution of the earlier system is unique.

Let $X(t)$ be any other solution in the vicinity of equilibrium solution X_*, then

$$X = X_* + \eta, \tag{9.4}$$

where $\eta = (\eta_1, \eta_2, \ldots, \eta_n)$ is a perturbation from the equilibrium solution. Then the perturbation vector can be written as

$$\frac{d\eta}{dt} = \frac{\partial G}{\partial X}\bigg|_{X=X_*} \eta \equiv A\eta, \tag{9.5}$$

where $A = (a_{ij})_{n \times n}$ is the variational matrix at the equilibrium solution X_*. Let $\eta(0)$ be the initial perturbation from the equilibrium solution X_*, then the formal matrix solution to (9.5) can be given by

$$\eta(t) = e^{At}\eta(0). \tag{9.6}$$

The system is stable about the equilibrium solution X_* if the perturbation $\eta(t)$ goes to zero as t tends to ∞. This is feasible only if the real parts of the characteristic values of the variational matrix A, namely, $Re\{\lambda_i\}$, are negative for each i. If $Re\{\lambda_i\} > 0$ for at least one value of i, then the equilibrium solution is unstable. Since $\eta(t)$ is the solution of linearized system (9.5), which is close to actual nonlinear system, the stability is referred to local/linear stability only.

Therefore, the characteristic values of the variational matrix decide whether the equilibrium solution is linearly stable or unstable. The characteristic equation for the variational matrix can be written as

$$\det(A - \lambda I) = a_0 \lambda^n + a_1 \lambda^{n-1} + \cdots + a_n = 0, \quad a_0 \neq 0. \tag{9.7}$$

The coefficients a_i, $i = 1, 2, \ldots, n$ of characteristic equation are all real. The system (9.2) is locally stable about the equilibrium point if all of the eigenvalues have negative real parts. On the other hand, the system is unstable if at least one of the eigenvalues has a positive real part. In other words, all the eigenvalues of Jacobian matrix must lie in the left half of the complex plane. Accordingly, the necessary condition (not sufficient) for all eigenvalues to have negative real part is

$$\text{Trace}\,(A) < 0.$$

In the special case, if Trace $(A) = 0$, then either at least one eigenvalue must lie in the right half plane, or all eigenvalues must be purely imaginary (the pathological case of neutral stability). Another necessary, but not sufficient, condition is

$$(-1)^n \det|A| > 0.$$

Routh-Hurwitz Criterion gives necessary and sufficient conditions to make certain that the real part of all characteristic roots are negative, that is it belongs to the left half complex plane. These conditions, collectively $a_n > 0$, are [32]

$$H_1 = a_1 > 0, \quad H_2 = \begin{vmatrix} a_1 & a_3 \\ 1 & a_2 \end{vmatrix} > 0, \quad H_3 = \begin{vmatrix} a_1 & a_3 & a_5 \\ 1 & a_2 & a_4 \\ 0 & a_1 & a_3 \end{vmatrix} > 0,$$

$$H_k = \begin{vmatrix} a_1 & a_3 & \cdot & \cdot & \cdot & \cdot \\ 1 & a_2 & a_4 & \cdot & \cdot & \cdot \\ 0 & a_1 & a_3 & \cdot & \cdot & \cdot \\ 0 & 1 & a_2 & \cdot & \cdot & \cdot \\ \cdot & \cdot & \cdot & \cdot & \cdot & \cdot \\ 0 & 0 & \cdot & \cdot & \cdot & a_k \end{vmatrix} > 0, \; k = 1, 2, 3, \ldots, n \tag{9.8}$$

If the values of parameters are such that the earlier restrictions are simultaneously satisfied, then the given system will be locally asymptotically stable at X_*.

9.2.2 Bifurcations

The theory of bifurcation is the mathematical study of sudden changes in the qualitative behavior of solutions of a nonlinear dynamical system. Bifurcation analysis shows the long-term dynamics of the interacting population depending on the system parameters. In particular, equilibrium point(s) can be created, destroyed, or their stability can change due to change of parameter values. The parameter values for which the bifurcation occurs are called bifurcating points. In this thesis, we particularly focused on local bifurcations that occur when a small change in the parameter value of a given dynamical system causes a sudden change in the qualitative behavior of the system in the neighborhood of a critical point of the system. Scientifically, they are important since they provide models of transitions and stabilities as the control parameter is varied. Some different types of local bifurcations are as follows:

- **Hopf bifurcation**: It is a type of bifurcation at which a stable equilibrium point loses its stability at a threshold value and gives birth to a limit cycle with the variation of the bifurcation parameter. The system experiences Hopf bifurcation when a purely complex conjugate crosses the boundary of stability. The Hopf bifurcation destroys the temporal symmetry of a system and gives rise to oscillations, which are uniform in space and periodic in time. Two types of Hopf bifurcation are observed: one is supercritical and other is subcritical. Supercritical Hopf bifurcation is a phenomenon in which the unstable limit cycle becomes stable at the bifurcation point. Subcritical Hopf bifurcation is a phenomenon in which the stable limit cycle becomes unstable at the bifurcation point.

- **Turing bifurcation**: It is the primary bifurcation that give rise to spatiotemporal patterns, and is crucial for almost all reaction-diffusion type mathematical systems for pattern formation in embryology, ecology, epidemiology, and to some other areas of biology, physics, and chemistry [2–4,33,34]. The primary concept of Turing bifurcation is that a uniform steady-state solution can be stable to uniform spatiotemporal perturbations but unstable to definite spatiotemporally changing perturbations, leading to the formation of patterns, that is, a spatial pattern. The mathematical model to describe spatial pattern formation in the theory of morphogenesis was given by Turning in 1952, viz., two reaction-diffusion type partial differential equations, the interacting chemicals having distinct coefficients of diffusion. For appropriate reaction kinetics, as the proportion of diffusivity increases (or decreases) from unity, e.g., there is a critical value at which the homogenous equilibrium solution becomes unstable to a particular spatiotemporal mode. Such kind of bifurcation is called the Turing bifurcation.

9.3 Diffusive Instability and Its Conditions

Spatial patterns are formed via the diffusive instability of the uniform equilibrium solution to small spatiotemporal perturbations. If the uniform equilibrium solution is stable, then small spatiotemporal perturbations from the equilibrium state will move towards back to an equilibrium state. In 1952, Alan Mathison Turing pointed out how a reaction-diffusion system, showing such instabilities, can form diffusive patterns [34].

Alan Turing, in 1952, demonstrated that the reaction-diffusion system may form the spatial pattern if the following two conditions hold:

- the coexistence steady state is linearly stable in the nonspatial (without diffusion) system, and

- after adding the diffusion term in system, the coexistence steady state is linearly unstable.

Proper mathematical analysis demonstrates that, at the beginning of instability, the model initially becomes unstable with regard to a spatiotemporally nonhomogenous perturbation with a definite wave number. Such type of instability is called diffusive instability (Turing instability).

The mathematical foundation of diffusive instability by considering two state variables, X_1 and X_2, are subject to one-dimensional space:

$$\frac{\partial X_1}{\partial t} = G_1(X_1, X_2) + d_1 \frac{\partial^2 X_1}{\partial x^2},$$

$$\frac{\partial X_2}{\partial t} = G_2(X_1, X_2) + d_2 \frac{\partial^2 X_2}{\partial x^2},$$

$$(9.9)$$

where "x" is the space coordinate and "t" is the time. d_1 and d_2 are diffusion coefficients of X_1 and X_2, respectively. $G_1(X_1, X_2)$ and $G_2(X_1, X_2)$ are the arbitrary interaction terms of X_1 and X_2, respectively.

To understand the effect of diffusion in pattern formation, we assume that, in the absence of diffusion, that is, when solutions are well mixed, the system has some positive spatially homogeneous steady state, $\left(X_1^*, X_2^*\right)$. Mathematically, this means that

$$\frac{\partial X_1^*}{\partial t} = 0 = \frac{\partial X_2^*}{\partial t},$$

$$\frac{\partial^2 X_1^*}{\partial x^2} = 0 = \frac{\partial^2 X_2^*}{\partial x^2}, \tag{9.10}$$

$$\Rightarrow G_1\left(X_1^*, X_2^*\right) = 0 = G_2\left(X_1^*, X_2^*\right). \tag{9.11}$$

Additionally, suppose that $\left(X_1^*, X_2^*\right)$ is stable with respect to spatially uniform perturbation, that is, the system is stable without diffusion.

To examine the effects of small nonhomogeneous perturbation on the stability of the system with respect to homogeneous steady state, we write

$$X_1(t, x) = X_1^* + X_1'(t, x),$$

$$X_2(t, x) = X_2^* + X_2'(t, x). \tag{9.12}$$

It is assumed that the perturbations are sufficiently small, that is, we analyze the local stability of system. Substituting (9.12) into (9.9) using (9.10) and linearizing the equations, we obtain

$$\frac{\partial X_1'}{\partial t} = a_{11} X_1' + a_{12} X_2' + d_1 \frac{\partial^2 X_1'}{\partial x^2},$$

$$\frac{\partial X_2'}{\partial t} = a_{21} X_1' + a_{22} X_2' + d_2 \frac{\partial^2 X_2'}{\partial x^2}, \tag{9.13}$$

where

$$a_{11} = \left.\frac{\partial G_1}{\partial X_1}\right|_{\left(X_1^*, X_2^*\right)}, \quad a_{12} = \left.\frac{\partial G_1}{\partial X_2}\right|_{\left(X_1^*, X_2^*\right)}$$

$$a_{21} = \left.\frac{\partial G_2}{\partial X_1}\right|_{\left(X_1^*, X_2^*\right)}, \quad a_{22} = \left.\frac{\partial G_2}{\partial X_2}\right|_{\left(X_1^*, X_2^*\right)} \tag{9.14}$$

and X_1' and X_2' are perturbations from X_1^* and X_2^*. Equation (9.13) can be written in compact matrix form as

$$X_t' = AX' + DX_{xx}', \tag{9.15}$$

where

$$X' = \begin{pmatrix} X_1'(t,x) \\ X_2'(t,x) \end{pmatrix} = \begin{pmatrix} X_1(t,x) - X_1^* \\ X_2(t,x) - X_2^* \end{pmatrix},$$

$$A = \begin{pmatrix} a_{11} & a_{12} \\ a_{21} & a_{22} \end{pmatrix},$$

$$D = \begin{pmatrix} d_1 & 0 \\ 0 & d_2 \end{pmatrix}.$$

For linear stability analysis, it is sufficient to assume solution of (9.13) in the form

$$X_1' = \exp(\mu t + ikx),$$
$$X_2' = \exp(\mu t + ikx),$$

(9.16)

where k and μ are the wave number and frequency, respectively. The corresponding characteristic equation is

$$\begin{vmatrix} a_{11} - d_1 k^2 - \mu & a_{12} \\ a_{21} & a_{22} - d_2 k^2 - \mu \end{vmatrix} = 0.$$

(9.17)

Solving for μ, we obtain

$$\mu = \frac{1}{2} \left(a_{11} + a_{22} - k^2 (d_1 + d_2) \right) \pm$$

$$\sqrt{\left(a_{11} + a_{22} - k^2 (d_1 + d_2) \right)^2 - 4 \left(\left(a_{11} - d_1 k^2 \right) \left(a_{22} - d_2 k^2 \right) - a_{12} a_{21} \right)}.$$

The condition $k = 0$ corresponds to the neglect of diffusion and, by definition, perturbations of zero wave number are stable when diffusive instability sets in. It is thus required that

$$a_{11} + a_{22} < 0,$$
$$a_{11} a_{22} - a_{12} a_{21} > 0.$$

(9.18)

Diffusive instability sets in when at least one of the following conditions is violated subject to the conditions (9.18)

$$\widetilde{a_{11}} + \widetilde{a_{22}} < 0,$$
$$\widetilde{a_{11}} \widetilde{a_{11}} - a_{12} a_{21} > 0.$$

(9.19)

It is seen that the first condition $\widetilde{a_{11}} + \widetilde{a_{22}} < 0$ is not violated when the requirement $a_{11} + a_{22} < 0$ is met. Hence, only a violation of the second condition $\widetilde{a_{11}}\widetilde{a_{11}} - a_{12}a_{21} > 0$ gives rise to diffusive instability. Reversal of the second inequality of (9.19) yields

$$Q\left(k^2\right) = d_1 d_2 k^4 - \left(d_1 a_{22} + d_2 a_{11}\right) k^2 + a_{11} a_{22} - a_{12} a_{21} < 0. \tag{9.20}$$

The minimum of $Q(k^2)$ occurs at $k^2 = k_m^2$, where

$$k_m^2 = \frac{d_1 a_{22} + d_2 a_{11}}{2 d_1 d_2} > 0. \tag{9.21}$$

Thus, a sufficient condition for instability is that $Q\left(k_m^{\,2}\right)$ be negative. Therefore,

$$\left(a_{11} a_{22} - a_{12} a_{21}\right) - \frac{\left(d_1 a_{22} + d_2 a_{11}\right)^2}{4 d_1 d_2} < 0. \tag{9.22}$$

Combination of (9.18), (9.21), and (9.22) leads to the following final criterion for diffusive instability:

$$d_1 a_{22} + d_2 a_{11} > 2\left(a_{11} a_{22} - a_{12} a_{21}\right)^{\frac{1}{2}} \left(d_1 d_2\right)^{\frac{1}{2}} > 0. \tag{9.23}$$

The critical conditions for the occurrence of the instability are obtained when the first inequality of (9.23) is an equality.

9.4 Model Description

Mathematical modeling methods are providing useful information for understanding the interplays of prey and predator in ecological models. Recently, a nonspatial prey–predator system is pondered with hunting cooperation and Allee effects in predators [22], due fundamentally to Ref. [22]:

$$\frac{dX}{dt'} - rX\left(1 - \frac{X}{K}\right) - (\lambda + aY)XY$$

$$\frac{dY}{dt'} = e(\lambda + aY)XY - mY \tag{9.24}$$

where $X(t')$ and $Y(t')$ are the densities of prey and predator population at time t', respectively. The parameter r is the rate of growth of prey, K is its holding efficiency, λ is the constant invasion rate, and a is the rate of predator hunting cooperation. The parameter e is the efficiency of conversion, and m is the natural death rate of predator. All parameters are positive.

On the other hand, we assume that the prey and predator populace densities spread without any order, and this random distribution of species is depicted by diffusion. Then, we propose a spatial prey–predator model with hunting cooperation and Allee effects in predators corresponding to (9.24) as follows:

$$\frac{\partial X}{\partial t'} = rX\left(1 - \frac{X}{K}\right) - (\lambda + aY)XY + d_1\nabla^2 X$$

$$\frac{\partial Y}{\partial t'} = e(\lambda + aY)XY - mY + d_2\nabla^2 Y$$

(9.25)

where the nonnegative constants d_1 and d_2 are the diffusion coefficients of prey and predator, respectively. ∇^2 is the usual Laplacian operator in $d \leq 3$ space dimensions.

A great number of research work are based on the nondimensional models of the nonlinear coupled partial differential equations, as they have less number of parameters. Following Ref. [22], the variables are scaled as

$$u = \frac{e\lambda}{m}X, v = \frac{\lambda}{m}Y, t = mt',$$

and dimensionless values of other parameters are given by

$$\sigma = \frac{r}{m}, N = \frac{e\lambda}{m}K, \alpha = \frac{am}{\lambda^2}.$$

With these changes, Eqs. (9.25) become

$$\frac{\partial u}{\partial t} = \sigma u\left(1 - \frac{u}{N}\right) - (1 + \alpha v)uv + d_1\nabla^2 u$$

$$\frac{\partial v}{\partial t} = (1 + \alpha v)uv - v + d_2\nabla^2 v.$$

(9.26)

This is the working spatial prey–predator model with hunting cooperation and Allee effects in predators. Generally, to make certain that spatial patterns are governed by reaction-diffusion method, model (9.26) is to be analyzed with the following initial conditions:

$$1D: u(x,0) > 0,\ v(x,0) > 0,\ x \in \Omega = [0,L]$$

$$2D: u(x,y,0) > 0,\ v(x,y,0) > 0,\ (x,y) \in \Omega = [0,L] \times [0,L]$$

(9.27)

and Neumann's boundary condition

$$\frac{\partial u}{\partial v} = \frac{\partial v}{\partial v} = 0,$$

(9.28)

where L is the size of the homogeneous spatial domain and v is the outward unit normal on the boundary $\partial\Omega$. The aim of our study in this chapter is to investigate the phenomena of diffusion-driven instability (spatial pattern) and higher-order instability analysis outside the Turing domain, in the predator–prey system with hunting cooperation and Allee effects in predator.

9.5 Spatial Dynamics of the Model

9.5.1 Initial Conditions

The spatial pattern generally onsets with a community interplay of species. The episodic initial conditions for model (9.26) should be stated by the mathematical compact support function, that is, within a definite domain, the initial distribution of prey and predator is nonzero and elsewhere is zero. The structure of the realm and the outlines of species densities can be dissimilar in different prey–predator models. For that purpose, we ponder the initial distribution of the populations in the shape of patch:

1-D Case:

$$u(x,0) = u_* \text{ for } (x - x_o)^2 < R_o, \quad \text{otherwise } u(x,0) = 0,$$

$$v(x,0) = v_* \text{ for } (x - x_o)^2 < R_o, \quad \text{otherwise } v(x,0) = 0,$$

(9.29)

2-D Case:

$$u(x,y,0) = u_o \text{ for } (x - x_o)^2 + (y - y_o)^2 < R_o,$$

$$\text{otherwise } u(x,y,0) = 0,$$

$$v(x,y,0) = v_o \text{ for } (x - x_o)^2 + (y - y_o)^2 < R_o,$$

$$\text{otherwise } v(x,y,0) = 0,$$

(9.30)

where $x_0, y_0, R_0, u_0,$ and v_0 are nonnegative parameters with clear interpretation.

9.5.2 Steady States

Further, we are concentrating on the spatial dynamics produced by the prey–predator model with hunting cooperation and Allee effects in predators. In the absence of diffusion [22], let us consider the equilibrium points of the system mentioned by (9.26). Explanations of ecological concern will have non-negative populace density u and v. We suppose that all the parameters are nonnegative. Information about the equilibrium points of (9.26) can be procured by scrutinizing the nullclines, that is the curves along which $du/dt = 0$ and $dv/dt = 0$. There are equilibria for the model at the intersections of these nullclines. There are two nullclines for the prey equation of (9.26). First is $u = 0$, and the other one is

$$u = \frac{N}{\sigma}(\sigma - (1 + \alpha v)v) \equiv g(v). \tag{9.31}$$

Also, there are two nullclines for predator equation (9.26). First is $v = 0$ and the other one is

$$u = \frac{1}{1 + \alpha v} \equiv h(v). \tag{9.32}$$

An equilibrium point with coordinates $(u, v) = (0, 0)$ is given by the intersection of $u = 0$ and $v = 0$ (both populations are extinct). The other equilibrium point with coordinates $(u, v) = (N, 0)$ is given by the intersection of $u = \dfrac{N}{\sigma}(\sigma - (1 + \alpha v)v)$ and $v = 0$ (the prey only survives). Putting $g(v)$ equal to $h(v)$, obtain a third-order polynomial for the v values of these equilibrium points:

$$A_3 v^3 + A_2 v^2 + A_1 v + A_0 = 0, \tag{9.33}$$

where

$$A_3 = N\alpha^2, \ A_2 = 2\alpha N, \ A_1 = N(1 - \sigma\alpha), \ A_0 = \sigma(1 - N).$$

The number of coexistence equilibrium points and their stability depend on the parameter values for the model (9.26).

9.5.3 Phase Space

From the ecological approach, we are interested in studying the qualitative properties of the interior equilibrium point. The phase portrait is shown in Figures 9.1 and 9.2 for different parameter values. When $0.8 < N < 1$ and $\alpha > 0.44$, there exist two real and positive equilibrium points (see in Ref. [22] for details), in which one is stable and the other one is unstable, see

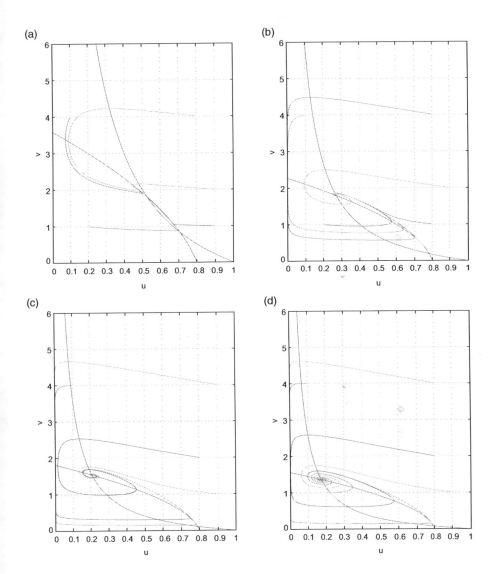

FIGURE 9.1
Phase plane diagrams of the system (9.26) for fixed parameters $\sigma = 10.0$ and $N = 0.8$ and different parameter values of α: (a) $\alpha = 0.5$; (b) $\alpha = 1.5$; (c) $\alpha = 2.5$; and (d) $\alpha = 3.5$.

in Figure 9.1. If $N > 1$, then there exists only one positive real equilibrium point, which is stable, see Figure 9.2. For more details, see reference [22]. Note that, the condition for the existence of nontrivial equilibrium points is that $N > 0.8$ and $\alpha > 0.44$ (see in Figures 9.1 and 9.2), and for more detailed analysis, see Ref. [22].

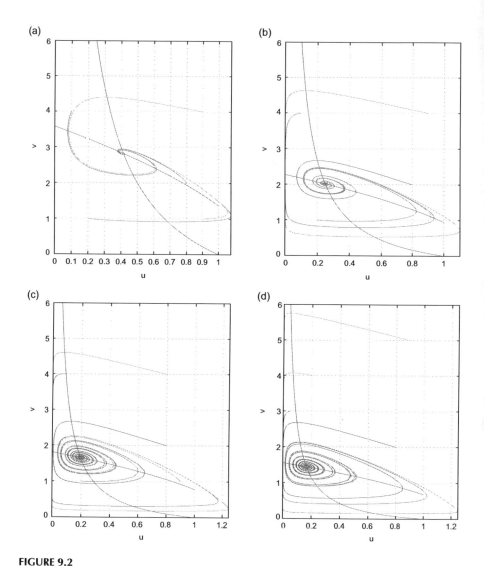

FIGURE 9.2
Phase plane diagrams of the system (9.26) for fixed parameters $\sigma = 10.0$ and $N = 1.3$ and different parameter values of α: (a) $\alpha = 0.5$; (b) $\alpha = 1.5$; (c) $\alpha = 2.5$; and (d) $\alpha = 3.5$.

9.5.4 Spatial Analysis

The Jacobian **J** of the nondiffusive model of (9.26) is as follows:

$$\mathbf{J} = \begin{bmatrix} \sigma - v - \alpha v^2 - \dfrac{2\sigma u}{N} & -u - 2\alpha uv \\[2mm] v + \alpha v^2 & u + 2\alpha uv - 1 \end{bmatrix} = \left(\Delta_{ij} \right)_{2 \times 2}. \tag{9.34}$$

The determinant $J_m = \det(J_m)$ and trace $J_m = \text{tr}(J_m)$, where $J_m = J$ evaluated at E_m, $m = 0,1,2,3,4$; $i = 1,2$; $j = 1,2$. Obviously, the equilibrium point $E_i(u^*, v^*)$, $i = 2,3,4$, for the model without diffusion is a spatially homogenous steady state for the reaction-diffusion model (9.26).

Here, we assume that $E_2(u^*, v^*)$ is stable in a nonspatial model of (9.26), which means that homogenous steady states are spatially stable with respect to homogenous spatial perturbations. Diffusion can make a spatially homogenous steady state linearly unstable with respect to nonhomogeneous perturbation [34,35]. Assume that interior steady-state point in the nonspatial homogenous point is stable with respect to spatial homogenous perturbation:

$$u(x,y,t) = u^* + \epsilon_1 \exp(\lambda_k t)\cos(k_x x)\cos(k_y y),$$

$$v(x,y,t) = v^* + \eta_1 \exp(\lambda_k t)\cos(k_x x)\cos(k_y y), \quad (9.35)$$

where ϵ_1 and η_1 are two nonzero reals and $k(\text{wave number}) = (k_x, k_y)$, such that $k^2 = (k_x^2 + k_y^2)$ is the wave number. On substitution of (9.35) into the model system (9.26), the solution of the linearized system is around the steady state $E_2(u^*, v^*)$ and suppose that at least one eigenvalue of the model's matrix (9.34) cuts the imaginary axis, the stability criteria will change if at least one condition is violated out of the following conditions:

$$a_1(k^2) = (d_1 + d_2)k^2 - \Delta_{11} - \Delta_{22} > 0, \quad (9.36)$$

$$a_2(k^2) = d_1 d_2 k^4 - (d_2 \Delta_{11} + d_1 \Delta_{22})k^2 + \Delta_{11}\Delta_{22} - \Delta_{12}\Delta_{21} > 0. \quad (9.37)$$

where Δ_{ij} are the elements of J (see Eq. 9.34). As d_1, d_2, and k^2 are both positive and $\Delta_{11} + \Delta_{22} < 0$ (stability of the temporal steady state), $a_1(k^2) > 0$ always holds. Only one possibility for instability after adding the diffusion term is that $a_2(k^2) < 0$. After some calculus, the minimum value of $a_2(k^2)$ is obtained at k^{cr}, where k^{cr} is defined by

$$k^{cr} = \sqrt{\frac{d_2 \Delta_{11} + d_1 \Delta_{22}}{2d_1 d_2}}. \quad (9.38)$$

As $\Delta_{11} + \Delta_{22} < 0$ and k^{cr} is real then we must have $\Delta_{11}\Delta_{22} < 0$. Hence, a sufficient condition for diffusion-driven instability is that

$$a_2\left((k^{cr})^2\right) = (\Delta_{11}\Delta_{22} - \Delta_{12}\Delta_{21}) - \frac{(d_2 \Delta_{11} + d_1 \Delta_{22})^2}{(4d_1 d_2)} < 0. \quad (9.39)$$

Therefore, from Eq. (9.39), we get the following condition for diffusion-driven instability:

$$\left(d_2\Delta_{11} + d_1\Delta_{22}\right)^2 > 4\left(\Delta_{11}\Delta_{22} - \Delta_{12}\Delta_{21}\right)d_1 d_2 > 0. \tag{9.40}$$

The critical wave number k^{cr} for the growing spatial patterns via diffusion-driven instability are obtained when Eq. (9.40) is an equality. The critical wave number k^{cr} for the growing spatial patterns is given by Eq. (9.38). After some algebra, the change of sign in $a_2(k^2)$ occurs when k^2 is in the range of $\left(k_1{}^2, k_2{}^2\right)$, where

$$k_1{}^2 = \frac{\Delta_{22}d_1 + \Delta_{11}d_2 - \sqrt{\left(\Delta_{22}d_1 + \Delta_{11}d_2\right)^2 - 4d_1 d_2\left(\Delta_{11}\Delta_{22} - \Delta_{12}\Delta_{21}\right)}}{2d_1 d_2} \tag{9.41}$$

and

$$k_2{}^2 = \frac{\Delta_{22}d_1 + \Delta_{11}d_2 + \sqrt{\left(\Delta_{22}d_1 + \Delta_{11}d_2\right)^2 - 4d_1 d_2\left(\Delta_{11}\Delta_{22} - \Delta_{12}\Delta_{21}\right)}}{2d_1 d_2}. \tag{9.42}$$

Especially, we have $a_2(k^2) < 0$, that is diffusion-driven instability occurs for $k_1{}^2 < k^2 < k_2{}^2$.

9.6 Outcomes of Numerical Simulations

The spatiotemporal mathematical model (9.26) is solved numerically in one- and two-dimensional spatial domains with the help of finite difference scheme for spatial derivatives. Euler method is employed for the reaction part of model (9.26) and general five-point finite difference scheme is employed for the diffusion part of the model (9.26). The numerical unification of the reaction-diffusion partial differential equations (9.26) is employed by using splitting method [36]. The numerical values of temporal and spatial step size selected have been adequately small for avoiding the numerical relics. We presented all the numerical simulations of system (9.26) over the zero-flux boundary condition with one- and two-dimensional spatial habitats [0, 2,000] and 200×200 domain size, respectively, and time step $\Delta t = 0.01$ and space step $\Delta x = \Delta y = 1$. For numerical simulation, we set σ at some hypothetical value as $\sigma = 10$ and consider α, N, and diffusion coefficients as controlling parameters. Selection of parameter values for spatial patterns (Turing patterns) in prey–predator model with hunting cooperation (9.26), (see Figure 9.3).

The ecological inspiration of this research work is perusal of the diverse (nonhomogeneous) patterns of prey and predator. The two-dimensional perusal gives more realistic phenomenon with much complexity than one-dimensional study, and that is why the main focus is to explore the solution of the two-dimensional system. Diffusion-driven instability (Turing's instability) conditions are satisfied in the system (9.26) (see in Figure 9.3); hence, Turing spatial patterns occur in system (9.26). Figure 9.3 shows the graph between $a_2(k^2)$ and k for the system (9.26) for fixed values of $\sigma = 10.0$, $N = 0.8$, $\alpha = 0.5$, and $d_2 = 0.3$ and for different values of d_1. Figure 9.3 shows the graph of $a_2(k^2)$ against k for the system (9.26) for fixed values of $\sigma = 10.0$, $N = 0.8$, $d_1 = 18.5$, and $d_2 = 0.3$ and for different values of cooperate rate α. Figure 9.3 shows the graph of $a_2(k^2)$ against k for the system (9.26) for fixed values of $\sigma = 10.0$, $\alpha = 0.5$, $d_1 = 21.5$, and $d_2 = 0.3$ and for different values of carrying capacity N. Figure 9.3 represents the plot of real part of largest eigenvalue $(Re(\lambda))$ against k of the system (9.26) for fixed values of $\sigma = 10.0$, $N = 0.8$, $\alpha = 0.5$, and $d_2 = 0.3$ and for different values of d_1. Figure 9.3 exhibits the plot of real part of largest eigenvalue $(Re(\lambda))$ against k of the system (9.26) for fixed values of $\sigma = 10.0$, $N = 0.8$, $d_1 = 18.5$, and $d_2 = 0.3$ and for different values of cooperate rate α, and finally Figure 9.3 exhibits the plot of real part of largest eigenvalue $(Re(\lambda))$ against k of the system (9.26) for fixed values of $\sigma = 10.0$, $\alpha = 0.5$, $d_1 = 21.5$, and $d_2 = 0.3$ and for different values of carrying capacity N. Clearly, we can see that the determinant of $[J_2 - Dk^2]$, that is, $a_2(k^2)$ is less than zero, and the real part of largest eigenvalue, that is, $Re(\lambda)$ is greater than zero from Figure 9.3, which is the condition for Turing's instability (diffusion-driven instability). From Figure 9.3, the parameter values for spatial patterns of the prey–predator system (9.26) are $\sigma = 10$, $N = 0.8$, and $\alpha = 0.5$, and diffusion coefficient value starts from $d_1 = 17.5$ to $d_2 = 0.3$.

Figure 9.4 shows that the one-dimensional stationary patterns of (9.26) for the prey and predator population at $T = 10{,}000$: black solid lines for prey and dotted lines for predator, for fixed values of $\sigma = 10.0$, $N = 0.8$, $\alpha = 0.5$, and $d_2 = 0.3$ and for different values of d_1, which is mentioned in figure (see in Figure 9.4). Figure 9.5 represents the one-dimensional stationary patterns of (9.26) for the prey and predator population at $T = 1{,}000$: black solid lines for prey and dotted lines for predator, for fixed values of $\sigma = 10.0$, $N = 0.8$, $d_1 = 18.5$, and $d_2 = 0.3$ and for different values of cooperate rate α, which is mentioned in figure (see Figure 9.5). Figure 9.6 shows the one-dimensional stationary patterns of (9.26) for the prey and predator population at different times T: black solid lines for prey and dotted lines for predator, for fixed values of $\sigma = 10.0$, $\alpha = 0.5$, $d_1 = 21.5$, and $d_2 = 0.3$ (see in Figure 9.6). Figure 9.7 shows the one-dimensional stationary patterns of (9.26) for the prey and predator

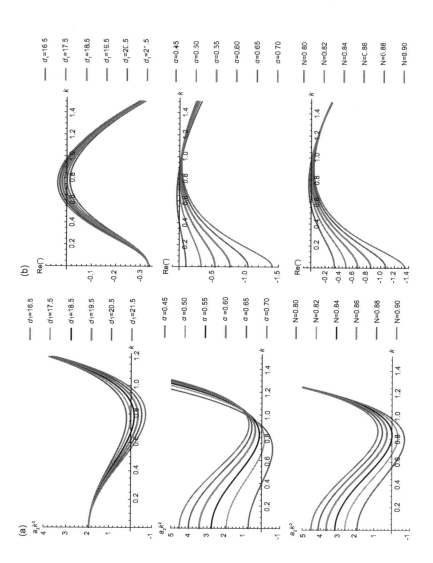

FIGURE 9.3

Plot of $a_2(k^2)$ (a) and real part of largest eigenvalue $Re(\lambda)$ (b) against wave number k for spatial mathematical system (9.26): for different values of diffusion coefficient d_1, α and N as explained in (a) and (b). Other parameters are $\sigma = 10.0$, $N = 0.8$, $\alpha = 0.5$, $d_1 = 18.5$, and $d_2 = 0.3$.

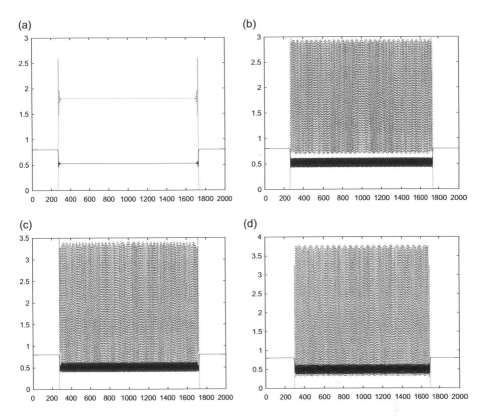

FIGURE 9.4
One-dimensional stationary patterns of (9.26) at $T = 10,000$: solid lines for prey and dotted lines for predator. Parameter values: $\sigma = 10.0$, $N = 0.8$, $\alpha = 0.5$, and $d_2 = 0.3$. (a) $d_1 = 17.5$; (b) $d_1 = 22.5$; (c) $d_1 = 27.5$; and (d) $d_1 = 32.5$.

population: black solid lines for prey and dotted lines for predator, for fixed values of $\sigma = 10.0$, $\alpha = 0.5$, $N = 0.8$, $d_1 = 27.5$, and $d_2 = 0.3$ and different times (see in Figure 9.7).

Figure 9.8 shows the two-dimensional stationary patterns of (9.26) for the prey population at $T = 500$, for fixed values of $\sigma = 10.0$, $N = 0.8$, $\alpha = 0.5$, and $d_2 = 0.3$ and for different values of diffusion coefficient d_1, which is mentioned in figure (see in Figure 9.8). It is easily seen that stripe patterns become separate patches and the size of circular patch also increases when the value of diffusion coefficients increases to 32.5. Figure 9.9 shows the two-dimensional stationary patterns of (9.26) for the predator population at $T = 500$, for fixed values of $\sigma = 10.0$, $N = 0.8$, $\alpha = 0.5$, and $d_2 = 0.3$ and for different values of diffusion coefficient d_1, which is mentioned in figure (see in Figure 9.9), and it is easily seen that stripe patterns become separate patches,

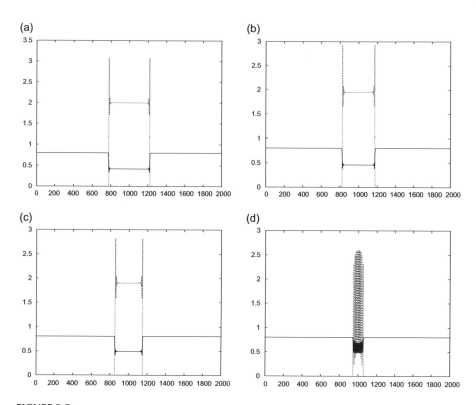

FIGURE 9.5
One-dimensional stationary patterns of (9.26) at $T = 1,000$: black solid lines for prey and dotted lines for predator. Parameter values: $\sigma = 10.0$, $N = 0.8$, $d_1 = 18.5$, and $d_2 = 0.3$. (a) $\alpha = 0.70$ ($u^* = 0.4171$, $v^* = 1.9963$), (b) $\alpha = 0.60$ ($u^* = 0.4602$, $v^* = 1.9546$), (c) $\alpha = 0.55$ ($u^* = 0.4888$, $v^* = 1.9014$), and (d) $\alpha = 0.45$ ($u^* = 0.5874$, $v^* = 1.5612$).

and the size of circular patch also increases when the value of diffusion coefficients increases to 32.5. Figure 9.9 shows the two-dimensional stationary patterns of (9.26) for the prey population at $T = 500$, for fixed values of $\sigma = 10.0$, $N = 0.8$, $d_1 = 18.5$, and $d_2 = 0.3$ and for different values of cooperate rate α (see in Figure 9.9). It can be clearly seen that large indistinct stripe circular patterns become sharp patches when the value of cooperate rate decreases to 0.45 from 0.60. Figure 9.10 shows the two-dimensional stationary patterns of (9.26) for the predator population at $T = 500$, for fixed values of $\sigma = 10.0$, $N = 0.8$, $d_1 = 18.5$, and $d_2 = 0.3$ and for different values of cooperate rate α (see in Figure 9.11). It can be clearly seen that large nondistinct stripe circular patterns become sharp patches when the value of cooperate rate decreases to 0.45 from 0.60. Figure 9.12 shows the two-dimensional stationary patterns of (9.26) for the prey population at time $T = 500$, for fixed values of $\sigma = 10.0$,

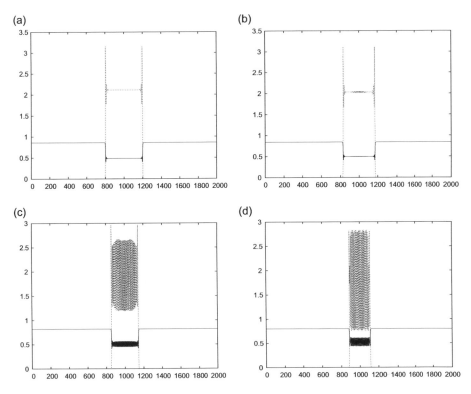

FIGURE 9.6

One-dimensional stationary patterns of (9.26) at $T = 1,000$: black solid lines for prey and dotted lines for predator. Parameter values: $\sigma = 10.0$, $\alpha = 0.5$, $d_1 = 21.5$, and $d_2 = 0.3$. (a) $N = 0.86$ ($u^* = 0.4862$, $v^* = 2.1132$), (b) $N = 0.84$ ($u^* = 0.4962$, $v^* = 2.0309$), (c) $N = 0.82$ ($u^* = 0.5088$, $v^* = 1.9310$), and (d) $N = 0.80$ ($u^* = 0.5874$, $v^* = 1.5612$).

$\alpha = 0.5$, $d_1 = 21.5$, and $d_2 = 0.3$ and for different values of carrying capacity N (see in Figure 9.12). Figure 9.13 shows the two-dimensional stationary patterns of (9.26) for the predator population at time $T = 500$, for fixed values of $\sigma = 10.0$, $\alpha = 0.5$, $d_1 = 21.5$, and $d_2 = 0.3$ and for different values of carrying capacity N (see in Figure 9.13). Figure 9.14 shows the two-dimensional stationary patterns of (9.26) for the prey population at different time values T, for fixed values of $\sigma = 10.0$, $\alpha = 0.5$, $N = 0.8$, $d_1 = 27.5$, and $d_2 = 0.3$ and at different times (see in Figure 9.14). Figure 9.15 shows the two-dimensional stationary patterns of (9.26) for the prey population at different time values T, for fixed values of $\sigma = 10.0$, $\alpha = 0.5$, $N = 0.8$, $d_1 = 27.5$, and $d_2 = 0.3$ and at different times (see in Figure 9.15). It is clearly seen that patterns become stationary patches and the size of circular patch also increases when the value of time increases (see in Figures 9.14 and 9.15).

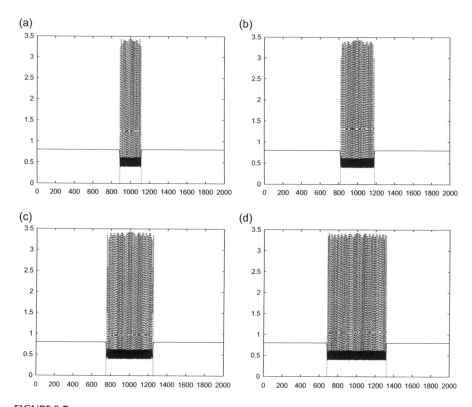

FIGURE 9.7
One-dimensional stationary patterns of (9.26) at different times: black solid lines for prey and dotted lines for predator. Parameter values: $\sigma = 10.0$, $\alpha = 0.50$, $N = 0.8$, $d_1 = 27.5$, and $d_2 = 0.3$. (a) Time = 1,000, (b) time = 2,000, (c) time = 3,000, and (d) time = 4,000.

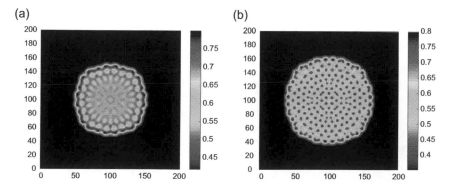

FIGURE 9.8
Two-dimensional stationary patterns of (9.26) for prey at $T = 500$. Parameter values: $\sigma = 10.0$, $N = 0.8$, $\alpha = 0.5$, and $d_2 = 0.3$. (a) $d_1 = 17.5$; (b) $d_1 = 22.5$; (c) $d_1 = 27.5$; and (d) $d_1 = 32.5$.

(*Continued*)

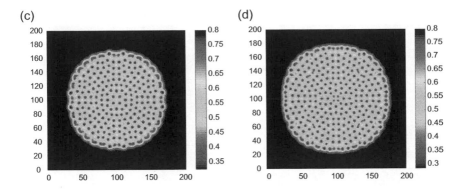

FIGURE 9.8 (CONTINUED)
Two-dimensional stationary patterns of (9.26) for prey at $T = 500$. Parameter values: $\sigma = 10.0$, $N = 0.8$, $\alpha = 0.5$, and $d_2 = 0.3$. (a) $d_1 = 17.5$; (b) $d_1 = 22.5$; (c) $d_1 = 27.5$; and (d) $d_1 = 32.5$.

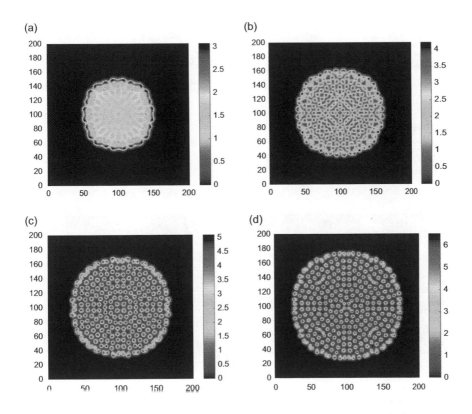

FIGURE 9.9
Two-dimensional stationary patterns of (9.26) for predator at $T = 500$. Parameter values: $\sigma = 10.0$, $N = 0.8$, $\alpha = 0.5$, and $d_2 = 0.3$. (a) $d_1 = 17.5$; (b) $d_1 = 22.5$; (c) $d_1 = 27.5$; and (d) $d_1 = 32.5$.

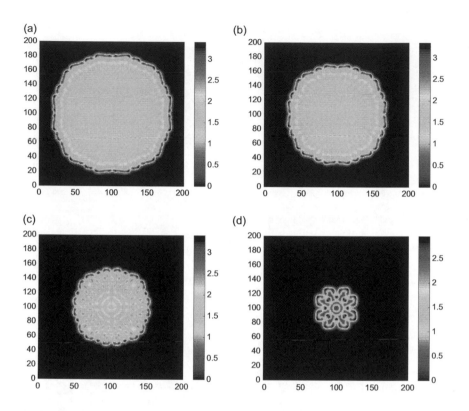

FIGURE 9.10
Two-dimensional stationary patterns of (9.26) for predator at $T = 500$. Parameter values: $\sigma = 10.0$, $N = 0.8$, $d_1 = 18.5$, and $d_2 = 0.3$. (a) $\alpha = 0.60$ ($u^* = 0.4602$, $v^* = 1.9546$), (b) $\alpha = 0.55$ ($u^* = 0.4888$, $v^* = 1.9014$), (c) $\alpha = 0.50$ ($u^* = 0.5262$, $v^* = 1.8009$), and (d) $\alpha = 0.45$ ($u^* = 0.5873$, $v^* = 1.5612$).

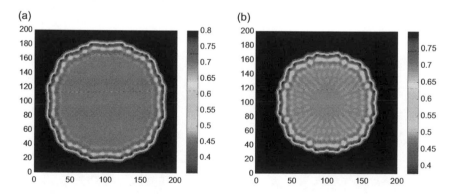

FIGURE 9.11
Two-dimensional stationary patterns of (9.26) for prey at $T = 500$. Parameter values: $\sigma = 10.0$, $N = 0.8$, $d_1 = 18.5$, and $d_2 = 0.3$. (a) $\alpha = 0.60$ ($u^* = 0.4602$, $v^* = 1.9546$), (b) $\alpha = 0.55$ ($u^* = 0.4888$, $v^* = 1.9014$), (c) $\alpha = 0.50$ ($u^* = 0.5262$, $v^* = 1.8009$), and (d) $\alpha = 0.45$ ($u^* = 0.5873$, $v^* = 1.5612$).

(Continued)

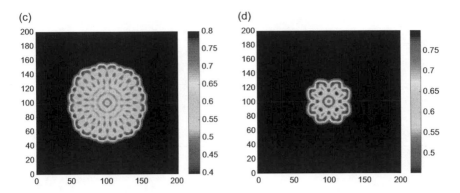

FIGURE 9.11 (CONTINUED)
Two-dimensional stationary patterns of (9.26) for prey at $T = 500$. Parameter values: $\sigma = 10.0$, $N = 0.8$, $d_1 = 18.5$, and $d_2 = 0.3$. (a) $\alpha = 0.60$ ($u^* = 0.4602$, $v^* = 1.9546$), (b) $\alpha = 0.55$ ($u^* = 0.4888$, $v^* = 1.9014$), (c) $\alpha = 0.50$ ($u^* = 0.5262$, $v^* = 1.8009$), and (d) $\alpha = 0.45$ ($u^* = 0.5873$, $v^* = 1.5612$).

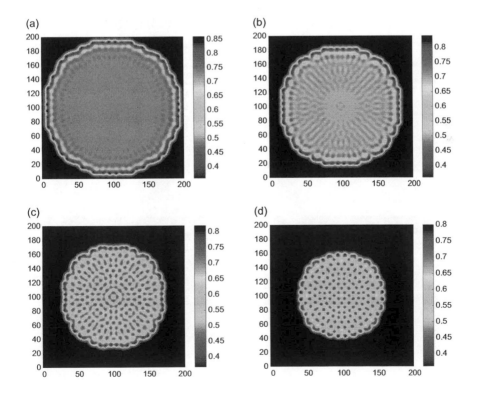

FIGURE 9.12
Two-dimensional stationary patterns of (9.26) for prey at $T = 500$. Parameter values: $\sigma = 10.0$, $\alpha = 0.5$, $d_1 = 21.5$, and $d_2 = 0.3$. (a) $N = 0.86$ ($u^* = 0.4862$, $v^* = 2.1132$), (b) $N = 0.84$ ($u^* = 0.4962$, $v^* = 2.0309$), (c) $N = 0.82$ ($u^* = 0.5088$, $v^* = 1.9310$), and (d) $N = 0.80$ ($u^* = 0.5262$, $v^* = 1.8009$).

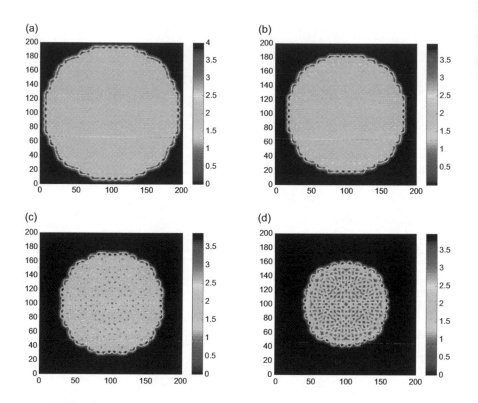

FIGURE 9.13

Two-dimensional stationary patterns of (9.26) for predator at $T = 500$. Parameter values: $\sigma = 10.0$, $\alpha = 0.5$, $d_1 = 27.5$, and $d_2 = 0.3$. (a) $N = 0.86$ ($u^* = 0.4862$, $v^* = 2.1132$), (b) $N = 0.84$ ($u^* = 0.4962$, $v^* = 2.0309$), (c) $N = 0.82$ ($u^* = 0.5088$, $v^* = 1.9310$), and (d) $N = 0.80$ ($u^* = 0.5262$, $v^* = 1.8009$).

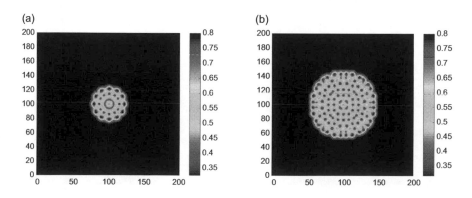

FIGURE 9.14

Two-dimensional stationary patterns of (9.26) for prey at different times. Parameter values: $\sigma = 10.0$, $N = 0.8$, $\alpha = 0.50$, $d_1 = 27.5$, and $d_2 = 0.3$. (a) time = 100, (b) time = 300, (c) time = 500, and (d) time = 700.

(Continued)

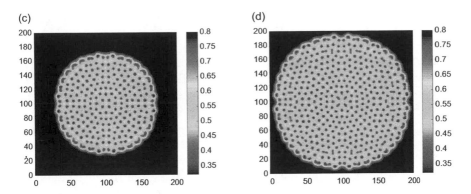

FIGURE 9.14 (CONTINUED)
Two-dimensional stationary patterns of (9.26) for prey at different times. Parameter values:
$\sigma = 10.0$, $N = 0.8$, $\alpha = 0.50$, $d_1 = 27.5$, and $d_2 = 0.3$. (a) time = 100, (b) time = 300, (c) time = 500, and
(d) time = 700.

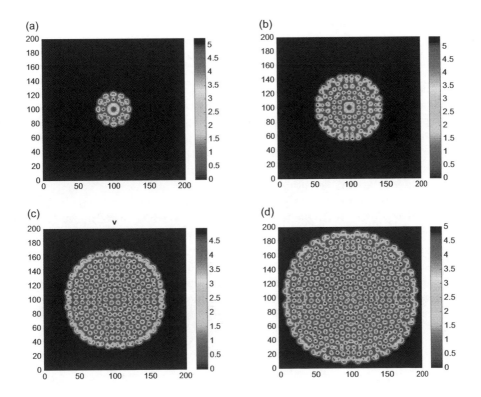

FIGURE 9.15
Two-dimensional stationary patterns of (9.26) for predator at different times. Parameter values:
$\sigma = 10.0$, $N = 0.8$, $\alpha = 0.50$, $d_1 = 27.5$, and $d_2 = 0.3$. (a) time = 100, (b) time = 300, (c) time = 500, and
(d) time = 700.

9.7 Discussion and Conclusion

Spatial pattern formation in predator–prey models plays an important role to understand real-world ecological models. It was shown that spatial patterns take place in prey–predator model through Turing's instability (diffusion-driven instability) [35]. It was shown that a nonconstant irregular spatial distribution of species is a rather usual natural phenomenon in prey–predator models [37,38]. This (spatial pattern formation) phenomenon of suppressing spatiotemporal chaos by increasing diffusivities is a sign that the formation of the "chaotic phase" in the system is possible only if the length of the domain exceeds a certain minimal value [39]. In Ref. [40], Yuan et al. show that prey–predator system with herd behavior can also cause patchiness through Turing instability. The qualitative variation in spatial patterns dynamics of prey–predator model is generally concerned with diffusion-driven instability.

In this chapter, we have considered a diffusive predator–prey model with hunting cooperation under uniform zero flux boundary conditions and circular initial conditions. We provided the elaborated analysis of spatial pattern in the prey–predator model with hunting cooperation through one mechanism named diffusion-driven instability and see the stripe and patchy spatial patterns. We analyzed the existence of the interior equilibrium point and phase space of the nonspatial model of the spatiotemporal model (9.26). Also, we pondered the diffusion-driven instability of the interior steady states point of the spatiotemporal model when the spatial domain is one- and two-dimensional bounded habitat, which generates spatial nonhomogeneous patterns. The sufficient conditions of diffusion-driven instability hold in prey–predator model with hunting cooperation, so there is a possibility of occurrence of the Turing patterns.

Consequently, our outcomes seem to ensure that spatial patterns should be connected with the Turing's instability (diffusion-driven instability). To this place we retrace that diffusion-driven instability cannot be recognized unless the diffusion coefficients are adequately sizable (see Figure 9.3). Particularly, the diffusion-driven instability cannot take place for $d_1 = d_2 = 1$. Diffusion-driven instability can fabricate the model dynamics more complex and the variety of growing spatial patterns generally. We have shown that, in a certain range of parameter, the model (9.26) develops a more complex wavelike pattern in one spatial dimension and two spatial dimensions in stripe and patchy patterns.

References

1. S.A. Levin and L.A. Segel. Hypothesis for origin of planktonic patchiness. *Nature*, 259(5545):659–659, 1976.

2. H. Malchow, S.V. Petrovskii, and E. Venturino. *Spatiotemporal Patterns in Ecology and Epidemiology: Theory, Models, and Simulation.* Boca Raton, FL: Chapman and Hall/CRC, 2008.

3. J.D. Murray. A pre-pattern formation mechanism for animal coat markings. *Journal of Theoretical Biology*, 88(1):161–199, 1981.

4. J.D. Murray. *Mathematical Biology II. Spatial Models and Biomedical Applications.* New York: Springer–Verlag, 2003.

5. A. Okubo & S.A. Levin. (2013). *Diffusion and Ecological Problems: Modern Perspectives* (Vol. 14). Heidelberg: Springer Science and Business Media.

6. A.B. Medvinsky, S.V. Petrovskii, I.A. Tikhonova, H. Malchow, and B.L. Li. Spatiotemporal complexity of plankton and fish dynamics. *SIAM Review*, 44(3):311–370, 2002.

7. M. Mimura and J.D. Murray. On a diffusive prey–predator model which exhibits patchiness. *Journal of Theoretical Biology*, 75(3):249–262, 1978.

8. A.Y. Morozov, S.V. Petrovskii, and B.L. Li. Spatiotemporal complexity of patchy invasion in a predator–prey system with the Allee effect. *Journal of Theoretical Biology*, 238(1):18–35, 2006.

9. S.V. Petrovskii, H. Malchow, F.M. Hilker, and E. Venturino. Patterns of patchy spread in deterministic and stochastic models of biological invasion and biological control. *Biological Invasions*, 7(5):771–793, 2005.

10. L. Edelstein-Keshet. *Mathematical Models in Biology.* Philadelphia, PA: SIAM, 1988.

11. M. Banerjee and S. Banerjee. Turing instabilities and spatiotemporal chaos in ratio-dependent Holling–Tanner model. *Mathematical Biosciences*, 236(1):64–76, 2012.

12. R.A. Barrio, C. Varea, J.L. Aragón, and P.K. Maini. A two-dimensional numerical study of spatial pattern formation in interacting turing systems. *Bulletin of Mathematical Biology*, 61(3):483–505, 1999.

13. B. Dubey, B. Das, and J. Hussain. A predator–prey interaction model with self and cross-diffusion. *Ecological Modelling*, 141(1):67–76, 2001.

14. H.P. Greenspan. Models for the growth of a solid tumor by diffusion. *Studies in Applied Mathematics*, 51(4):317–340, 1972.

15. P.K. Maini, D.L. Benson, and J.A. Sherratt. Pattern formation in reaction–diffusion models with spatially inhomogeneous diffusion coefficients. *Mathematical Medicine and Biology: A Journal of the IMA*, 9(3):197–213, 1992.

16. A. Matzavinos, M.A. Chaplain, and V.A. Kuznetsov. Mathematical modeling of the spatiotemporal response of cytotoxic T-lymphocytes to a solid tumour. *Mathematical Medicine and Biology*, 21(1):1–34, 2004.

17. M. Papadogiorgaki, P. Koliou, X. Kotsiakis, and M.E. Zervakis. Mathematical modelling of spatiotemporal glioma evolution. *Theoretical Biology and Medical Modelling*, 10(1):1–47, 2013.

18. X.C. Zhang, G.Q. Sun, and Z. Jin. Spatial dynamics in a predator–prey model with Beddington-DeAngelis functional response. *Physical Review E*, 85(2):021924–021938, 2012.

19. Q.Q. Zheng and J.W. Shen. Dynamics and pattern formation in a cancer network with diffusion. *Communications in Nonlinear Science and Numerical Simulation*, 27(1):93–109, 2015.

20. M. Banerjee and S.V. Petrovskii. Self-organised spatial patterns and chaos in a ratio-dependent predator–prey system. *Theoretical Ecology*, 4(1):37–53, 2011.

21. W. Wang, Q.X. Liu, and Z. Jin. Spatiotemporal complexity of a ratio–dependent predator–prey system. *Physical Review E*, 75(5):051913–051922, 2007.

22. M.T. Alves and F.M. Hilker. Hunting cooperation and Allee effects in predators. *Journal of Theoretical Biology*, 419(1):13–22, 2017.
23. A. Bompard, I. Amat, X. Fauvergue, and T. Spataro. Host-parasitoid dynamics and the success of biological control when parasitoids are prone to Allee effects. *Bulletin of Mathematical Biology*, 8(10):076768–076779, 2013.
24. F. Courchamp, T. Clutton-Brock, and B. Grenfell. Inverse density dependence and the Allee effect. *Trends in Ecology and Evolution*, 14(10):405–410, 1999.
25. A. DeRoss, L. Persson, and H.R. Thieme. Emergent Allee effects in top predators feeding on structured prey populations. *Proceedings of the Royal Society of London B: Biological Sciences*, 270(1515):611–618, 2003.
26. E. González-Olivares, H. Meneses-Alcay, B. González-Yañez, J. Mena-Lorca, A. Rojas-Palma, and R. Ramos-Jiliberto. Multiple stability and uniqueness of the limit cycle in a Gause-type predator–prey model considering the Allee effect on prey. *Nonlinear Analysis: Real World Applications*, 12(6):2931–2942, 2011.
27. A.Y. Morozov, S.V. Petrovskii, and B.L. Li. Bifurcations and chaos in a predator–prey system with the Allee effect. *Proceedings of the Royal Society of London B: Biological Sciences*, 271(1546):1407–1414, 2004.
28. F. Rao and Y. Kang. The complex dynamics of a diffusive prey–predator model with an Allee effect in prey. *Ecological Complexity*, 28(1):123–144, 2016.
29. A. Verdy. Modulation of predator–prey interactions by the Allee effect. *Ecological Modelling*, 221(8):1098–1107, 2010.
30. J. Wang, J. Shi, and J. Wei. Dynamics and pattern formation in a diffusive predator–prey system with strong Allee effect in prey. *Journal of Differential Equations*, 251(5):1276–1304, 2011.
31. S.R. Zhou, Y.F. Liu, and G. Wang. The stability of predator–prey systems subject to the Allee effects. *Theoretical Population Biology*, 67(1):23–31, 2005.
32. J.D. Murray. *Mathematical Biology I. An Introduction*. New York: Springer–Verlag, 2002.
33. V. Ardizzone, P. Lewandowski, M.H. Luk, Y.C. Tse, N.H. Kwong, A. Lücke, M. Abbarchi, E. Baudin, E. Galopin, J. Bloch, A. Lemaitre, P.T. Leung, P. Roussignol, R. Binder, J. Tignon, and S. Schumacher. Formation and control of Turing patterns in a coherent quantum fluid. *Scientific Reports*, 3(1):1–6, 2013.
34. A.M. Turing. The chemical basis of morphogenesis. *Philosophical Transactions of the Royal Society of London B: Biological Sciences*, 237(641):37–72, 1952.
35. L.A. Segel and J.L. Jackson. Dissipative structure: An explanation and an ecological example. *Journal of Theoretical Biology*, 37(3):545–559, 1972.
36. I. Kozlova, M. Singh, A. Easton, and P. Ridland. Twospotted spider mite predator–prey model. *Mathematical and Computer Modelling*, 42(11):1287–1298, 2005.
37. S.V. Petrovskii and H. Malchow. Wave of chaos: New mechanism of pattern formation in spatiotemporal population dynamics. *Theoretical Population Biology*, 59(2):157–174, 2001.
38. F. Yi, J. Wei, and J. Shi. Bifurcation and spatiotemporal patterns in a homogeneous diffusive predator–prey system. *Journal of Differential Equations*, 246(5):1944–1977, 2009.
39. S.V. Petrovskii and H. Malchow. A minimal model of pattern formation in a prey–predator system. *Mathematical and Computer Modelling*, 29(8):49–63, 1999.
40. S. Yuan, C. Xu, and T. Zhang. Spatial dynamics in a predator–prey model with herd behavior. *Chaos: An Interdisciplinary Journal of Nonlinear Science*, 23(3):033102–033113, 2013.

10

Solving Multi-objective Transportation Problem under Cost Reliability Using Utility Function

Gurupada Maity, Dharmadas Mardanya, and Sankar Kumar Roy

Vidyasagar University

CONTENTS

10.1 Introduction and Preliminaries

Operations research (OR), a branch of applied mathematics plays important roles for solving real-life decision-making problems related to optimization. The study of transportation problem (TP) is a special case for solving decision-making problem in OR. In a broad sense, TP can be considered as a special case of linear programming problem (LPP), and its model is used to find how many units of items to be transported from each origin to various destinations, satisfying source availabilities and destination demands by a minimum transportation cost.

Hitchcock [1] developed the basic mathematical model of TP, and later on Koopmans [2] formulated the TP model in a more compact form. Kanntorovich [3] studied mathematical methods of organizing and planning production. Hamzehee et al. [4] introduced linear programming with rough interval coefficients. James et al. [5] improved transportation service quality based on information fusion. Kumar and Bierlaire [6] solved multi-objective airport gate assignment problem. Luathep et al. [7] discussed global optimization method for mixed transportation network design problem. Realizing the present competitive market scenario, it is easy to recall that the single-objective TP is inadequate to handle real-life decision-making situations. In this regard, we introduce the multi-objective concept in TP, where the objective functions are conflicting to each other. In the area of multi-objective decision-making problems, Charnes et al. [8] first discussed various approaches on the solution for managerial-level problems involving multiple conflicting objective functions. Vincent et al. [9] studied about an interactive approach for multi-objective TP with interval parameters. Maity and Roy [10] solved a multi-objective TP with non-linear cost and multichoice demand. Maity and Roy [11] solved multichoice multi-objective TP using utility function approach. Again Maity and Roy [12] solved multichoice multi-objective TP with interval goal using utility function approach. Multi-objective interval valued transportation problem involving log normal is solved by Mahapatra and Roy [13]. Multi-choice stochastic TP involving Weibul distribution has been studied by Roy [14]. Roy et al. [15] presented a study on fixed-charge TP under rough environment and Midya and Roy [16] analysis the interval programming and its application to fixed-charge TP. Recently, Roy and Maity [17] studied minimizing cost and time through single objective function in multi-choice interval valued TP, and Roy et al. [18] introduced a new approach for solving the intuitionistic fuzzy multi-objective TP.

Time is an important factor in every transportation system. In most of the real-life transportation, the customers wish to receive ordered goods in their schedule time. But, most of the times, it is impossible to complete transportation in schedule time by the sellers. As a result, the customers suffer due to their loss of profit. To remove this factor, we introduce the reliability in cost parameters of the TP. Again, in the study of multi-objective TP (MOTP) in which the objective functions are conflicting type, the reliability in cost penalty may cause to increase the cost of transportation or increase the values of objective functions which are minimization type and to decrease the profit of decision maker (DM) or decrease the value of objective functions which are maximization type. When the cost reliability factor is introduced as the rate of damage goods which creates a diverse relation for both maximization and minimization types of objective functions. Also, the amount of loss of profit or it is increased the cost of transportation caused by the cost reliability factors are independent. So, it is clear that the reliability in cost parameter preserves the conflicting nature to the objective functions of MOTP. Recently, Maity et al. [19] developed a study on MOTP using cost reliability. Goal programming (GP), an analytical approach, is devised to address the decision-making problem, where targets have been assigned to all objective functions that are conflicting and commensurable to each other, and DM interests to maximize the achievement level of the corresponding goals. Charnes et al. [8] introduced the concept of GP that is further developed by several researchers such as Lee [20], Zeleny [21], Ignizio [22], Tamiz et al. [23,24], Romero [25], Maity et al. [26], Roy et al. [27] and many others. Earlier, the main concept of GP was to minimize the deviation between the achievement goals and achievement levels. The mathematical model of multi-objective decision making (MODM) can be considered in the following form:

- **Model GP**

$$\text{minimize} \sum_{t=1}^{r} w_t \left| Z^t(x) - g_t \right|$$

subject to $x \in F$,

where F is the feasible set and w_t are the weights attached to the deviation of the achievement function. $Z_t(x)$ is the ith objective function of the tth goal, and g_i is the aspiration level of the ith goal. $\left| Z^t(x) - g_t \right|$ represents the deviation of the tth goal. Later on, a modification of GP is provided, and it is denoted as weighted GP (WGP), which can be displayed in the following form:

- **Model WGP**

$$\text{minimize} \sum_{t=1}^{r} w_t \left(d_t^+ + d_t^- \right)$$

subject to $Z^t(x) - d_t^+ + d_t^- = g_t$,

$d_t^+ \geq 0, d_t^- \geq 0 \ (t = 1, 2, \ldots, r)$,

$x \in F$,

where F is the feasible set, d_t^+ and d_t^- are over and underachievements of the tth goal, respectively.

However, the conflicts of resources and incompleteness of available information make these almost impossible for DMs to set the specific aspiration levels and choose a better decision. To overcome this situation, multichoice GP (MCGP) approach has been presented by Chang [28] with a new direction to solve the MODM problem. In the next year, Chang [29] proposed the revised form of MCGP and defined as revised multichoice GP (RMCGP) to solve MODM. Mahapatra et al. [30] studied about TP involving multichoice costs. Paksoy and Chang [31] used revised MCGP for multiperiod, multistage inventory-controlled supply chain model. The mathematical model of MODM using RMCGP is defined as follows:

- **Model RMCGP**

$$\text{minimize} \sum_{t=1}^{r} \left[w_t \left(d_t^+ + d_t^- \right) + \alpha_t \left(e_t^+ + e_t^- \right) \right]$$

subject to $Z^t(x) - d_t^+ + d_t^- = y_t$ $(t = 1, 2, \ldots, r)$,

$y_t - e_t^+ + e_t^- = g_{t,\max}$ or $g_{t,\min}$ $(t = 1, 2, \ldots, r)$,

$g_{t,\min} \leq y_t \leq g_{t,\max}$ $(t = 1, 2, \ldots, r)$,

$d_t^+, d_t^-, e_t^+, e_t^- \geq 0, (t = 1, 2, \ldots, r)$,

$x \in F$.

Here, F being the feasible set, y_t is the continuous variable associated with t^{th} goal which restricted between the upper $(g_{t,\max})$ and lower $(g_{t,\min})$ bounds and e_t^+ and e_t^- are positive and negative deviations attached to the tth goal of $|y_t - g_{t,\max}|$ and α_t is the weight attached to the sum of deviations of $|y_t - g_{t,\max}|$, and other variables are defined as in WGP.

In recent years, the notion of utility function is introduced by several researchers such as Al-Nowaihi et al. [32], Yu et al. [33], and Podinovski [34]. Recently, Maity and Roy [12] solved MOTP under multichoice programming using utility function approach.

The rest of this chapter is organized as follows: In Sections 10.2 and 10.3, the problem background and mathematical models are described for MOTP. Section 10.4 contains the solution procedure. We demonstrate the usefulness of the proposed model with a realistic example in MOTP in Section 10.5. Finally, the conclusion and future study are discussed in Section 10.6.

10.2 Problem Background

The study of TP is highly connected with the time of transportation. Especially, transportation time in the transportation of perishable items (fishes, vegetables, etc.) is very much essential to transport in schedule time. In this regard, transportation of amount of goods should be done within a specified time; otherwise, there may be damage of items or storing problems, and/or the customer may reject the ordered item. Therefore, the transportation penalties such as transportation cost, profit, etc. may not be taken as crisp value.

To accommodate the situations, transportation cost, profit, etc. are computed under a function that depends on scheduled time of transportation, and a reliability approach is employed over the transportation parameters. Then the selection of goals for the objective functions or the solution of MOTP cannot be made in a usual way. To overcome this difficulty by selecting the proper goals to the objective functions, we incorporate the concept of reliability for the cost parameters in TP. In that situation, we introduce a new term called "cost reliability" for the transportation cost in the proposed study.

Reliability (Richard et al. [35]) is considered as the probability of a machine operating its scheduled purpose adequately for the period of time desired under the prescribed operating conditions. More precisely, reliability is the probability with which a device will not fail to perform its desired operation for a certain period of time.

Definition 10.1

(Cost Reliability): Cost reliability refers the probability that the transportation of goods will not fail to transport goods in the schedule time. This fact creates a probabilistic cost in the TP. This probabilistic cost in the TP increases the original value of transportation of goods and simultaneously makes a difference in profit margin. More conveniently, the values of transportation parameters of minimizing type are increased and maximizing type are decreased due to the occurrence of this probabilistic cost.

Let us assume that T be the estimated or schedule time of transportation of goods in a transportation system. The transportation of goods before or after the scheduled time (error time) creates lots of problems to the customer. As the customer sells goods after purchasing, he finds less profit due to the error time. Here it is considered that the transportation is maintained by the source person. So, the error time enlarges the maintenance cost, which is taken as transportation cost here. Let us consider $\delta\tau$ be the error time in the transportation of goods; Again, we assume the failure rate λ, which is defined by $\lambda = \dfrac{\delta\tau}{T}$, where τ represents the time.

Now, the cost reliability $R(\tau)$, a function of time (τ), is defined as follows:

$$R(\tau) = \frac{\text{Amount of goods remains in good condition due to error time}}{\text{Total amount of goods}}$$

Furthermore, the probability of failure, $Q(\tau)$ can be expressed in the following way:

$$Q(\tau) = \frac{\text{Amount of goods damaged due to error time}}{\text{Total amount of goods}}$$

By definition of $R(\tau)$ and $Q(\tau)$, it is observed that, $R(\tau)+Q(\tau)=1$, then, $Q(\tau)=1-R(\tau)$. Furthermore, assuming the failure rate λ as a constant with respect to time (τ), we have $R(\tau)=e^{-\lambda\tau}$. And finally, $Q(\tau)$ can be obtained as $Q(\tau)=1-R(\tau)=1-e^{-\lambda\tau}$. To analyze our study under the light of real-life situations, we take the parameter λ as the ratio of a function of decision variable x_{ij}, and a_i, i.e., $\lambda=\dfrac{x_{ij}}{a_i}$. It is clear that, if the value of λ increases, then the value of reliability $R(\tau)$ decreases, which means that if the amount of transported goods becomes larger, then the amount of items may be defective or damaged in a bigger rate. Again, $R(\tau)=e^{-\lambda\tau}$ depends on time. If the transportation is made in time, then $\tau=0$, so the reliability value is maximized, i.e., $R(\tau)=1$. Furthermore, it is obvious that, in the solution of TP, some variables may take value as zero. It means no item is transported in the respective routes. So the reliability is again equal to 1 for this path in the proposed model, which does not create any complexity to take the decision for the DM.

10.3 Model Formulation

The classical TP, a way of scheduling a homogenous product, is to be transported from several origins (or sources) to numerous destinations in such a way that the total transportation cost is minimized. Suppose there are m origins ($i=1,2,\ldots,m$) and n destinations ($j=1,2,\ldots,n$). The sources may be production facilities, warehouses etc., and these are characterized by available supplies $a_1,a_2,\ldots,a_m$. The destinations may be warehouses, sales outlets etc., and these are also characterized by demand levels $b_1,b_2,\ldots,b_n$. The transportation cost C_{ij} is associated with transporting a unit of product from the origin i to the destination j. A variable x_{ij} is used to represent the unknown quantity to be transported from the origin O_i to the destination D_j. The mathematical model of a classical TP is as follows:

- **Model 1**

$$\text{minimize/maximize } Z = \sum_{i=1}^{m}\sum_{j=1}^{n}C_{ij}x_{ij},$$

$$\text{subject to } \sum_{j=1}^{n}x_{ij}\le a_i, \quad i=1,2,\ldots,m$$

$$\sum_{i=1}^{m}x_{ij}\ge b_j, \quad j=1,2,\ldots,n$$

$$x_{ij}\ge 0 \; \forall \; i \text{ and } j.$$

The feasibility condition is considered as $\sum_{i=1}^{m} a_i \geq \sum_{j=1}^{n} b_j$. According to the challenging competitive market scenario, several objective functions are related to a TP like minimizing the transportation cost, maximizing the transportation profit, minimizing the toll tax, etc. are required. Again, there are no connections between the cost parameters in the different objective functions of TP, so, they are considered as conflicting and commensurable to each other. In a multi-objective environment of TP, the MOTP can be defined as follows:

- **Model 2**

$$\text{minimize/maximize } Z^t = \sum_{i=1}^{m} \sum_{j=1}^{n} C_{ij}^t x_{ij}, t = 1, 2, \ldots, r$$

$$\text{subject to } \sum_{j=1}^{n} x_{ij} \leq a_i, i = 1, 2, \ldots, m \qquad (10.1)$$

$$\sum_{i=1}^{m} x_{ij} \geq b_j, j = 1, 2, \ldots, n \qquad (10.2)$$

$$x_{ij} \geq 0, \forall \ i \text{ and } j. \qquad (10.3)$$

Here C_{ij}^t, a_i, b_j are the cost, supply, and demand parameters of tth objective function in MOTP, respectively, and $\sum_{i=1}^{m} a_i \geq \sum_{j=1}^{n} b_j$ is the feasibility condition.

Here, we introduce the reliability by considering due or early transportation time of delivering the goods. Due to the late or early reach of the transporting goods, the customer or the storekeeper fails to manage it. So, DM should consider to his mind the matter, and as a whole, the optimum value of the objective functions is affected. Considering this situation, we introduce time in the cost parameter, which reduces the cost parameter of MOTP to the cost parameter with reliability. When the cost reliability is considered for all objective functions, then the time is taken as independent to each other, and the conflicting nature of the objective functions are preserved in the MOTP.

- **Model 3**

$$\text{minimize/maximize } Z^t = \sum_{i=1}^{m} \sum_{j=1}^{n} R\left(C_{ij}^t\right) x_{ij}, t = 1, 2, \ldots, r$$

subject to the constraints (10.1) – (10.3).

In Model 3, $R\left(C_{ij}^t\right)$ is considered as cost parameter under reliability. The delay of supply of items causes damage to the items, and in this case, the value of profit function (maximization type) decreases, and the penalty cost due to loss of time is considered when the objective function is of minimization type (like transportation cost) which increases the value of the respective objective function. Then, $R\left(C_{ij}^t\right)$ takes the following form of cost parameter as

$$R\left(C_{ij}^t\right) = C_{ij}^t + C_{ij}^t\left(1 - R\left(C_{ij}^t\right)\right)$$

for the objective function is of minimization type and

$$R\left(C_{ij}^t\right) = C_{ij}^t - C_{ij}^t\left(1 - R\left(C_{ij}^t\right)\right)$$

for the objective function is of maximization type. Here, the reliability function of the tth objective function for the (i, j)-th node depends on fixed time (τ), decision variable (x_{ij}), and demand (a_j). For consistency of reliability in each node, the DM measures the time (τ) in a unit scale; otherwise, there may occur large deviations in the cost values and produce an optimal solution, which is not significantly a good result.

10.4 Solution Procedure

In order to solve MOTP, we use GP, revised MCGP, and utility function, and they are defined in the following way:

- **Goal programming:**
  Let d_t^+ and d_t^- be positive and negative deviations corresponding to the tth goal of the objective function. Then the mathematical model using GP is depicted as follows:
  - **Model 1A**

$$\text{minimize} \sum_{t=1}^{r} w_t\left(d_t^+ + d_t^-\right)$$

subject to $Z^t(X) - d_t^+ + d_t^- = y_t \ (t = 1, 2, \ldots, r),$

$g_{t,\min} \le y_t \le g_{t,\max} \ (t = 1, 2, \ldots, r),$

$d_t^+, d_t^- \ge 0 \ (t = 1, 2, \ldots, r),$

the constraints $(10.1) - (10.3).$

- **Revised MCGP:**
  The mathematical model of RMCGP to solve MOTP is presented in the following model (see **Model 1B**):
  - **Model 1B**

$$\text{minimize} \sum_{t=1}^{r} \left[w_t \left(d_t^+ + d_t^- \right) + \alpha_t \left(e_t^+ + e_t^- \right) \right]$$

$$\text{subject to } Z^t(X) - d_t^+ + d_t^- = y_t \ (t = 1, 2, \ldots, r),$$

$$y_t - e_t^+ + e_t^- = g_{t,\max} \text{ or } g_{t,\min} \ (t = 1, 2, \ldots, r),$$

$$g_{t,\min} \le y_t \le g_{t,\max} \ (t = 1, 2, \ldots, r),$$

$$d_t^+, d_t^-, e_t^+, e_t^- \ge 0 \ (t = 1, 2, \ldots, r),$$

$$\text{the constraints (10.1)} - (10.3).$$

where t^{th} aspiration level is defined as y_t which lies between upper ($g_{t,\max}$) and lower ($g_{t,\min}$) bounds. Again e_t^+ and e_t^- are positive and negative deviations corresponding to t^{th} goal of $|y_t - g_{t,\max}|$, and α_t is the weight connected with the sum of deviations $|y_t - g_{t,\max}|$.

- **Utility function approach:**
  Here, the concept of utility function is addressed to solve MOTP. At first a short introduction of utility function is presented here, and then we discuss the methodology for solving MOTP using utility function.

  In this chapter, the introduction of utility is taken to be correlative to "Desire" or "Want." It has been already argued that desire cannot be measured directly, but only indirectly, by the outward phenomena in which the context is presented.

Definition 10.2

The utility function describes a function $U : X \to \mathbb{R}$ which assigns a real number to every outcome in such a way that it captures the DM's preferences over the desired goals of objectives, where X is the set of feasible points and $\mathbb{R}$ is the set of real numbers.

The purpose of this study is to derive the achievement function of MOTP under the light of utility function for DM according to the priority of goals. In our proposed approach, the DM wants to maximize his/her expected utility. For the sake of simplicity, two popular utility functions (linear and S-shaped) are considered as follows:

Linear utility function $u_i(y_i)$ for decision-making (management) problems can be found in Lai and Hwang [36], and S-shaped utility function (for the same purpose) has been proposed by Chang [31]. The utility function is generally considered in three cases as follows:

- **Case 1: Left Linear Utility Function (LLUF)**

$$u_i(y_i) = \begin{cases} 1, & \text{if } y_i \leq g_{i,\min} \\ \dfrac{g_{i,\max} - y_i}{g_{i,\max} - g_{i,\min}}, & \text{if } g_{i,\min} \leq y_i \leq g_{i,\max}, \quad i = 1,2,\dots,r \\ 0, & \text{if } y_i \geq g_{i,\max} \end{cases}$$

- **Case 2: Right Linear Utility Function (RLUF)**

$$u_i(y_i) = \begin{cases} 1, & \text{if } y_i \geq g_{i,\max} \\ \dfrac{y_i - g_{i,\min}}{g_{i,\max} - g_{i,\min}}, & \text{if } g_{i,\min} \leq y_i \leq g_{i,\max}, \quad i = 1,2,\dots,r \\ 0, & \text{if } y_i \leq g_{i,\min} \end{cases}$$

- **Case 3: S-shaped Utility Function**

$$u_i(y_i) = \begin{cases} 0, & \text{if } y_i \leq g_{i2} \\ \dfrac{y_i - g_{i2}}{g_{i8} - g_{i2}}, & \text{if } g_{i2} \leq y_i \leq g_{i4} \\ \dfrac{y_i - g_{i3}}{g_{i6} - g_{i3}}, & \text{if } g_{i4} \leq y_i \leq g_{i5} \\ \dfrac{y_i - g_{i1}}{g_{i7} - g_{i1}}, & \text{if } g_{i5} \leq y_i \leq g_{i7} \end{cases} \quad , i = 1,2,\dots,r$$

where $g_{i,\min}$ and $g_{i,\max}$ are lower and upper bounds corresponding to the ith goal, respectively. The graphs of the earlier utility functions are drawn in Figures 10.1–10.3: Using the utility function on the GP model, we can formulate the following mathematical models for different cases like LLUF, RLUF, and S-shaped utility function.

- **Model formulation for LLUF**
  The DM would like to increase the utility value $u_t(y_t)$ as much as possible in the case of LLUF (Figure 10.1). In order to achieve this goal, the value of y_t should be as close to the target value $g_{t,\min}$ as possible. The MOTP from **Model 1A** can be reformulated using the proposed LLUF as follows:

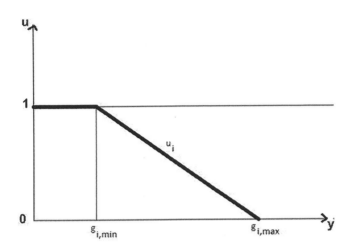

FIGURE 10.1
LLUF.

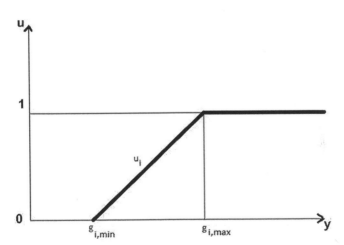

FIGURE 10.2
RLUF.

- **Model 2A**

$$\text{minimize} \sum_{t=1}^{r} \left[w_t \left(d_t^+ + d_t^- \right) + \beta_t f_t^- \right]$$

$$\text{subject to } Z^t(X) - d_t^+ + d_t^- = y_t, \ t = 1, 2, \ldots, r \tag{10.4}$$

$$g_{t,\min} \leq y_t \leq g_{t,\max}, \ t = 1, 2, \ldots, r \tag{10.5}$$

$$u_t \leq \frac{g_{t,\max} - y_t}{g_{t,\max} - g_{t,\min}}, \ t = 1, 2, \ldots, r \tag{10.6}$$

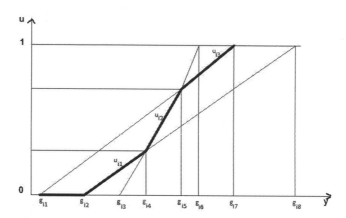

FIGURE 10.3
S-shaped utility function.

$$u_t + f_t^- = 1, t = 1, 2, \ldots, r \tag{10.7}$$

$$u_t, f_t^- \geq 0, t = 1, 2, \ldots, r \tag{10.8}$$

the constraints $(10.1) - (10.3)$.

where β_t is the weight attached to deviation f_t^-. The role of weight β_t can be seen as the preferential component for the utility value u_t.

Let us discuss how the mathematical model (see **Model 2A**) produces optimal solution connecting with the utility value in the following proposition.

Proposition 1

The optimal solution of **Model 2A** is equivalent with the achievement of optimal utility in LLUF (Figure 10.1).

Proof: The deviation $f_t^- \to 0$ when u_t approaches to the highest value 1. Again, f_t^- should be minimized in the objective function to obtain the optimal solution of **Model 2A**. This forces y_t tend to $g_{t,\min}$ [from Eq. 10.6]. Then $Z_t(X)$ is closer to $g_{t,\min}$ [from Eq. 10.4] because d_t^+ and d_t^- should also be minimized in the objective function. It is clear that the behavior of **Model 2A** and the level of utility were achieved. This completes the proof.

- **Model formulation for RLUF**
  The DM would like to increase the utility value $u_t(y_t)$ as much as possible in the case of RLUF (Figure 10.2). In order to achieve this goal, the value of y_t should be as close to the target value $g_{t,\max}$ as possible. The MOTP from **Model 1A** can be reformulated using the proposed RLUF as follows:

- **Model 2B**

$$\text{minimize} \sum_{t=1}^{r} \left[w_t \left(d_t^+ + d_t^- \right) + \beta_t f_t^- \right]$$

$$\text{subject to } Z^t(X) - d_t^+ + d_t^- = y_t, \ t = 1, 2, \ldots, r \qquad (10.9)$$

$$g_{t,\min} \le y_t \le g_{t,\max}, \ t = 1, 2, \ldots, r \qquad (10.10)$$

$$u_t \le \frac{y_t - g_{t,\min}}{g_{t,\max} - g_{t,\min}} \qquad (10.11)$$

$$u_t + f_t^- = 1 \qquad (10.12)$$

$$u_t, f_t^- \ge 0 \qquad (10.13)$$

the constraints (10.1) – (10.3).

where β_t is weight attached to the deviation f_t^-. The role of weight β_t can be seen as a preferential component for the utility value u_t.

Here, we present the proposition that confirms that the mathematical model (see **Model 2B**) produces optimal solution.

Proposition 2

The optimal solution of **Model 2B** is equivalent with the achievement of optimal utility in the RLUF (Figure 10.2).

Proof: The proof is similar to the proof of **Proposition 1**.

The advantages using LLUF and RLUF in the decision-making problems are as follows:

1. DM can easily formulate the MOTP by taking into account their preference mappings with utility functions in real-life situation and
2. The two linear utility models represented as linear form can be easily solved using software.

Due to variation of deviation variables d_t^+, d_t^-, and f^- in different ranges, biasness may be occurred towards the objective functions with larger magnitude. Normalization technique may help to remove the biasness. Several normalization approaches such as Percentage, Euclidean, Summation, and Zero-one notarizations (Tamiz et al. [24], Kettani et al. [37]) are available to execute this. According to the normalization technique proposed by Tamiz et al. [24], **Model 2A** can be redesigned as follows:

- **Model 2AN**

$$\text{minimize} \sum_{t=1}^{r} \left[\frac{w_t \left(d_t^+ + d_t^- \right) + \beta_t f_t^-}{\phi_t} \right]$$

subject to the constraints (11.10) − (11.14) and

the constraints (11.2) − (11.4).

where ϕ_t is the normalization constant for t^{th} goal. In order to solve this problem, utility normalization concept is introduced as follows: Let $d_t^+, d_t^- \in [0, \overline{u}_t]$ and $f_t^- \in [0,1]$, where $\overline{u}_t$ is the upper bound of d_t^+ and d_t^-. The normalized weights w_t and β_t can be easily obtained as $w_t = \dfrac{1}{1 + \overline{u}_t}$ and $\beta_t = \dfrac{\overline{u}_t}{1 + \overline{u}_t}$. This technique of normalization ensures that deviation variables d_t^+, d_t^-, and f_t^- approximated the same magnitude.

Similarly, the same methodology can be applied to **Model 2B**.

- **Model formulation for S-shaped utility function**
  Utility value for S-shaped utility function can be expressed as a sum of linear utility functions (RLUF or LLUF) by introducing binary variables [38]. But Chang [31] proposed in his paper that the utility value for S-shaped utility function can be considered without using the binary variables, and this is shown in the following model (see **Model 2C**).

  - **Model 2C**

$$\text{minimize} \sum_{t=1}^{r} w_t \left[\left[p_{t1} + p_{t2} + p_{t3} \right] + \beta_t f_t^- \right]$$

subject to $Z^t(X) - d_t^+ + d_t^- = y_t, \quad t = 1, 2, \ldots, r$ (10.14)

$$g_{t,\min} \leq y_t \leq g_{t,\max}, \quad t = 1, 2, \ldots, r \tag{10.15}$$

$$\left(u_t \left(g_{t4} \right) - u_t \left(g_{t2} \right) \right) \left(\frac{p_{t1} - p_{t2}}{g_{t4} - g_{t2}} \right) + \left(u_t \left(g_{t5} \right) - u_t \left(g_{t4} \right) \right) \left(\frac{p_{t2} - p_{t3}}{g_{t5} - g_{t4}} \right)$$

$$+ \left(u_t \left(g_{t7} \right) - u_t \left(g_{t5} \right) \right) \left(\frac{p_{t3}}{g_{t7} - g_{t5}} \right), \quad t = 1, 2, \ldots, r \tag{10.16}$$

$$y_t - p_{t1} + n_{t1} = g_{t2}, \quad t = 1, 2, \ldots, r \tag{10.17}$$

$$y_t - p_{t2} + n_{t2} = g_{t4}, \ t = 1, 2, \ldots, r \tag{10.18}$$

$$y_t - p_{t3} + n_{t3} = g_{t5}, \ t = 1, 2, \ldots, r \tag{10.19}$$

$$u_t + f_t^- = 1, \ t = 1, 2, \ldots, r \tag{10.20}$$

$$u_t, p_{tl}, n_{tl} \geq 0, \ t = 1, 2, \ldots, r, l = 1, 2, 3 \tag{10.21}$$

the constraints (10.1) – (10.3).

Using the formulated models such as **Models 2A, 2B,** and **2C,** we may formulate that the MOTP occurred in real-life situations with interval goals under the consideration of utility functions related to these goals. Solving the formulated problem, the DM obtains the satisfactory solution.

10.5 Numerical Example

To test the efficiency of our proposed study, we consider a fish TP, which mainly refers to the MOTP with interval goals to the objective functions. The MOTP is designed based on the concept of reliability to the cost parameter of the TP. DM plans to distribute the fishes from three fish stores F_1, F_2, and F_3 to four markets that are situated in the cities, C_1, C_2, C_3, and C_4. During planning, we wished to optimize the following objective functions as

- minimize the transportation cost (Z^1),
- maximize the profit (Z^2),
- minimize the toll tax (Z^3).

DM wishes to choose the optimal goals for the objective functions, where Z^1, Z^2, and Z^3 are interval goals. The goals corresponding to the objectives lie within the intervals [17,500, 21,000]; [2,300, 2,600]; [420, 700], respectively. The demands in the cities C_1, C_2, C_3, and C_4 are 9, 8, 10, and 7 quintals, respectively. The supply from fish stores F_1, F_2, and F_3 are 18, 15, and 20 quintals, respectively.

- The cost C_{ij}^1 for transporting 1 quintal of fishes from the resource i to the destination j and the approximate loss of time of delivery from the sources to the destinations are also known to the DM, for $i = 1, 2, 3$ and $j = 1, 2, 3, 4$, respectively.

- The profit C_{ij}^2 for transporting 1 quintal of fishes from the resource i to the destination j and DM makes a prediction of approximate loss of time of delivery from the sources to the destinations, for $i = 1, 2, 3$ and $j = 1, 2, 3, 4$, respectively.
- The toll tax cost C_{ij}^3 for transporting 1 quintal of fishes from the resource i to the destination j, and it is fixed value for $i = 1, 2, 3$ and $j = 1, 2, 3, 4$, respectively.

The data for the transportation costs C_{ij}^t for $t = 1, 2, 3$ are represented in Tables 10.1–10.3. Assume that each goal is equally important to the company. Goal of the second objective function in the given problem keeps the demand high of the fish TP in the city. Obviously, company tries to earn more money and he wants more profit during the transportation, so Goal 2 tends to reach a maximum value. When Goal 2 tended to its maximum value of the specified problem, sometimes customers face problems for not getting their goods in the scheduled time. Then to keep the good reputation in the market, company makes the profit in such a way that the Goal 2 tends to a value surrounding 2,400, and the priority of goal value follows an S-shaped utility

TABLE 10.1

Transportation Cost $\left(C_{ij}^1\right)$ (in Rupees) and Loss of Time (in Weeks)

	C_1	C_2	C_3	C_4
F_1	(500,0.3)	(600,0.2)	(620,0.2)	(550,0)
F_2	(600,0.2)	(500,0.1)	(550,0.1)	(570,0.1)
F_3	(650,0)	(600,0.3)	(580,0.2)	(620,0)

TABLE 10.2

Cost Parameters $\left(C_{ij}^2\right)$ Related to Profit (in Rupees) and Loss of Time (in Weeks)

	C_1	C_2	C_3	C_4
F_1	(70,0.15)	(80,0.1)	(78,1)	(75,1.5)
F_2	(80,0.2)	(72,0.2)	(84,0.23)	(82,0.1)
F_3	(90,0.1)	(80,0.3)	(76,0.2)	(73,0)

TABLE 10.3

Toll Tax Cost $\left(C_{ij}^3\right)$ (in Rupees) for Transportation of Goods

	C_1	C_2	C_3	C_4
F_1	10	15	25	12
F_1	18	17	15	14
F_1	20	16	22	25

function (see Figure 10.4). Using the data provided in Tables 10.1–10.3, we formulate the following MOTP model:

- **Model 3**

$$\min Z^1 = \sum_{i=1}^{3}\sum_{j=1}^{4} R\left(C_{ij}^1\right)x_{ij}, \ Z^1 \text{ has the interval goal } [17,500, \ 21,000]$$

Z^1 follows LLUF

$$\max z^2 = \sum_{i=1}^{3}\sum_{j=1}^{4} R\left(C_{ij}^2\right)x_{ij}, \ Z^2 \text{ has the interval goal } [2,300, \ 2,600]$$

Z^2 follows RLUF

$$\min z^3 = \sum_{i=1}^{3}\sum_{j=1}^{4} C_{ij}^3 x_{ij}, \ Z^3 \text{ has the interval goal } [420, 700]$$

Z^3 follows S-shaped utility function

$$\text{subject to } x_{11} + x_{12} + x_{13} + x_{14} \leq 18 \qquad (10.22)$$

$$x_{21} + x_{22} + x_{23} + x_{24} \leq 15 \qquad (10.23)$$

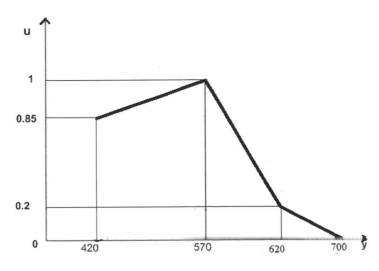

FIGURE 10.4
S-shaped utility function for Goal 3 of numerical example.

$$x_{31} + x_{32} + x_{33} + x_{34} \leq 20 \tag{10.24}$$

$$x_{11} + x_{21} + x_{31} \geq 9 \tag{10.25}$$

$$x_{12} + x_{22} + x_{32} \geq 8 \tag{10.26}$$

$$x_{13} + x_{23} + x_{33} \geq 10 \tag{10.27}$$

$$x_{14} + x_{24} + x_{34} \geq 7 \tag{10.28}$$

$$x_{ij} \geq 0, \quad i,j = 1,2,3. \tag{10.29}$$

The Model 3 can be reduced in the following form as

- **Model 4**

$$\min Z^1 = \sum_{i=1}^{3}\sum_{j=1}^{4}\left(C_{ij}^1 + C_{ij}^1\left(1 - R_{ij}^1\right)\right)x_{ij}, Z^1 \text{ has the interval goal } [17,500, \ 21,000]$$

$$\max Z^2 = \sum_{i=1}^{3}\sum_{j=1}^{4}\left(C_{ij}^2 - C_{ij}^2\left(1 - R_{ij}^2\right)\right)x_{ij}, Z^2 \text{ has the interval goal } [2,300, \ 2,600]$$

$$\min Z^3 = \sum_{i=1}^{3}\sum_{j=1}^{4}C_{ij}^3 x_{ij}, Z^3 \text{ has the interval goal } [420, 700]$$

subject to the constraints $(10.22) - (10.29)$.

Here, R_{ij}^1 and R_{ij}^2 are the reliability of completing the job of transportation in time. These are taken as function of the decision variables and time deviation for completing the work. To achieve the goals in the proposed problem (see **Model 3**), we may formulate the following models:

In the proposed problem, the deviations of goals 1, 2, and 3 are 3,500, 300, and 280, respectively. By considering the weights $w_1 = \dfrac{1}{3,500}, w_2 = \dfrac{1}{300}$, and $w_3 = \dfrac{1}{280}$ for **Model 1A**, the earlier **Model 4** reduces to the model (see **Model 4A**) as

- **Model 4A**

$$\text{minimize } \frac{1}{3,500}\left(d_1^+ + d_1^-\right) + \frac{1}{300}\left(d_2^+ + d_2^-\right) + \frac{1}{280}\left(d_3^+ + d_3^-\right)$$

subject to $\left(1,000 - 500e^{-\frac{0.3}{18}}\right)x_{11} + \left(1,200 - 600e^{-\frac{0.2}{18}}\right)x_{12} + \left(1,240 - 620e^{-\frac{0.2}{18}}\right)x_{13}$

$+\left(1,100 - 550e^{-\frac{0}{18}}\right)x_{14} + \left(1,200 - 600e^{-\frac{0.2}{15}}\right)x_{21} + \left(1,000 - 500e^{-\frac{0.1}{15}}\right)x_{22}$

$+\left(1,100 - 550e^{-\frac{0.1}{15}}\right)x_{23} + \left(1,140 - 570e^{-\frac{0.1}{15}}\right)x_{24} + \left(1,300 - 650e^{-\frac{0}{20}}\right)x_{31}$

$+\left(1,200 - 600e^{-\frac{0.3}{20}}\right)x_{32} + \left(1,160 - 580e^{-\frac{0.2}{20}}\right)x_{33} + \left(1,240 - 620e^{-\frac{0}{20}}\right)x_{34}$

$$-d_1^+ + d_1^- = y_1 \qquad (10.30)$$

$$17,500 \leq y_1 \leq 21,000 \qquad (10.31)$$

$70e^{-\frac{0.15}{18}}x_{11} + 80e^{-\frac{0.1}{18}}x_{12} + 78e^{-\frac{1}{18}}x_{13} + 75e^{-\frac{1.5}{18}}x_{14} + 80e^{-\frac{0.2}{15}}x_{21}$

$+72e^{-\frac{0.2}{15}}x_{22} + 84e^{-\frac{0.23}{15}}x_{23} + 82e^{-\frac{0.1}{15}}x_{24} + 90e^{-\frac{0.1}{20}}x_{31} + 80e^{-\frac{0.3}{20}}x_{32}$

$+76e^{-\frac{0.2}{20}}x_{33} + 73e^{-\frac{0}{20}}x_{34} - d_2^+ + d_2^- = y_2 \qquad (10.32)$

$$2,300 \leq y_2 \leq 2,600 \qquad (10.33)$$

$10x_{11} + 15x_{12} + 25x_{13} + 12x_{14} + 18x_{21} + 17x_{22} + 15x_{23} + 14x_{24}$

$+ 20x_{31} + 16x_{32} + 22x_{33} + 25x_{34} - d_3^+ + d_3^- = y_3 \qquad (10.34)$

$$420 \leq y_3 \leq 700 \qquad (10.35)$$

$$d_t^+, d_t^- \geq 0, t = 1, 2, 3$$

the constraints (10.22) – (10.29).

Again, considering the same weights w_t as used in **Model 4A** for $t = 1, 2, 3$ and setting $\alpha_t = w_t$ for $t = 1, 2, 3$ for deviation of goals and using **Model 1B**, then **Model 4** reduces to the model (see **Model 4B**) as

- **Model 4B**

$$\text{minimize } \frac{1}{3,500}\left(d_1^+ + d_1^-\right) + \frac{1}{300}\left(d_2^+ + d_2^-\right) + \frac{1}{280}\left(d_3^+ + d_3^-\right)$$

$$+ \frac{1}{3,500}\left(e_1^+ + e_1^-\right) + \frac{1}{300}\left(e_2^+ + e_2^-\right) + \frac{1}{280}\left(e_3^+ + e_3^-\right)$$

$$\text{subject to } y_1 - e_1^+ + e_1^- = 17,500$$

$$y_2 - e_2^+ + e_2^- = 2,600$$

$$y_3 - e_3^+ + e_3^- = 700$$

$$d_t^+, d_t^-, e_t^+, e_t^- \geq 0, \ t = 1,2,3$$

the constraints $(10.22) - (10.35)$.

Using the concept of utility function described in Section 4, **Model 4** can be reformulated as follows:

The consideration of utility function depends on the DM. Here, we assume that Goals 1, 2, and 3 follow the utility function LLUF (Figure 10.1), RLUF (Figure 10.2), and S-shaped utility function as given in Figure 10.4, respectively. In the given example, the upper bound of variations $d_1^+, d_1^-, d_2^+, d_2^-, d_3^+, d_3^-$ are 3,500, 3,500, 300, 300, 280, and 280, respectively, and the upper bounds of $f_1^-, f_2^-,$ and f_3^- are 1. We find the weights as described in Section 10.4 as follows:

$$w_1 = \frac{1}{3,501}, w_2 = \frac{1}{301}, w_3 = \frac{1}{281}, \beta_1 = \frac{3,500}{3,501}, \beta_2 = \frac{300}{301}, \text{ and } \beta_3 = \frac{280}{281}.$$

With the supplied data, **Model 4** can be reformulated as follows (see **Model 4C**):

- **Model 4C**

$$\text{minimize } w_1\left(d_1^+ + d_1^-\right) + \beta_1 f_1^- + w_2\left(d_2^+ + d_2^-\right) + \beta_2 f_2^- + w_3\left(d_{31}^+ + d_{32}^+ + d_{33}^+\right) + \beta_3 f_3^-$$

$$\text{subject to } u_1 \leq \frac{21,000 - y_1}{3,500}$$

$$f_1^- + u_1 = 1$$

$$u_2 \leq \frac{y_2 - 2,300}{300}$$

$$f_2^- + u_2 = 1$$

$$u_3 = \frac{(1 - 0.85)\left(d_{31}^+ - d_{32}^+\right)}{150} + \frac{(0.2 - 1)\left(d_{32}^+ - d_{33}^+\right)}{50} + \frac{(0 - 0.2)d_{33}^+}{50}$$

$$y_3 - d_{31}^+ + dn_{31} = 570, y_3 - d_{32}^+ + dn_{32} = 620, y_3 - d_{33}^+ + dn_{33} = 700$$

$$d_{31}^+ dn_{31} = 0, d_{32}^+ dn_{32} = 0, d_{33}^+ dn_{33} = 0, \ f_3^- + u_3 = 1$$

$$u_t \geq 0, f_t^- \geq 0, t = 1,2,3$$

the constraints (11.28) – (11.41).

10.5.1 Results and Discussion

Using LINGO software, solve Models 4A, 4B, and 4C, and the optimal solution is obtained by different methods as mentioned in Table 10.4. From the solution, we observe that the optimal solution by RMCGP is better than the solution of GP. The optimal solution is derived by GP, optimize the profit as the best but unable to minimize the other costs accordingly as DM wishes. The RMCGP finds optimal values that tend to the value of objective function as the upper bound of goal for maximization problem and lower bound of goals for minimization problem. The optimal solution is calculated by utility function approach as per the utility values corresponding to the objective function. Sometimes, the maximum profits cause to worse relationship between the customer and the supplier. In that case, seller would like to think the future and fixed the best utility value corresponding to an objective function, which is not its original optimal value. In this regard, the solution by utility function is better than GP and RMCGP. Each objective function in the numerical example on MOTP has been considered with interval goals. Moreover, we incorporate the situation of cost reliability with the cost parameters due to the delay of delivery of goods before/after the scheduled time. Usually, the MOTP having the objective functions are of either maximization type or minimization type, but here, the aim of DM is not likely to the solution of traditional MOTP. Again, the proposed problem has been solved by GP, RMCGP, and utility function approach. From the obtained optimal solution, it is observed that RMCGP provides better solution than GP. Furthermore, utility function indicates the solution which is according to the respective utility of the objective preferred by DM.

TABLE 10.4

Optimal Solution of the Proposed Models by Different Techniques

Method	Optimal Solution	Optimal Value
GP	$x_{11} = 9, x_{14} = 7, x_{32} = 8, x_{33} = 10,$	$Z^1 = 19,153.55, Z^2 = 2,490.70.76,$ $Z^3 = 522.$
RMCGP	$x_{11} = 2, x_{21} = 7, x_{22} = 8, x_{33} = 10, x_{34} = 7,$ and other variables are zero.	$Z^1 = 17,500, Z^2 = 2,523.23,$ $Z^3 = 677.$
Utility approach	$x_{14} = 7, x_{21} = 3.43, x_{22} = 8, x_{23} = 3.56, x_{31} = 5.56,$ $x_{33} = 6.43,$ and other variables are zero.	$Z^1 = 19,325.34, Z^2 = 2,600,$ $Z^3 = 570.$

Basically, the work of S-shaped utility function for the third objective function establishes a solution for that objective function, which is impossible to find by GP and RMCGP. In this aspect, the utility approach is a better technique to solve MOTP when compared with GP and RMCGP. On the other hand, time is very much important for transporting the goods to real-life TPs, so the decision making under time consideration and cost reliability provides a better way of selecting the goals for objective functions.

10.6 Conclusion and Future Study

In this chapter, we have analyzed the real-life MOTP using the concept of cost reliability. Realizing the importance of time of transportation in TP, we have incorporated time in the transportation parameter and defined the concept of cost reliability in MOTP. We have presented GP, revised MCGP, and utility function approach for solving MOTP with cost reliability. A real-life example has been encountered to establish the usefulness of the proposed study.

In future investigation, the proposed study can be implemented as a decision-making tool for multimodal transportation planning, multicriterion decision-making problems to cover all real-life problems occurring in the field of economical, agricultural, industrial management, military, etc. In the future, the reliability and utility function approach can be used for the fix-up cost of goods and can be applicable in supply chain management to get better results in real-life MODM problems.

References

1. Hitchcock, F.L. (1941). The distribution of a product from several sources to numerous localities. *Journal of Mathematical Physics*, 20, 224–230.
2. Koopmans, T.C. (1949). Optimum utilization of the transportation System. *Econo-metrica*, 17, 136–146.
3. Kantorovich, L.V. (1960). Mathematical methods of organizing and planning production. *Management Science*, 6(4), 366–422.
4. Hamzehee, A., Yaghoobi, M.A., Mashinchi, M. (2014). Linear programming with rough interval coefficients. *Journal of Intelligent & Fuzzy Systems*, 26(3), 1179–1189.
5. James, J.H.L., Hsu, C.C., Chen, Y.S. (2014). Improving transportation service quality based on information fusion. *Transportation Research Part A: Policy and Practice*, 67, 225–239.

6. Kumar, V.P., Bierlaire, M. (2014). Multi-objective airport gate assignment problem in planning and operations. *Journal of Advanced Transportation*, 48(7), 902–926.
7. Luathep, P., Sumalee, A., Lam, W.H.K., Li, Z.C., Lo, H.K. (2011). Global optimization method for mixed transportation network design problem: A mixed-integer linear programming approach. *Transportation Research Part B: Methodological*, 45(5), 808–827.
8. Charnes, A., Cooper, W.W., Ferguson, R. (1955). Optimal estimation of executive compensation by linear programming. *Management Science*, 1, 138–151.
9. Yu, V.F., Hu, K.J., Chang, A.Y. (2015). An interactive approach for the multi-objective transportation problem with interval parameters. *International Journal of Production Research*, 53(4), 1051–1064.
10. Maity, G., Roy, S.K. (2016). Solving multi-objective transportation problem with nonlinear cost and multi-choice demand. *International Journal of Management Science and Engineering Management*, 11(1), 62–70.
11. Maity, G., Roy, S.K. (2014). Solving multi-choice multi-objective transportation problem: A utility function approach. *Journal of Uncertainty Analysis and Applications*. doi:10.1186/2195-5468-2-11.
12. Maity, G., Roy, S.K. (2016). Solving multi-objective transportation problem with interval goal using utility function approach. *International Journal of Operational Research*, 27(4), 513–529.
13. Mahapatra, D.R., Roy, S.K. (2011). Multi-objective interval-valued transportation probabilistic problem involving log-normal. *International Journal of Mathematics and Scientific Computing*, 1(2), 14–21.
14. Roy, S.K. (2014). Multi-choice stochastic transportation problem involving Weibul distribution. *International Journal of Operational Research*, 21(1), 38–58.
15. Roy, S.K., Midya, S., Yu, V.F. (2018). Multi-objective fixed charge transportation problem with random rough variables. *International Journal of uncertainty, Fuzziness and Knowledge-Based-Syatems*, 26(6), 971–996.
16. Midya, S., Roy, S.K. (2017). Analysis of interval programming in different environments and its application to fixed-charge transportation problem. *Discrete Mathematics, Algorithms and Applications*, 9(03), 1750040.
17. Roy, S.K., Maity, G. (2017). Minimizing cost and time through single objective function in multi-choice interval valued transportation problem. *Journal of Intelligent & Fuzzy Systems*, 32(3), 1697–1709.
18. Roy, S.K., Ebrahimnejad, A., Verdegay J.L., Das, S. (2018). New approach for solving intuitionistic fuzzy multi-objective transportation problem. *Sadhana*, 43(1), 3. doi:10.1007/s12046-017-0777-7.
19. Maity, G., Roy, S.K., Verdegay, J.L. (2016). Multi-objective transportation problem with cost reliability under uncertain environment, *International Journal of Computational Intelligence Systems*. 9(5), 839–849.
20. Lee, S.M. (1972). *Goal Programming for Decision Analysis*. Auerbach Publishers, Philadelphia, PA: 252–260.
21. Zeleny, M. (1982). The pros and cons of goal programming. *Computers and Operations Research*, 8, 357–359.
22. Ignizio, J.P. (1985). An algorithm for solving the linear goal programming problem by solving its dual. *Journal of the Operational Research Society*, 36(6), 507–515.

23. Tamiz, M., Jones, D.F. (1997). Interactive frameworks for investigation of goal programming models: Theory and practice. *Journal of Multi-Criteria Decision Analysis*, 6(1), 52–60.
24. Tamiz, M., Jones, D., Romero, C. (1998). Goal programming for decision making: An overview of the current state-of-the-art. *European Journal of Operational Research*, 111(3), 567–581.
25. Romero, C. (2004). A general structure of achievement function for a goal programming model. *European Journal of Operational Research*, 153(3), 675–686.
26. Maity, G., Mardanya, D., Roy, S.K. (2019). A new approach for solving dual hesitant fuzzy transportation problem with restrictions. *Sadhana*, 44, 75. doi:10.1007/s12046-018-1045-1.
27. Roy, S.K., Maity, G., Weber, G.W., Gok, S.Z.A. (2017). Conic scalarization approach to solve multi-choice multi-objective transportation problem with interval goal. *Annals of Operations Research*, 253(1), 599–620.
28. Chang, C.T. (2007). Multi-choice goal programming. *Omega*, 35(4), 389–396.
29. Chang, C.T. (2008). Revised multi-choice goal programming. *Applied Mathematical Modelling*, 32(12), 2587–2595.
30. Mahapatra, D.R., Roy, S.K., Biswal, M.P., (2013). Multi-choice stochastic transportation problem involving extreme value distribution. *Applied Mathematical Modelling*, 37(4), 2230–2240.
31. Chang, C.-T. (2010). An approximation approach for representing S-shaped membership functions. *IEEE Transactions on Fuzzy Systems*, 18(2), 412–424.
32. Al-Nowaihi, A., Bradley, I., Dhami, S. (2008). A note on the utility function under prospect theory. *Economics letters*, 99(2), 337–339.
33. Yu, B.W.T., Pang, W.K., Troutt, M.D., Hou, S.H. (2009). Objective comparisons of the optimal portfolios corresponding to different utility functions. *European Journal of Operational Research*, 199(2), 604–610.
34. Podinovski, V.V. (2010). Set choice problems with incomplete information about the preferences of the decision maker. *European Journal of Operational Research*, 207(1), 371–379.
35. Richard, E.B., Frank, P., Larry, C.H. (1967). *Mathematical Theory of Reliability*. John Wiley and Sons, Inc., New York-London-Sydney, Third edition.
36. Lai, Y.-J., Hwang, C.-L. (1994). *Fuzzy Multiple Objective Decision Making: Methods and Applications*. Springer-Verlag, Berlin.
37. Kettani, O., Aouni, B., Martel, J.-M. (2004). The double role of the weight factor in the goal programming model. *Computers and Operations Research*, 31(4), 1833–1845.
38. Chang, C.T. (2006). Mixed binary interval goal programming. *Journal of the Operational Research Society*, 57(4), 469–473.

11

Effect of Environmental Pollutants on Rain due to Stakeholders

Nita H. Shah, Moksha H. Satia, and Foram A. Thakkar

Gujarat University

CONTENTS

11.1 Introduction

"Population" is the word used for plants, animals, or other organisms making up a group that stays together and procreates. These populations make the Earth very unique. The impact of population is a complex key issue for environmental problem on Earth. Discrete populations like animals, plants, humans have different impacts on environment. Animals and plants are organisms that are not as active as human beings, but that also contaminate the environment mostly after completing their lifespan. Moreover, the human population plays a leading role during their lifespans to vitiate the environment in several ways by constructing industries, driving vehicles, depleting natural resources, and other damaging actions and their consequences. Each population contributes environmental pollutants by decreasing water quality, causing pollution, increasing greenhouse gas emissions, creating global climate change, etc. There is a direct relation between environment pollution and rain intensity. If environment is polluted, then there are more chances to have polluted rain (Traistaa et al., 2003). This may be because rain contains some amount of atmospheric water droplets, and some of the water droplets dissolve with the gaseous pollutants. It is also observed by some scientists that an increase in environmental pollution causes a decrease in rainfall. Some researchers have studied this problem but cannot give satisfactory proof of what drives other researchers to think in this direction.

Khemani and Murty measured rainfall variation of an urban industrial region in 1973. In 2008a,b, Shukla et al. studied the effect of rain on the removal of a gaseous pollutant and two different particulate matters from the atmosphere of a city. An extended model was plotted with the effect of cloud density, later in the same year. The impact of rain on population exposed to air pollution modeling was set in 2017 by Sharma and Kumari. Study on environmental pollution was done by some researchers. In 1992, Pandey et al. introduced air pollutant concentrations in Varanasi, India. In 2000, the interaction of two biological species in a polluted environment through modeling was proposed by Dubey and Hussain. Naresh has prepared a qualitative analysis for removal of air pollutants using a nonlinear model in 2003. Shukla et al. declared a model that consists of effects for primary and secondary toxicants on renewable resources in 2003. In 2010, a model was developed by Dubey for the effect of pollutant on human population dependent on a resource with environmental and health policy. Shah et al. proposed a model for optimum control for spread of pollutants through forest resources in 2017. Also, they approached household waste causing environmental pollutants due to landfill and treatments mathematically in 2018. To revive environment, some models based on environment are developed using mathematical concepts by some researchers. In 2017, optimal control on the depletion of green belt due to industries was deliberated by Shah et al.; Kademi et al. expressed the dynamical model of toxicity related with aflatoxins in foods and feeds in 2017. In 2018, Shah et al. calculated stability of "GO-CLEAN" model through graphs for betterment of environment.

In this chapter, to study the effect of environmental pollutants and rain intensity, mathematical models are formulated for different populations in Section 11.2. Section 11.3 comprises local and global stability of each equilibrium points. Sensitivity of each model parameter is illustrated in Section 11.4. In Section 11.5, numerical simulation is displayed to compare the effect of stakeholders on environmental pollutants and rain intensity.

11.2 Formulation of the Model

We are studying the effect of environmental pollutants (E_P) on rain intensity (R) due to animal population (A), plant population (P), and human

TABLE 11.1

Notation, Description, and Parametric Values

Notation	Description	Parametric Value
B_A	The growth rate of the respective population	0.1
B_P		0.3
B_H		0.5
β_A	The rate of environmental pollutants caused by the dead population	0.25
β_P		0.1
β_H	The rate of environmental pollutants caused by human activities	0.4
δ_A	The rate of environmental pollutants absorbed by stakeholders	0.15
δ_P		0.3
δ_H		0.1
ε_1	The rate at which environmental pollutants decrease rain intensity due to stakeholders	0.05
γ_1		
η_1		
ε_2	The rate at which rain dissolves environmental pollutants caused due to stakeholders	0.3
γ_2		
η_2		
μ_A	The rate of unobserved activity through the respective population	0.4
μ_P		0.4
μ_H		0.5

population (H). The models are developed based on the following assumptions and parametric values given in the Table 11.1.

- **Assumptions:**
  1. The models deal with three different populations.
  2. The rate of environmental pollution created by human is maximum and that created by plant is minimum.
  3. Environmental pollutants are absorbed by all stakeholders. This may be in terms of inhaling carbon dioxide or poisonous gases.
  4. Rain dissolves gaseous pollutants.
  5. Rain guarantees environmental pollutants free equilibrium point.
  6. Unobserved rate from each compartment is taken as maximum in human population.

- **Animal population model**
  Figure 11.1 shows the transmission diagram of the environmental pollutants caused due to animal population.
  The following system of ordinary differential equations represents the model for animal population:

$$\frac{dA}{dt} = B_A - \beta_A A E_P + \delta_A A E_P - \mu_A A$$

$$\frac{dE_P}{dt} = \beta_A A E_P - \delta_A A E_P - \varepsilon_1 E_P + \varepsilon_2 R - \mu_A E_P \qquad (11.1)$$

$$\frac{dR}{dt} = \varepsilon_1 E_P - \varepsilon_2 R - \mu_A R$$

Adding all differential equations of system (11.1), we get

$$\frac{d}{dt}(A + E_P + R) = B_A - \mu_A(A + E_P + R) \geq 0$$

Therefore, the positive invariant set for system (11.1) is

$$\Lambda_A = \left\{ (A, E_P, R) : A + E_P + R \leq \frac{B_A}{\mu_A} ; A > 0, E_P \geq 0, R \geq 0 \right\}$$

- **Plant population model**
  The transmission diagram of the environmental pollutants caused by plant population is shown in Figure 11.2.
  The system of ordinary differential equations that represents the model for the plant population is

$$\frac{dP}{dt} = B_P - \beta_P P E_P + \delta_P P E_P - \mu_P P$$

$$\frac{dE_P}{dt} = \beta_P P E_P - \delta_P P E_P - \gamma_1 E_P + \gamma_2 R - \mu_P E_P \qquad (11.2)$$

$$\frac{dR}{dt} = \gamma_1 E_P - \gamma_2 R - \mu_P R$$

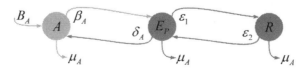

FIGURE 11.1
Transmission diagram of an animal model.

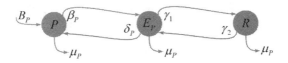

FIGURE 11.2
Transmission diagram of a plant model.

From differential equations of system (11.2), we get

$$\frac{d}{dt}(P + E_P + R) = B_P - \mu_P(P + E_P + R) \geq 0$$

Therefore, the positive invariant set for system (11.2) is

$$\Lambda_P = \left\{ (P, E_P, R) : P + E_P + R \leq \frac{B_P}{\mu_P}; P > 0, E_P \geq 0, R \geq 0 \right\}$$

- **Human population model**
  Figure 11.3 represents the transmission diagram of the environmental pollutants caused by human population.
  The following system of ordinary differential equations represents the model for human population:

$$\frac{dH}{dt} = B_H - \beta_H H E_P + \delta_H H E_P - \mu_H H$$

$$\frac{dE_P}{dt} = \beta_H H E_P - \delta_H H E_P - \eta_1 E_P + \eta_2 R - \mu_H E_P \qquad (11.3)$$

$$\frac{dR}{dt} = \eta_1 E_P - \eta_2 R - \mu_H R$$

Adding the aforementioned differential equations of system (11.3), we get

$$\frac{d}{dt}(H + E_P + R) = B_H - \mu_H(H + E_P + R) \geq 0$$

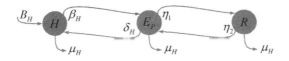

FIGURE 11.3
Transmission diagram of a human model.

Therefore, the positive invariant set for system (11.3) is

$$\Lambda_H = \left\{ (H, E_P, R) : H + E_P + R \le \frac{B_H}{\mu_H} ; H > 0, E_P \ge 0, R \ge 0 \right\}$$

11.2.1 Equilibrium Points

The equilibrium points of the systems are given as follows:

- **Equilibrium points for animal population model**
  Solving the system (11.1) of animal population, we get two equilibrium points.

  i. *Animal equilibrium point:*

  $$E_0(A) = \left(\frac{B_A}{\mu_A}, 0, 0 \right)$$

  ii. *Interior equilibrium point due to animal population:*

  $$E^*(A) = (A^*, E_P^*, R^*)$$

  where

  $$A^* = \frac{(\varepsilon_1 + \varepsilon_2 + \mu_A)\mu_A}{(\beta_A - \delta_A)(\varepsilon_2 + \mu_A)}$$

  $$E_P^* = \frac{(\beta_A - \delta_A)(\varepsilon_2 + \mu_A)B_A - (\varepsilon_1 + \varepsilon_2 + \mu_A)\mu_A{}^2}{(\beta_A - \delta_A)(\varepsilon_1 + \varepsilon_2 + \mu_A)\mu_A}$$

  $$R^* = \frac{\left((\beta_A - \delta_A)(\varepsilon_2 + \mu_A)B_A - (\varepsilon_1 + \varepsilon_2 + \mu_A)\mu_A{}^2 \right)\varepsilon_1}{(\beta_A - \delta_A)(\varepsilon_2 + \mu_A)(\varepsilon_1 + \varepsilon_2 + \mu_A)\mu_A}$$

- **Equilibrium points for plant population model**
  We get two equilibrium points by solving the system of plant population (11.2).

  i. *Plant equilibrium point:*

  $$E_0(P) = \left(\frac{B_P}{\mu_P}, 0, 0 \right)$$

  ii. *Interior equilibrium point due to plant population:*

  $$E^*(P) = (P^*, E_P^*, R^*)$$

where

$$P* = \frac{(\gamma_1 + \gamma_2 + \mu_P)\mu_P}{(\beta_P - \delta_P)(\gamma_2 + \mu_P)}$$

$$E_P* = \frac{(\beta_P - \delta_P)(\gamma_2 + \mu_P)B_P - (\gamma_1 + \gamma_2 + \mu_P)\mu_P^2}{(\beta_P - \delta_P)(\gamma_1 + \gamma_2 + \mu_P)\mu_P}$$

$$R* = \frac{((\beta_P - \delta_P)(\gamma_2 + \mu_P)B_P - (\gamma_1 + \gamma_2 + \mu_P)\mu_P^2)\gamma_1}{(\beta_P - \delta_P)(\gamma_2 + \mu_P)(\gamma_1 + \gamma_2 + \mu_P)\mu_P}$$

- **Equilibrium points for human population model**
  The system (11.3) of human population gives two equilibrium points.
  i. *Human equilibrium point:*

$$E_0(H) = \left(\frac{B_H}{\mu_H}, 0, 0\right)$$

  ii. *Interior equilibrium point due to human population:*

$$E*(H) = (H*, E_P*, R*)$$

where

$$H* = \frac{(\eta_1 + \eta_2 + \mu_H)\mu_H}{(\beta_H - \delta_H)(\eta_2 + \mu_H)}$$

$$E_P* = \frac{(\beta_H - \delta_H)(\eta_2 + \mu_H)B_H - (\eta_1 + \eta_2 + \mu_H)\mu_H^2}{(\beta_H - \delta_H)(\eta_1 + \eta_2 + \mu_H)\mu_H}$$

$$R* = \frac{((\beta_H - \delta_H)(\eta_2 + \mu_H)B_H - (\eta_1 + \eta_2 + \mu_H)\mu_H^2)\eta_1}{(\beta_H - \delta_H)(\eta_2 + \mu_H)(\eta_1 + \eta_2 + \mu_H)\mu_H}$$

11.2.2 Computation of Threshold Quantity

In this section, we calculate the threshold quantity of each model using the next generation matrix method (Diekmann et al., 2010) to study the effect of environmental pollutants on rain due to stakeholders. Here, if the threshold for any model is greater than one, that is, if the quantity of pollutants in environment due to any stakeholders exceeds the carrying capacity, then it creates an imbalanced environment. The system is under control if the threshold is less than one.

- **Threshold R_{0_A} for animal population**
  The threshold of animal model R_{0_A} is an average number of secondary infectious pollutants caused by a single infected pollutant in a completely susceptible animal population. We find threshold as follows:

$$
F_A = \begin{bmatrix} \beta_A A & 0 & \beta_A E_P \\ 0 & 0 & 0 \\ \delta_A A & 0 & \delta_A E_p \end{bmatrix} \text{ and}
$$

$$
V_A = \begin{bmatrix} \delta_A A + \varepsilon_1 + \mu_A & -\varepsilon_2 & \delta_A E_P \\ -\varepsilon_1 & \varepsilon_2 + \mu_A & 0 \\ \beta_A A & 0 & \beta_A E_P + \mu_A \end{bmatrix}
$$

Finding F_A and V_A about $E_0(A)$, we get

$$
F_A(E_0(A)) = \begin{bmatrix} \dfrac{\beta_A B_A}{\mu_A} & 0 & 0 \\ 0 & 0 & 0 \\ \dfrac{\delta_A B_A}{\mu_A} & 0 & 0 \end{bmatrix} \text{ and}
$$

$$
V_A(E_0(A)) = \begin{bmatrix} \dfrac{\delta_A B_A}{\mu_A} + \varepsilon_1 + \mu_A & -\varepsilon_2 & 0 \\ -\varepsilon_1 & \varepsilon_2 + \mu_A & 0 \\ \dfrac{\beta_A B_A}{\mu_A} & 0 & \mu_A \end{bmatrix}
$$

Here, V_A is nonsingular matrix.
  The threshold of an animal model is the spectral radius of matrix $F_A V_A^{-1}$

$$
R_{0A} = \frac{B_A \beta_A (\varepsilon_2 + \mu_A)}{\varepsilon_1 \mu_A{}^2 + (B_A \delta_A + \mu_A{}^2)(\varepsilon_2 + \mu_A)} \tag{11.4}
$$

- **Threshold R_{0_P} for plant population**
  The threshold of plant model R_{0_P} is an average number of secondary infectious pollutants caused by a single infected pollutant in

a completely susceptible plant population. Arguing as (11.4), the threshold R_{0p} is given by

$$R_{0p} = \frac{B_P \beta_P (\gamma_2 + \mu_P)}{\gamma_1 \mu_P{}^2 + (B_P \delta_P + \mu_P{}^2)(\gamma_2 + \mu_P)} \tag{11.5}$$

- **Threshold R_{0_H} for human population**
  The threshold of human model R_{0_H} is an average number of secondary infectious pollutants caused by a single infected pollutant in a completely susceptible human population. Similarly, threshold R_{0_H} is given by

$$R_{0_H} = \frac{B_H \beta_H (\eta_2 + \mu_H)}{\eta_1 \mu_H{}^2 + (B_H \delta_H + \mu_H{}^2)(\eta_2 + \mu_H)} \tag{11.6}$$

11.3 Stability Analysis of the Equilibrium

In this section, the local and global stability analysis for each model will be derived for the transmission of environmental pollutants due to different populations.

11.3.1 Local Stability Analysis

In this section, we carry out the local stability for each model. For this, Jacobian matrices are calculated, and if all the eigenvalues about each equilibrium point of that Jacobian are negative, then the equilibrium is locally asymptotically stable.

Using the system (11.1), we formulate the following Jacobian

$$J(A) = \begin{bmatrix} -\beta_A E_P + \delta_A E_P - \mu_A & -\beta_A A + \delta_A A & 0 \\ \beta_A E_P - \delta_A E_P & \beta_A A - \delta_A A - \varepsilon_1 - \mu_A & \varepsilon_2 \\ 0 & \varepsilon_1 & -\varepsilon_2 - \mu_A \end{bmatrix} \tag{11.7}$$

The Jacobian matrix calculated through system (11.2) is

$$J(P) = \begin{bmatrix} -\beta_P E_P + \delta_P E_P - \mu_P & \beta_P P + \delta_P P & 0 \\ \beta_P E_P - \delta_P E_P & \beta_P P - \delta_P P - \gamma_1 - \mu_P & \gamma_2 \\ 0 & \gamma_1 & -\gamma_2 - \mu_P \end{bmatrix} \tag{11.8}$$

From the system (11.3), the following Jacobian is formulated

$$J(H) = \begin{bmatrix} -\beta_H E_P + \delta_H E_P - \mu_H & -\beta_H H + \delta_H H & 0 \\ \beta_H E_P - \delta_H E_P & \beta_H H - \delta_H H - \eta_1 - \mu_H & \eta_2 \\ 0 & \eta_1 & -\eta_2 - \mu_H \end{bmatrix} \quad (11.9)$$

Theorem 11.1: Local stability of $E_0(A)$

The animal equilibrium point $E_0(A)$ is locally asymptotically stable.
   Proof: From (11.7), we have

$$J_0(A) = \begin{bmatrix} -\mu_A & -\dfrac{B_A(\beta_A - \delta_A)}{\mu_A} & 0 \\ 0 & \dfrac{B_A(\beta_A - \delta_A)}{\mu_A} - \varepsilon_1 - \mu_A & \varepsilon_2 \\ 0 & \varepsilon_1 & -\varepsilon_2 - \mu_A \end{bmatrix}$$

The eigenvalues of $J_0(A)$ are given by

$$\omega_1 = -\mu_A$$

$$\omega_2 = \frac{-1}{2\mu_A}\left(\left(\varepsilon_1 + \varepsilon_2 + 2\mu_A\right) - B_A\left(\beta_A - \delta_A\right)\right.$$

$$\left. - \sqrt{\left(B_A\left(\beta_A - \delta_A\right) + \left(\varepsilon_1 + \varepsilon_2\right)\mu_A\right)^2 - 2B_A\varepsilon_1\mu_A\left(\beta_A - \delta_A\right)}\right)$$

$$\omega_3 = \frac{-1}{2\mu_A}\left(\left(\varepsilon_1 + \varepsilon_2 + 2\mu_A\right) - B_A\left(\beta_A - \delta_A\right)\right.$$

$$\left. + \sqrt{\left(B_A\left(\beta_A - \delta_A\right) + \left(\varepsilon_1 + \varepsilon_2\right)\mu_A\right)^2 - 2B_A\varepsilon_1\mu_A\left(\beta_A - \delta_A\right)}\right)$$

Clearly, $\omega_1, \omega_2, \omega_3 < 0$.
Hence, $E_0(A)$ is locally asymptotically stable.

Theorem 11.2: Local stability of $E_0(P)$

The plant equilibrium point $E_0(P)$ is locally asymptotically stable.
   Proof: From (11.8), we have $J_0(P)$

$$J_0(P) = \begin{bmatrix} -\mu_P & -\dfrac{B_P(\beta_P - \delta_P)}{\mu_P} & 0 \\[2ex] 0 & \dfrac{B_P(\beta_P - \delta_P)}{\mu_P} - \gamma_1 - \mu_P & \gamma_2 \\[2ex] 0 & \gamma_1 & -\gamma_2 - \mu_P \end{bmatrix}$$

The eigenvalues of $J_0(P)$ are given by

$$\omega_4 = -\mu_P$$

$$\omega_5 = \frac{-1}{2\mu_P}\left((\gamma_1 + \gamma_2 + 2\mu_P) - B_P(\beta_P - \delta_P) \right.$$

$$\left. - \sqrt{\left(B_P(\beta_P - \delta_P) + (\gamma_1 + \gamma_2)\mu_P\right)^2 - 2B_P\gamma_1\mu_P(\beta_P - \delta_P)} \right)$$

$$\omega_6 = \frac{-1}{2\mu_P}\left((\gamma_1 + \gamma_2 + 2\mu_P) - B_P(\beta_P - \delta_P) \right.$$

$$\left. + \sqrt{\left(B_P(\beta_P - \delta_P) + (\gamma_1 + \gamma_2)\mu_P\right)^2 - 2B_P\gamma_1\mu_P(\beta_P - \delta_P)} \right)$$

Clearly, $\omega_4, \omega_5, \omega_6 < 0$
Hence, $E_0(P)$ is locally asymptotically stable.

Theorem 11.3: Local stability of $E_0(H)$

The human equilibrium point $E_0(H)$ is locally asymptotically stable.
   Proof: From (11.9), we get the following $J_0(H)$

$$J_0(H) = \begin{bmatrix} -\mu_H & -\dfrac{B_H(\beta_H - \delta_H)}{\mu_H} & 0 \\[2ex] 0 & \dfrac{B_H(\beta_H - \delta_H)}{\mu_H} - \eta_1 - \mu_H & \eta_2 \\[2ex] 0 & \eta_1 & -\eta_2 - \mu_H \end{bmatrix}$$

The eigenvalues of $J_0(H)$ are given by

$$\omega_7 = -\mu_H$$

$$\omega_8 = \frac{-1}{2\mu_H}\left(\left(\eta_1 + \eta_2 + 2\mu_H\right) - B_H\left(\beta_H - \delta_H\right)\right.$$

$$\left. -\sqrt{\left(B_H\left(\beta_H - \delta_H\right) + \left(\eta_1 + \eta_2\right)\mu_H\right)^2 - 2B_H\eta_1\mu_H\left(\beta_H - \delta_H\right)}\right)$$

$$\omega_9 = \frac{-1}{2\mu_H}\left(\left(\eta_1 + \eta_2 + 2\mu_H\right) - B_H\left(\beta_H - \delta_H\right)\right.$$

$$\left. +\sqrt{\left(B_H\left(\beta_H - \delta_H\right) + \left(\eta_1 + \eta_2\right)\mu_H\right)^2 - 2B_H\eta_1\mu_H\left(\beta_H - \delta_H\right)}\right)$$

Clearly, $\omega_7, \omega_8, \omega_9 < 0$
Hence, $E_0(H)$ is locally asymptotically stable.

Theorem 11.4: Local stability of $E^*(A)$

Interior equilibrium point for animal population $E^*(A)$ is locally asymptotically stable.
   Proof: From (11.7), we get the following

$$J^*(A) = \begin{bmatrix} -\beta_A E_P{}^* + \delta_A E_P{}^* - \mu_A & -\beta_A A^* + \delta_A A^* & 0 \\ \beta_A E_P{}^* - \delta_A E_P{}^* & \beta_A A^* - \delta_A A^* - \varepsilon_1 - \mu_A & \varepsilon_2 \\ 0 & \varepsilon_1 & -\varepsilon_2 - \mu_A \end{bmatrix}$$

The eigenvalues of $J^*(A)$ are given by

$$\omega_{10} = -\mu_A$$

$$\omega_{11} = \frac{-1}{2}\left(\left(\varepsilon_1 + \varepsilon_2 + 2\mu_A\right) + \left(\beta_A - \delta_A\right)\left(A^* - E_P{}^*\right)\right.$$

$$\left. -\sqrt{\left(\left(\beta_A - \delta_A\right)\left(A^* - E_P{}^*\right) + \varepsilon_1 + \varepsilon_2\right)^2 - 4\varepsilon_1 A^*}\right)$$

$$\omega_{12} = \frac{-1}{2}\left(\left(\varepsilon_1 + \varepsilon_2 + 2\mu_A\right) + \left(\beta_A - \delta_A\right)\left(A^* - E_P{}^*\right)\right.$$

$$\left. +\sqrt{\left(\left(\beta_A - \delta_A\right)\left(A^* - E_P{}^*\right) + \varepsilon_1 + \varepsilon_2\right)^2 - 4\varepsilon_1 A^*}\right)$$

Clearly, $\omega_{10}, \omega_{11}, \omega_{12} < 0$.
Hence, $E^*(A)$ is locally asymptotically stable.

Theorem 11.5: Local stability of $E*(P)$

Interior equilibrium point for plant population $E*(P)$ is locally asymptotically stable.

Proof: From (11.8), we get the following matrix

$$J*(P) = \begin{bmatrix} -\beta_P E_P * + \delta_P E_P * - \mu_P & -\beta_P P * + \delta_P P * & 0 \\ \beta_P E_P * - \delta_P E_P * & \beta_P P * - \delta_P P * - \gamma_1 - \mu_P & \gamma_2 \\ 0 & \gamma_1 & -\gamma_2 - \mu_P \end{bmatrix}$$

The eigenvalues of $J*(P)$ are given by

$$\omega_{13} = -\mu_P$$

$$\omega_{14} = \frac{-1}{2}\left((\gamma_1 + \gamma_2 + 2\mu_P) + (\beta_P - \delta_P)(P* - E_P *) \right.$$

$$\left. -\sqrt{\left((\beta_P - \delta_P)(P* - E_P *) + \gamma_1 + \gamma_2 \right)^2 - 4\gamma_1 P*} \right)$$

$$\omega_{15} = \frac{-1}{2}\left((\gamma_1 + \gamma_2 + 2\mu_P) + (\beta_P - \delta_P)(P* - E_P *) \right.$$

$$\left. +\sqrt{\left((\beta_P - \delta_P)(P* - E_P *) + \gamma_1 + \gamma_2 \right)^2 - 4\gamma_1 P*} \right)$$

Clearly, $\omega_{13}, \omega_{14}, \omega_{15} < 0$
Hence, $E*(P)$ is locally asymptotically stable.

Theorem 11.6: Local stability of $E*(H)$

Interior equilibrium point for human population $E*(H)$ is locally asymptotically stable.

Proof: From (11.9), the matrix $J*(H)$ is given by

$$J*(H) = \begin{bmatrix} -\beta_H E_P * + \delta_H E_P * - \mu_H & -\beta_H P * + \delta_H P * & 0 \\ \beta_H E_P * - \delta_H E_P * & \beta_H P * - \delta_H P * - \eta_1 - \mu_H & \eta_2 \\ 0 & \eta_1 & -\eta_2 - \mu_H \end{bmatrix}$$

The eigenvalues of $J*(H)$ are given by

$$\omega_{16} = -\mu_H$$

$$\omega_{17} = \frac{-1}{2}\left((\eta_1 + \eta_2 + 2\mu_H) + (\beta_H - \delta_H)(H^* - E_P^*)\right.$$

$$\left. -\sqrt{\left((\beta_H - \delta_H)(H^* - E_P^*) + \eta_1 + \eta_2\right)^2 - 4\eta_1 H^*}\right)$$

$$\omega_{18} = \frac{-1}{2}\left((\eta_1 + \eta_2 + 2\mu_H) + (\beta_H - \delta_H)(H^* - E_P^*)\right.$$

$$\left. +\sqrt{\left((\beta_H - \delta_H)(H^* - E_P^*) + \eta_1 + \eta_2\right)^2 - 4\eta_1 H^*}\right)$$

Clearly, $\omega_{16}, \omega_{17}, \omega_{18} < 0$
Hence, $E^*(H)$ is locally asymptotically stable.

11.3.2 Global Stability Analysis

In this section, we study the global stability analysis. For each animal, plant, and human equilibrium point, stability will be proved by LaSalle's Invariance Principle (La Salle, 1976), and for each interior point, it will be analyzed using the theorem given in Busenberg et al. (1990), Cai et al. (2009), and Awan et al. (2017).

Theorem 11.7: Global stability of $E_0(A)$

The animal equilibrium point $E_0(A)$ is globally asymptotically stable.
Proof: Consider Lyapunov function for animal population as given later:

$$L_A(t) = E_p(t) + R(t)$$

then

$$L_A'(t) = E_p'(t) + R'(t)$$

$$= \beta_A AE_P - \delta_A AE_P - \varepsilon_1 E_P + \varepsilon_2 R - \mu_A E_P + \varepsilon_1 E_P - \varepsilon_2 R - \mu_A R$$

$$= B_A - \mu_A A - \mu_A(E_P + R)$$

$$\leq B_A - \mu_A\left(\frac{B_A}{\mu_A}\right) - \mu_A(E_P + R)$$

$$= -\mu_A(E_P + R) \leq 0$$

We get $\dfrac{dL_A}{dt} < 0$, whereas $\dfrac{dL_A}{dt} = 0$ only if $E_p = R = 0$. So that, by LaSalle's Invariance Principle, all the roots of the equations of the system (11.1) are having initial conditions that approach to $E_0(A)$ as t goes to infinity. Hence, in the result of it, the unique positive equilibrium point $E_0(A) = \left(\dfrac{B_A}{\mu_A}, 0, 0 \right)$ is globally asymptotically stable.

Theorem 11.8: Global stability of $E_0(P)$

The plant equilibrium point $E_0(P)$ is globally asymptotically stable.
Proof: For plant population, consider Lyapunov function as given later

$$L_P(t) = E_p(t) + R(t)$$

then

$$L'_P(t) = E'_p(t) + R'(t)$$

$$= \beta_P P E_P - \delta_P P E_P - \gamma_1 E_P + \gamma_2 R - \mu_P E_P + \gamma_1 E_P - \gamma_2 R - \mu_P R$$

$$= B_P - \mu_P P - \mu_P (E_P + R)$$

$$\leq B_P - \mu_P \left(\dfrac{B_P}{\mu_P} \right) - \mu_P (E_P + R)$$

$$= -\mu_P (E_P + R) \leq 0$$

We get $\dfrac{dL_P}{dt} < 0$, whereas $\dfrac{dL_P}{dt} = 0$ only if $E_P = R = 0$. So that, by LaSalle's Invariance Principle, all the roots of the equations of the system (11.2) are having initial conditions that approach to $E_0(P)$ as t goes to infinity. Hence, in the results, the unique positive equilibrium point $E_0(P) = \left(\dfrac{B_P}{\mu_P}, 0, 0 \right)$ is globally asymptotically stable.

Theorem 11.9: Global stability of $E_0(H)$

The human equilibrium point $E_0(H)$ is globally asymptotically stable.
Proof: Consider Lyapunov function for human population as given later

$$L_H(t) = E_p(t) + R(t)$$

then

$$L'_H(t) = E'_P(t) + R'(t)$$

$$= \beta_H HE_P - \delta_H HE_P - \eta_1 E_P + \eta_2 R - \mu_H E_P + \eta_1 E_P - \eta_2 R - \mu_H R$$

$$= B_H - \mu_H H - \mu_H (E_P + R)$$

$$\leq B_H - \mu_H \left(\frac{B_H}{\mu_H} \right) - \mu_H (E_P + R)$$

$$= -\mu_H (E_P + R) \leq 0$$

We get $\dfrac{dL_H}{dt} < 0$, whereas $\dfrac{dL_H}{dt} = 0$ only if $E_P = R = 0$. So that, by LaSalle's Invariance Principle, all the roots of the equations of the system (11.3) are having initial conditions that approach to $E_0(H)$ as t goes to infinity. Hence, in the result of it, the unique positive equilibrium point $E_0(H) = \left(\dfrac{B_H}{\mu_H}, 0, 0 \right)$ is globally asymptotically stable.

Theorem 11.10: Global stability of $E^*(A)$

Consider a piecewise smooth vector field $g_A(A, E_P, R) = \{g_1(A, E_P, R), g_2(A, E_P, R), g_3(A, E_P, R)\}$ on Λ^*_A, where $f_A(f_1, f_2, f_3)$ is a Lipschitz continuous field inside Λ^*_A and $\overrightarrow{n^*_A}$ is the normal vector to Λ^*_A. Then, the system of differential equations (11.1) where $A = f_1$, $E_P = f_2$, and $R = f_3$ has no homoclinic loops, periodic solutions, and oriented phase polygons inside Λ^*_A.

Proof: Let $\Lambda^*_A = \{(A, E_P, R): A + E_P + R = 1, \quad A > 0, E_P \geq 0, R \geq 0\}$. It can easily prove that Λ^*_A is a subset of Λ_A, Λ^*_A is positively invariant, and endemic equilibrium $E^*(A)$ goes to Λ^*_A. Let f_1, f_2, f_3 denote the right-hand side of the equation in the model, respectively. Using $A + E_P + R = 1$ to write f_1, f_2, f_3 in the equivalent forms, we get

$$f_1(A, E_P) = B_A - \beta_A AE_P + \delta_A AE_P - \mu_A A$$

$$f_1(A, R) = B_A - \beta_A A(1 - A - R) + \delta_A A(1 - A - R) - \mu_A A$$

$$f_2(A, E_P) = \beta_A AE_P - \delta_A AE_P - \varepsilon_1 E_P + \varepsilon_2 (1 - A - E_P) - \mu_A E_P$$

$$f_2(E_P, R) = \beta_A (1 - E_P - R)E_P - \delta_A (1 - E_P - R)E_P - \varepsilon_1 E_P + \varepsilon_2 R - \mu_A E_P$$

$$f_3(A, R) = \varepsilon_1 (1 - A - R) - \varepsilon_2 R - \mu_A R$$

$$f_3(E_P, R) = \varepsilon_1 E_P - \varepsilon_2 R - \mu_A R$$

Assume, $g_A = (g_1, g_2, g_3)$ be a vector field such that

$$g_1 = \frac{f_3(A,R)}{AR} - \frac{f_2(A,E_P)}{AE_P}$$

$$= \varepsilon_1 \left(\frac{1}{AR} - \frac{1}{R} \right) - \beta_A + \delta_A - \varepsilon_2 \left(\frac{1}{AE_P} - \frac{1}{E_P} \right)$$

$$g_2 = \frac{f_1(A,E_P)}{AE_P} - \frac{f_3(E_P,R)}{E_P R}$$

$$= \frac{B_A}{AE_P} - \beta_A + \delta_A - \frac{\varepsilon_1}{R} + \frac{\varepsilon_2}{E_P}$$

$$g_3 = \frac{f_2(E_P,R)}{E_P R} - \frac{f_1(A,R)}{AR}$$

$$= 2\beta_A \left(\frac{1}{R} - 1 \right) - 2\delta_A \left(\frac{1}{R} - 1 \right) - \frac{\beta_A E_P}{R} + \frac{\delta_A E_P}{R} - \frac{\beta_A A}{R} + \frac{\delta_A A}{R} - \frac{\varepsilon_1}{R} + \frac{\varepsilon_2}{E_P} - \frac{B_A}{AR}$$

Here, $f_1, f_2,$ and f_3 are equivalent in Λ_A^*. So,

$$g_A \cdot f_A = g_1 \cdot f_1 + g_2 \cdot f_2 + g_3 \cdot f_3 = 0$$

Now, let's find *curl* g_A,

$$curl\ g_A = \begin{vmatrix} \hat{i} & \hat{j} & \hat{k} \\ \dfrac{\partial}{\partial A} & \dfrac{\partial}{\partial E_P} & \dfrac{\partial}{\partial R} \\ g_1 & g_2 & g_3 \end{vmatrix}$$

$$= \left(\frac{-\beta_A}{R} + \frac{\delta_A}{R} - \frac{\varepsilon_1}{R^2} - \frac{\varepsilon_2}{E_P{}^2} \right) \hat{i} - \left(\frac{-\beta_A}{R} + \frac{\delta_A}{R} + \frac{B_A}{A^2 R} - \varepsilon_1 \left(\frac{-1}{AR^2} + \frac{1}{R^2} \right) \right) \hat{j}$$

$$+ \left(-\frac{B_A}{A^2 E_P} - \varepsilon_2 \left(\frac{1}{AE_P{}^2} - \frac{1}{E_P{}^2} \right) \right) \hat{k}$$

Now, normal vector is $\vec{n_A}$, $\vec{n_A}$, where for Λ_A^*, we have

$$(curl\ g_A) \cdot \vec{n_A} = -\frac{B_A}{A^2 R} - \frac{B_A}{A^2 E_P} - \frac{\varepsilon_1}{AR^2} - \frac{\varepsilon_2}{AE_P{}^2} < 0$$

Therefore, the system has no homoclinic loops, periodic solutions, and oriented phase polygons inside Λ_A^*. Thus, $E^*(A)$ is globally asymptotically stable.

Theorem 11.11: Global stability of $E^*(P)$

Consider a piecewise smooth vector field $g_P(P, E_P, R) = \{g_4(P, E_P, R),$ $g_5(P, E_P, R), g_6(P, E_P, R)\}$ on Λ_P^*, where $f_P(f_4, f_5, f_6)$ is a Lipschitz continuous field inside Λ_P^* and $\vec{n}_P$ is the normal vector to Λ_P^*. Then, the system of differential equations (11.2) where $P = f_4$, $E_P = f_5$, and $R = f_6$ has no homoclinic loops, periodic solutions, and oriented phase polygons inside Λ_P^*.

Proof: Let $\Lambda_P^* = \{(P, E_P, R): P + E_P + R = 1, \quad P > 0, E_P \geq 0, R \geq 0\}$. It can easily prove that Λ_P^* is a subset of Λ_P, Λ_P^* is positively invariant, and endemic equilibrium $E^*(P)$ goes to Λ_P^*. Let f_4, f_5, f_6 denote the right-hand side of the equation in the model, respectively. Using $P + E_P + R = 1$ to write f_4, f_5, f_6 in the equivalent forms, we get

$$f_4(P, E_P) = B_P - \beta_P P E_P + \delta_P P E_P - \mu_P P$$

$$f_4(P, R) = B_P - \beta_P P(1 - P - R) + \delta_P P(1 - P - R) - \mu_P P$$

$$f_5(P, E_P) = \beta_P P E_P - \delta_P P E_P - \gamma_1 E_P + \gamma_2(1 - P - E_P) - \mu_P E_P$$

$$f_5(E_P, R) = \beta_P(1 - E_P - R)E_P - \delta_P(1 - E_P - R)E_P - \gamma_1 E_P + \gamma_2 R - \mu_P E_P$$

$$f_6(P, R) = \gamma_1(1 - P - R) - \gamma_2 R - \mu_P R$$

$$f_6(E_P, R) = \gamma_1 E_P - \gamma_2 R - \mu_P R$$

Assume, $g_P = (g_4, g_5, g_6)$ be a vector field such that

$$g_4 = \frac{f_6(P, R)}{PR} - \frac{f_5(P, E_P)}{PE_P}$$

$$-\gamma_1\left(\frac{1}{PR} - \frac{1}{R}\right) - \beta_P + \delta_P - \gamma_2\left(\frac{1}{PE_P} - \frac{1}{E_P}\right)$$

$$g_5 = \frac{f_4(P, E_P)}{PE_P} - \frac{f_6(E_P, R)}{E_P R}$$

$$= \frac{B_P}{PE_P} - \beta_P + \delta_P - \frac{\gamma_1}{R} + \frac{\gamma_2}{E_P}$$

$$g_6 = \frac{f_5(E_P, R)}{E_P R} - \frac{f_4(P, R)}{PR}$$

$$= 2\beta_P\left(\frac{1}{R} - 1\right) - 2\delta_P\left(\frac{1}{R} - 1\right) - \frac{\beta_P E_P}{R} + \frac{\delta_P E_P}{R} - \frac{\beta_P P}{R} + \frac{\delta_P P}{R} - \frac{\gamma_1}{R} + \frac{\gamma_2}{E_P} - \frac{B_P}{PR}$$

Here, $f_4, f_5,$ and f_6 are equivalent in Λ_P^*. So,

$$g_P \cdot f_P = g_4 \cdot f_4 + g_5 \cdot f_5 + g_6 \cdot f_6 = 0$$

Now, let's find $curl\ g_P$,

$$curl\ g_P = \begin{vmatrix} \hat{i} & \hat{j} & \hat{k} \\ \dfrac{\partial}{\partial P} & \dfrac{\partial}{\partial E_P} & \dfrac{\partial}{\partial R} \\ g_4 & g_5 & g_6 \end{vmatrix}$$

$$= \left(\frac{-\beta_P}{R} + \frac{\delta_P}{R} - \frac{\gamma_1}{R^2} - \frac{\gamma_2}{E_P{}^2} \right)\hat{i} - \left(\frac{-\beta_P}{R} + \frac{\delta_P}{R} + \frac{B_P}{P^2 R} - \gamma_1 \left(\frac{-1}{PR^2} + \frac{1}{R^2} \right) \right)\hat{j}$$

$$+ \left(-\frac{B_P}{P^2 E_P} - \gamma_2 \left(\frac{1}{PE_P{}^2} - \frac{1}{E_P{}^2} \right) \right)\hat{k}$$

Now, normal vector is $\vec{n_P}, \vec{n_P}$, where for Λ_P^*, we have

$$\left(curl\ g_P \right) \cdot \vec{n_P} = -\frac{B_P}{P^2 R} - \frac{B_P}{P^2 E_P} - \frac{\gamma_1}{PR^2} - \frac{\gamma_2}{PE^2} < 0$$

Therefore, the system has no homoclinic loops, periodic solutions, and oriented phase polygons inside Λ_P^*. Thus, $E^*(P)$ is globally asymptotically stable.

Theorem 11.12: Global stability of $E^*(H)$

Consider a piecewise smooth vector field $g_H(H, E_P, R) = \{ g_7(H, E_P, R), g_8(H, E_P, R), g_9(H, E_P, R) \}$ on Λ_H^*, where $f_H(f_7, f_8, f_9)$ is a Lipschitz continuous field inside Λ_H^* and $\vec{n_H}$ is the normal vector to Λ_H^*. Then, the system of differential equations (11.3) where $H = f_7$, $E_P = f_8$, and $R = f_9$ has no homoclinic loops, periodic solutions, and oriented phase polygons inside Λ_H^*.

Proof: Let $\Lambda_H^* = \{ (H, E_P, R) : H + E_P + R = 1, \quad H > 0, E_P \geq 0, R \geq 0 \}$. It can easily prove that Λ_H^* is a subset of Λ_H, Λ_H^* is positively invariant, and endemic equilibrium $E^*(H)$ goes to Λ_H^*. Let f_7, f_8, f_9 denote the right-hand side of the equation in the model, respectively. Using $H + E_P + R = 1$ to write f_7, f_8, f_9 in the equivalent form, we get

$$f_7(H, E_P) = B_H - \beta_H H E_P + \delta_H H E_P - \mu_H H$$

$$f_7(H,R) = B_H - \beta_H H(1-H-R) + \delta_H H(1-H-R) - \mu_H H$$

$$f_8(H,E_P) = \beta_H H E_P - \delta_H H E_P - \eta_1 E_P + \eta_2(1-H-E_P) - \mu_H E_P$$

$$f_8(E_P,R) = \beta_H(1-E_P-R)E_P - \delta_H(1-E_P-R)E_P - \eta_1 E_P + \eta_2 R - \mu_H E_P$$

$$f_9(H,R) = \eta_1(1-H-R) - \eta_2 R - \mu_H R$$

$$f_9(E_P,R) = \eta_1 E_P - \eta_2 R - \mu_H R$$

Assume $g_H = (g_7, g_8, g_9)$ be a vector field such that

$$g_7 = \frac{f_9(H,R)}{HR} - \frac{f_8(H,E_P)}{HE_P}$$

$$= \eta_1\left(\frac{1}{HR} - \frac{1}{R}\right) - \beta_H + \delta_H - \eta_2\left(\frac{1}{HE_P} - \frac{1}{E_P}\right)$$

$$g_8 = \frac{f_7(H,E_P)}{PE_P} - \frac{f_9(E_P,R)}{E_P R}$$

$$= \frac{B_H}{HE_P} - \beta_H + \delta_H - \frac{\eta_1}{R} + \frac{\eta_2}{E_P}$$

$$g_9 = \frac{f_8(E_P,R)}{E_P R} - \frac{f_7(H,R)}{HR}$$

$$= 2\beta_H\left(\frac{1}{R} - 1\right) - 2\delta_H\left(\frac{1}{R} - 1\right) - \frac{\beta_H E_P}{R} + \frac{\delta_H E_P}{R} - \frac{\beta_H H}{R} - + \frac{\delta_H H}{R} - \frac{\eta_1}{R} + \frac{\eta_2}{E_P} - \frac{B_H}{HR}$$

Here, f_7, f_8, and f_9 are equivalent in Λ_H^*. So,

$$g_H \cdot f_H = g_7 \cdot f_7 + g_8 \cdot f_8 + g_9 \cdot f_9 = 0$$

Now, let's find *curl* g_H,

$$curl\; g_H = \begin{vmatrix} \hat{i} & \hat{j} & \hat{k} \\ \dfrac{\partial}{\partial H} & \dfrac{\partial}{\partial E_P} & \dfrac{\partial}{\partial R} \\ g_7 & g_8 & g_9 \end{vmatrix}$$

$$= \left(\frac{-\beta_H}{R} + \frac{\delta_H}{R} - \frac{\eta_1}{R^2} - \frac{\eta_2}{E_P^{\,2}}\right)\hat{i} - \left(\frac{-\beta_H}{R} + \frac{\delta_H}{R} + \frac{B_H}{H^2 R} - \eta_1\left(\frac{-1}{HR^2} + \frac{1}{R^2}\right)\right)\hat{j}$$

$$+ \left(-\frac{B_H}{H^2 E_P} - \eta_2\left(\frac{1}{HE_P^{\,2}} - \frac{1}{E_P^{\,2}}\right)\right)\hat{k}$$

Now, normal vector is $\vec{n}_H$, $\vec{n}_H$, where for Λ_H^*, we have

$$\left(curl\ g_H \right)\cdot \vec{n}_H = -\frac{B_H}{H^2 R} - \frac{B_H}{H^2 E_P} - \frac{\eta_1}{HR^2} - \frac{\eta_2}{HE_P^2} < 0$$

Therefore, the system has no homoclinic loops, periodic solutions, and oriented phase polygons inside Λ_H^*. Thus, $E^*(H)$ is globally asymptotically stable.

11.4 Sensitivity Analysis

In this section, the sensitivity analysis for all parameters is discussed for the effect of environmental pollutants on rain due to stakeholders. The normalized sensitivity of the parameters is computed using the Christoffel formula

$$\Gamma_{\alpha}^{R_0} = \frac{\partial R_0}{\partial \alpha} \cdot \frac{\alpha}{R_0}$$

where $\Gamma_{\alpha}^{R_0}$ represents the change in threshold of the system R_0 with respect to the parameter α.

- **Sensitivity analysis of R_{0A}**

$$\Gamma_{B_A}^{R_{0A}} = \frac{(\varepsilon_1 + \varepsilon_2 + \mu_A)\mu_A^2}{B_A \delta_A (\varepsilon_2 + \mu_A) + (\varepsilon_1 + \varepsilon_2 + \mu_A)\mu_A^2} > 0$$

$$\Gamma_{\delta_A}^{R_{0A}} = \frac{-B_A \delta_A (\varepsilon_2 + \mu_A)}{B_A \delta_A (\varepsilon_2 + \mu_A) + (\varepsilon_1 + \varepsilon_2 + \mu_A)\mu_A^2} < 0$$

$$\Gamma_{\varepsilon_1}^{R_{0A}} = \frac{-\varepsilon_1 \mu_A^2}{B_A \delta_A (\varepsilon_2 + \mu_A) + (\varepsilon_1 + \varepsilon_2 + \mu_A)\mu_A^2} < 0$$

$$\Gamma_{\varepsilon_2}^{R_{0A}} = \frac{\varepsilon_1 \varepsilon_2 \mu_A^2}{(\varepsilon_2 + \mu_A)\left(B_A \delta_A (\varepsilon_2 + \mu_A) + (\varepsilon_1 + \varepsilon_2 + \mu_A)\mu_A^2\right)} > 0$$

- **Sensitivity analysis of R_{0P}**

$$\Gamma_{B_P}^{R_{0P}} = \frac{(\gamma_1 + \gamma_2 + \mu_P)\mu_P^2}{B_P \delta_P (\gamma_2 + \mu_P) + (\gamma_1 + \gamma_2 + \mu_P)\mu_P^2} > 0$$

$$\Gamma_{\delta_P}^{R_{0P}} = \frac{-B_P \delta_P (\gamma_2 + \mu_P)}{B_P \delta_P (\gamma_2 + \mu_P) + (\gamma_1 + \gamma_2 + \mu_P)\mu_P^2} < 0$$

$$\Gamma_{\gamma_1}^{R_{0P}} = \frac{-\gamma_1 \mu_P^2}{B_P \delta_P (\gamma_2 + \mu_P) + (\gamma_1 + \gamma_2 + \mu_P)\mu_P^2} < 0$$

$$\Gamma_{\gamma_2}^{R_{0P}} = \frac{\gamma_1 \gamma_2 \mu_P{}^2}{(\gamma_2 + \mu_P)\left(B_P \delta_P (\gamma_2 + \mu_P) + (\gamma_1 + \gamma_2 + \mu_P)\mu_P{}^2\right)} > 0$$

- **Sensitivity analysis of R_{0_H}**

$$\Gamma_{B_H}^{R_{0H}} = \frac{(\eta_1 + \eta_2 + \mu_H)\mu_H{}^2}{B_H \delta_H (\eta_2 + \mu_H) + (\eta_1 + \eta_2 + \mu_H)\mu_H{}^2} > 0$$

$$\Gamma_{\delta_H}^{R_{0H}} = \frac{-B_H \delta_H (\eta_2 + \mu_H)}{B_H \delta_H (\eta_2 + \mu_H) + (\eta_1 + \eta_2 + \mu_H)\mu_H{}^2} < 0$$

$$\Gamma_{\eta_1}^{R_{0H}} = \frac{-\eta_1 \mu_H{}^2}{B_H \delta_H (\eta_2 + \mu_H) + (\eta_1 + \eta_2 + \mu_H)\mu_H{}^2} < 0$$

$$\Gamma_{\eta_2}^{R_{0H}} = \frac{\eta_1 \eta_2 \mu_H{}^2}{(\eta_2 + \mu_H)\left(B_H \delta_H (\eta_2 + \mu_H) + (\eta_1 + \eta_2 + \mu_H)\mu_H{}^2\right)} > 0$$

11.4.1 Observations

Threshold governing that growth rate B_i (where i represents A, P, H i.e. B_A, B_P, B_H) and the dissolving rate of environmental pollutants due to rain #$_2$ (where # denotes $\varepsilon, \gamma, \eta$ i.e. $\varepsilon_2, \gamma_2, \eta_2$) have positive impact, while other model parameters have a negative impact.

The recruitment rate of all the stakeholders increases environmental pollutants. The rate of absorption of environmental pollutants decreases the lifespan of population density of stakeholders as it causes inheritance of life-threatening diseases. Rain intensity decreases because of environmental pollutants, which is a major cause of global warming. Rain dissolves some of the gaseous pollutants that are in favor of the population in focus.

11.5 Numerical Simulation

Here, numerical simulation is given to study how models are working and what is the impact of environmental pollutants through stakeholders and rain.

Figure 11.4 represents the impact of stakeholders on environmental pollutants. It is observed that animal and human population increase environmental pollutants up to 96 and 150 ppm, respectively, whereas plant population helps to reduce environmental pollutants within 0.3 years by inhaling.

Figure 11.5 shows the depletion in rain intensity due to stakeholders. Here, animal and human populations decrease rain intensity at a higher rate compared with the plant population.

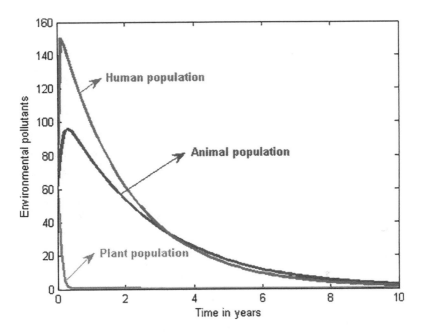

FIGURE 11.4
Transmission of environmental pollutants through different populations.

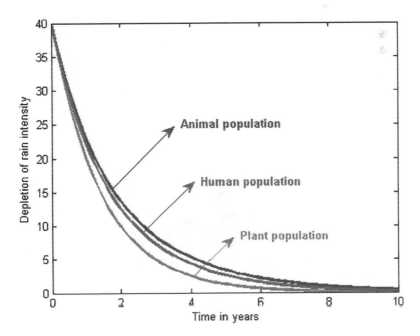

FIGURE 11.5
Depletion of rain intensity through different populations.

The animal population is a major source of environmental pollution. 51% of the animal population creates environmental pollution in terms of dead animals, discharge from animals, fungal and bacterial infection of animals, and much more. Though human beings are considered to be intelligent creatures, 47% environmental pollutants are created by them. These environmental pollutants are excreted in terms of household waste, industrial waste, biowaste, etc. Only 2% of the environmental pollution is due to plants. This pollution is due to forest fires (Figure 11.6).

Depletion of rain intensity is highest by human activities. Unfavorable human activities need to be curtailed to maintain rain intensity. The animal population decreases rain intensity only by 29%, while the plant population decreases rain intensity by 22%. This suggests that more and more plant/forest should be grown to increase rain intensity (Figure 11.7).

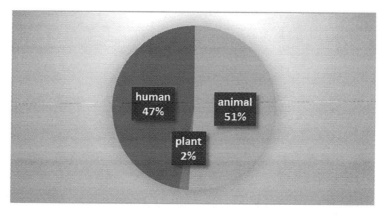

FIGURE 11.6
Existence of environmental pollutants.

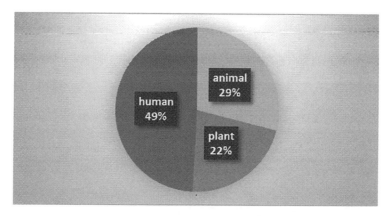

FIGURE 11.7
Survival of rain intensity.

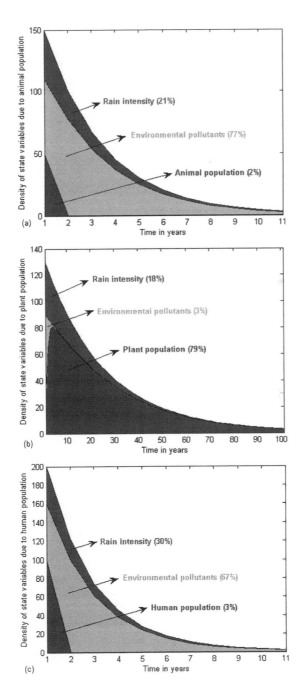

FIGURE 11.8
Density of state variables due to stakeholders. (a) Density of compartments due to animal population; (b) density of compartments due to plant population; (c) density of compartments due to human population.

Human population can use advance techniques to reduce environmental pollutants when compared with animal population. The activity performed by humans reduces rain intensity by 30% and that is the maximum among all three stakeholders. This suggests that there is a dire need to control environmental pollutants and increase rain intensity at the human end. We can also have more greenery to increase rain intensity. The minimum environmental pollution due to plant population is because the plant itself absorbs some of the pollutants and rain dissolves gaseous pollutants (Figure 11.8).

11.6 Summary and Conclusions

In the proposed chapter, the effect of environmental pollutants on rain due to stakeholders is studied. Here, the animal equilibrium point, plant equilibrium point, human equilibrium point, and interior equilibrium points due to stakeholders are acquired for different population models. These equilibrium points are locally and globally asymptotically stable. Sensitivity analysis has been performed to illustrate the nature of each model parameter. Numerical simulation results in a comparison between all three stakeholders. Among them, the human population is creating the highest risk for environment that diminishes rain intensity.

This study has shown that 13.41%, 11.48%, and 63.37% environmental pollutants are spread through the animal, plant, and human populations, respectively. This suggests that efforts should be made to limit human activities, give proper treatment to ill/dead animals and plants, and try to reduce forest fires. The results also advocate for green surroundings so that more pollutants can be absorbed from air and hence the depletion of rain intensity can be decreased.

Acknowledgement

The authors thank DST-FIST file #MSI-097 for technical support to the department.

References

Awan, A. U., et al. Smoking model with cravings to smoke. *Advanced Studies in Biology*, 2017, 9(1), 31–41.

Busenberg, S., Van den Driessche, P. Analysis of a disease transmission model in a population with varying size. *Journal of Mathematical Biology*, 1990, 28(3), 257–270.

Cai, L., Li, X., Ghosh, M., Guo, B. Stability analysis of an HIV/AIDS epidemic model with treatment. *Journal of Computational and Applied Mathematics*, 2009, 229(1), 313–323.

Diekmann, O., Heesterback, J. A. P., Roberts, M. G. The construction of next generation matrices for compartmental epidemic models. *Journal of the Royal Society Interface*, 2010, 7(47), 873–885.

Dubey, B. A model for the effect of pollutant on human population dependent on a resource with environmental and health policy. *Journal of Biological Systems*, 2010, 18(03), 571–592.

Dubey, B., Hussain, J. Modelling the interaction of two biological species in a polluted environment. *Journal of Mathematical Analysis and Applications*, 2000, 246(1), 58–79.

Kademi, H. I., Baba, I. A., Saad, F. T. Modelling the dynamics of toxicity associated with aflatoxins in foods and feeds. *Toxicology Reports*, 2017, 4, 358–363.

Khemani, L. T., Murty, B. V. R. Rainfall variations in an urban industrial region. *Journal of Applied Meteorology*, 1973, 12(1), 187–194.

La Salle, J. P. *The Stability of Dynamical Systems*. Regional Conference Series in Applied Mathematics, Philadelphia, PA: Society for Industrial and Applied Mathematics, 1976.

Naresh, R. Qualitative analysis of a nonlinear model for removal of air pollutants. *International Journal of Nonlinear Sciences and Numerical Simulation*, 2003, 4(4), 379–386.

Pandey, J., Agrawal, M., Khanam, N., Narayan, D., Rao, D. N. Air pollutant concentrations in Varanasi, India. *Atmospheric Environment. Part B. Urban Atmosphere*, 1992, 26(1), 91–98.

Shah, N. H., Satia, M. H., Thakkar, F. A. Mathematical approach on household waste causing environmental pollutants due to landfill and treatments. *International Journal of Engineering Technologies and Management Research*, 2018, 5(2), 266–282.

Shah, N. H., Satia, M. H., Yeolekar, B. M. Optimal control on depletion of green belt due to industries. *Advances in Dynamical Systems and Applications*, 2017, 12(3), 217–232.

Shah, N. H., Satia, M. H., Yeolekar, B. M. Optimum control for spread of pollutants through forest resources. *Applied Mathematics*, 2017, 8(05), 607–620.

Shah, N. H., Satia, M. H., Yeolekar, B. M. Stability of 'GO-CLEAN' model through graphs. *Journal of Computer and Mathematical Sciences*, 2018, 9(2), 79–93.

Sharma, S. & Kumari, N. *Arxiv preprint arxiv:1705.01895*. Modeling the impact of rain on population exposed to air pollution, 2017.

Shukla, J. B., Agrawal, A. K., Sinha, P., Dubey, B. Modeling effects of primary and secondary toxicants on renewable resources. *Natural Resource Modeling*, 2003, 16(1), 99–120.

Shukla, J. B., Misra, A. K., Sundar, S., Naresh, R. Effect of rain on removal of a gaseous pollutant and two different particulate matters from the atmosphere of a city. *Mathematical and Computer Modelling*, 2008a, 48(5–6), 832–844.

Shukla, J. B., Sundar, S., Misra, A. K., Naresh, R. Modelling the removal of gaseous pollutants and particulate matters from the atmosphere of a city by rain: Effect of cloud density. *Environmental Modeling & Assessment*, 2008b, 13(2), 255–263.

Traistaa, E., Ionicaa, M., Codrea, V., Barbu, O. Correlations between the air pollution and the rainfall composition in Jiului Valley area. *Studia UBB Geologia*, 2003, 48(2), 95–100.

12

Sliding Mode Control Approaches for Robust Control of Quadruple Tank System

Dipesh Shah and Dhruv Patel
Sardar Vallabhbhai Patel Institute of Technology

Axaykumar Mehta
Institute of Infrastructure Technology Research and Management

CONTENTS

12.1 Introduction

In process industries, the processes are complex as it possesses multiple interacting and noninteracting loops [1]. Hence, there is a demand to design robust controllers for such processes and plants that take care of uncertainties and disturbances. The characteristics of process variables in complex system depends on various parameters such as accuracy, precision, resolution span adjustment, placing of the sensors, and unwanted internal or external disturbances. Generally, in control system domain, the systems are classified based on the number of inputs and outputs such as single input single output (SISO), single input multiple output (SIMO), multiple input single output (MISO), and multiple input multiple output (MIMO) systems. Among all these systems, MIMO is the most complex multivariable control system that possesses nonlinear characteristics and multiple interactions between input and output variables [2]. In the last few decades, quadruple tank process (QTP) has become a popular benchmark for the researchers as it involves four highly interconnected tanks having nonlinear character- istics [3]. Also, the QTP system has a wide range of applications in process industries such as petrochemicals, wastewater treatment and purification, biochemical, spray coating, filling-disposal plants, beverages, and pharma- ceutical industries. The liquid level control of QTP in the presence of inter- nal or external disturbances is one of the essential problems in the process industries. Various control methodologies have been developed to improve the performance of the quadruple tank system. Johansson [4] developed a decentralized proportional integral (PI) controller for the quadruple tank process in the absence of disturbances. Azam and Jorgensen [5] proposed a mathematical model of a modified quadruple tank system in which matched uncertainties were introduced in the upper two tanks. Grebeck [6] tested the effectiveness of centralized PI controller for quadruple tank process in the absence of matched uncertainty. Multivariable controller tuning for qua- druple tank process based on relay feedback approach was investigated by researchers in Refs. [7,8]. Tunyasrirut et al. [9] designed a proportional inte- gral derivative (PID) controller with auto-tuning approach for interacting- type water-level process. Shehu and Wahab [10] designed and compared a model predictive control with PI controller for coupled tank systems. The results proved that model predictive control (MPC) algorithm shows better performance than PI controller in the presence of system uncertainty. Gouta et al. [11,12] designed an observer-based backstepping controller and a model- based predictive controller for quadruple tank system with bounded control inputs in the absence of disturbances. Basci and Derdiyok [13] designed an adaptive fuzzy controller (AFC) to realize the position control of two coupled water tank systems. The fuzzy control system includes an adaptive model identifier and controller. The parameters of the fuzzy identifier model are adjusted online by using recursive least square algorithm. The effectiveness of the proposed control algorithm was tested in the presence of disturbances.

With rapid advancement in control technology in recent years, sliding mode controller (SMC) has received much attention in process applications due to its robust properties and simple structure. SMC is a nonlinear control technique preferred for the systems where the performance depends on model uncer- tainties and external disturbances. Various researchers [14–19] have tried to design SMCs for controlling the liquid level of coupled and quadruple tank system in the presence of matched and unmatched disturbances. The sliding mode control involves the design of sliding surface and reaching law. Presently, constant-rate, power-rate, and exponential reaching laws are widely used in continuous-time domain to derive the sliding mode control, as they provide faster convergence in the presence of model uncertainties and disturbances.

In all earlier literatures [14–19], researchers have tried to design SMCs using these mentioned reaching laws for quadruple and coupled tank process. The major drawback of derived control algorithms is, despite providing faster convergence, it generates more chattering with an increase in the amplitude of control signal, which leads to the degradation of pumps and performance of the system. This motivates the authors to develop a robust nonswitching control algorithm that provides faster convergence and generates less chattering without increasing the amplitude of the control signal. The main advantage of the proposed technique is that it improves the efficiency of pumps and the performance of the system in the presence of matched disturbances. The results of the proposed control algorithm are compared with the exponential reaching law, which proves to be better than constant-rate and power-rate reaching laws.

The layout of this chapter is as follows: The mathematical model of QTP with matched uncertainty and problem statement are presented in Section 12.2. Section 12.3 describes the design of SMC with exponential and nonswitching reaching laws. The stability analysis for both the cases is presented in Section 12.4. The simulation results are discussed in Section 12.5 followed by a conclusion in Section 12.6.

12.2 Quadruple Tank System

Figure 12.1 shows the schematic diagram of a quadruple tank system [4] with matched disturbances. The disturbance d_1 and d_2 applied at the input side of the system are deterministic in nature. As shown in Figure 12.1, the liquid from the reservoir is fed to tank-3 and tank-4 through a cross connection of inlet pipes from pump-1 and pump-2, respectively. The main advantage of cross connection in the inlet pipes for tank-3 and tank-4 is to mix the liquids in the lower two tanks in the absence of stirrer, resulting in two different products. The main target is to control the level in the lower two tanks with two pumps. The process inputs (v_1, v_2) are used to drive the pumps, and the outputs (y_1, y_2) are measured from the lower two tanks. The measured level signals are assumed to be proportional to the true levels, i.e. $y_1 = k_{m_1} h_1$ and $y_2 = k_{m_2} h_2$. The level sensors are calibrated as $k_{m_1} = k_{m_2} = 1$. The differential equations representing the mass balances in the quadruple tank process are given by

$$\frac{dh_1}{dt} = -\frac{a_1}{A_1}\sqrt{2gh_1} + \frac{a_3}{A_1}\sqrt{2gh_3} + \frac{\gamma_1 k_1}{A_1} v_1. \tag{12.1}$$

$$\frac{dh_2}{dt} = -\frac{a_2}{A_2}\sqrt{2gh_2} + \frac{a_4}{A_2}\sqrt{2gh_4} + \frac{\gamma_2 k_2}{A_2} v_2. \tag{12.2}$$

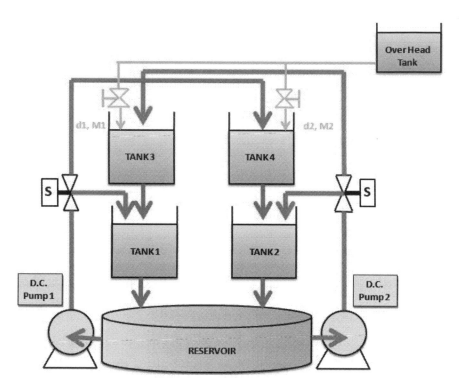

FIGURE 12.1
Schematic diagram of a quadruple tank system with disturbances.

$$\frac{dh_3}{dt} = -\frac{a_3}{A_3}\sqrt{2gh_3} + \frac{(1-\gamma_2)}{A_3}k_2v_2 - \frac{k_{d_1}}{A_3}d_1. \tag{12.3}$$

$$\frac{dh_4}{dt} = -\frac{a_4}{A_4}\sqrt{2gh_4} + \frac{(1-\gamma_1)}{A_4}k_1v_1 - \frac{k_{d_2}}{A_4}d_2. \tag{12.4}$$

where
A_i is the crosssection of the tank i, a_i is the crosssection of the outlet hole, h_i is the water level, and d_1 and d_2 are the flow disturbances applied to tank-3 and tank-4, respectively.

This process exhibits interacting multivariable dynamics, because each of the pumps affects both outputs. γ_1 is the ratio of water diverted from tank-1 to tank-4 and γ_2 is the corresponding ratio diverted from tank-2 to tank-3. The nonlinear model presented in Eqs. (12.1)–(12.4) is linearized at the operating point $\delta h_i = h_i - h_i{}^0$.

Thus, the state space model of the earlier MIMO system in linear form is represented as

$$\dot{x} = Ax(t) + B[u(t) + Dd(t)], \tag{12.5}$$

$$y(t) = Cx(t). \tag{12.6}$$

where $x(t) \in R^{n\times1}$ is system state vector in terms of height, $u(t) \in R^{m\times1}$ is control input vector in terms of voltage, $y(t) \in R^{r\times1}$ is system output vector, $A \in R^{n\times n}$, $B \in R^{n\times m}$, $C \in R^{r\times n}$, and $D \in R^{p\times n}$ are the matrices of appropriate dimensions, and $d(t) \in R^{p\times1}$ is a match-bounded disturbance vector applied at the input side of the channel with $|d(t)| \le d_{\max}$.

Using Eqs. (12.1)–(12.4) the matrices A, B, C, and D are written as

$$A = \begin{bmatrix} -\dfrac{1}{T_1} & 0 & \dfrac{A_3}{A_1 T_3} & 0 \\ 0 & -\dfrac{1}{T_2} & 0 & \dfrac{A_4}{A_2 T_4} \\ 0 & 0 & -\dfrac{1}{T_3} & 0 \\ 0 & 0 & 0 & -\dfrac{1}{T_4} \end{bmatrix}, \quad B = \begin{bmatrix} \dfrac{\gamma_1 k_1}{A_1} & 0 \\ 0 & \dfrac{\gamma_2 k_2}{A_2} \\ 0 & \dfrac{(1-\gamma_2)k_2}{A_3} \\ \dfrac{(1-\gamma_1)k_1}{A_4} & 0 \end{bmatrix},$$

$$C = \begin{bmatrix} 1 & 0 & 0 & 0 \\ 0 & 1 & 0 & 0 \end{bmatrix}, \quad D = \begin{bmatrix} 0 & 0 \\ 0 & 0 \\ 0 & -\dfrac{k_{d_1}}{A_3} \\ -\dfrac{k_{d_2}}{A_4} & 0 \end{bmatrix}, \quad d_1(t) = d_2(t) = 0.2\sin 0.086t.$$

where the time constants are given by

$$T_i = \dfrac{A_i}{a_i}\sqrt{\dfrac{2h_i^0}{g}}, \quad i = 1, 2, 3, 4. \tag{12.7}$$

where

$$A_i = \pi \dfrac{D_i^2}{4}, a_i = \pi \dfrac{d_i^2}{4}, h_i^0 = \dfrac{h_{max}}{2}, k_1 = \dfrac{Q_1}{\upsilon_1}, k_2 = \dfrac{Q_2}{\upsilon_2}, k_{d_1} = \dfrac{Q_{d_1}}{\upsilon_{d_1}}, k_{d_2} = \dfrac{Q_{d_2}}{\upsilon_{d_2}}.$$

D_i is the diameter of tank i, d_i is the diameter of outlet hole, and Q_1 and Q_2 are flow rates in tank-1 and tank-2, respectively.

Assumption 1: The process delays and the computational delays are negligible in nature as the driving capacity of each pump is very high compared to the distance traveled by the water through the pipes.

Problem Statement: The main objective is to design and compare the robustness of sliding mode control, derived using nonswitching type reaching law, with the exponential reaching law for quadruple tank MIMO system in Eq. (12.5) in the presence of matched disturbances applied to tank-3 and tank-4, respectively.

12.3 Design of SMC with Different Approaches

In this section, sliding mode control law is derived for a quadruple tank system using two different approaches, namely (i) exponential reaching law and (ii) nonswitching reaching law. The exponential reaching law provides faster convergence than power-rate and constant-rate reaching laws [19].However, nonswitching reaching law provides faster convergence with negligible chattering without increasing the amplitude of the control signal when compared with the exponential reaching law [20]. It is further observed in the Results and Discussion section.

12.3.1 Exponential Reaching Law

Theorem 12.1

The system states in Eq. (12.5) would reach in a finite time onto the sliding surface in Eq. (12.9) in the presence of matched uncertainty, provided the control law is designed as

$$u(t) = -(C_sB)^{-1}\left[C_sAx(t) + \frac{k}{N(s)}sign(s)\right] - Dd(t). \qquad (12.8)$$

where C_s is the sliding gain.

Proof: Consider the sliding surface as

$$s(t) = C_sx(t), \qquad (12.9)$$

where $s(t) = [s_1 \ s_2]^T$ is the sliding surface vector and C_s is the sliding gain vector designed using linear quadratic regulator (LQR) approach.

The sliding surface in Eq. (12.9) should satisfy following "Δ" reaching condition that ensures the finite time convergence to $s = 0$.

$$\dot{s}s \langle -\Delta |s|. \ \Delta \rangle 0, \ \forall t. \tag{12.10}$$

Thus, to satisfy the condition in Eq. (12.10), the exponential reaching law proposed in Ref. [21] is given by

$$\dot{s} = -\frac{k}{N(s)} sign(s), \quad k > 0 \tag{12.11}$$

where

$$N(s) = \delta_0 + (1 - \delta_0) e^{-\alpha |s|^p}.$$

δ_0 is a strictly positive offset that is less than 1 and p and α are strictly positive integers.

From the reaching law stated in Eq. (12.11), it can be observed that if $|s|$ increases, $N(s)$ approaches to δ_0, and therefore $\frac{k}{N(s)}$ will approach to $\frac{k}{\delta_0}$, which is greater than k. This means that $\frac{k}{N(s)}$ increases in the reaching law, and consequently, the attraction of the sliding surface is faster. If $|s|$ decreases, then $N(s)$ approaches to 1, and therefore $\frac{k}{N(s)}$ will approach to k. Thus, it means that the system approaches to sliding surface $\frac{k}{N(s)}$ gradually decreases to limit chattering.

Integrating Eq. (12.11) with respect to time, the convergence time (t_c) for the system state variables to reach $s = 0$ is given by Ref. [19]:

$$t_c = \frac{|s(0)|}{k}. \tag{12.12}$$

From Eq. (12.9), Eq. (12.11) can be written as

$$C_s \dot{x}(t) = -\frac{k}{N(s)} sign(s), \tag{12.13}$$

Substituting $\dot{x}(t)$ from Eq. (12.5), Eq. (12.13) is written as

$$C_s [Ax(t) + B(u(t) + Dd(t))] = -\frac{k}{N(s)} sign(s), \tag{12.14}$$

Further simplification gives

$$C_s Ax(t) + C_s B(u(t) + Dd(t)) = -\frac{k}{N(s)} sign(s), \tag{12.15}$$

Further, solving earlier Eq. (12.15), the control law can be expressed as

$$u(t) = -(C_s B)^{-1} \left[C_s A x(t) + \frac{k}{N(s)} sign(s) \right] - Dd(t). \tag{12.16}$$

This completes the **Proof**.

12.3.2 Nonswitching Reaching Law

Theorem 12.2

The system states in Eq. (12.5) would reach in a finite time onto the sliding surface mentioned in Eq. (12.9) in the presence of matched uncertainty, provided the control law is designed as

$$u(t) = -(C_s B)^{-1} \left[C_s A x(t) + \frac{k_1 s_0}{s_0 + |s(t)|} s(t) \right] - Dd(t). \tag{12.17}$$

Proof: In order to satisfy the condition in Eq. (12.10) the nonswitching reaching law [20] in continuous-time domain is given as

$$\dot{s} = -\frac{k_1 s_0}{s_0 + |s(t)|} s(t), \tag{12.18}$$

where
   k_1 and s_0 are positive real constants.
   The convergence time (t_c) for the system state variables to reach $s = 0$ is given by Ref. [20]

$$t_c = \frac{1}{k_1} \left\{ \frac{|s(0) - s(t)|}{s_0} + \ln \left[\frac{s(0)}{s(t)} \right] \right\}. \tag{12.19}$$

Referring to Eq. (12.9), Eq. (12.18) can be written as

$$C_s \dot{x}(t) = -\frac{k_1 s_0}{s_0 + |s(t)|} s(t), \tag{12.20}$$

Substituting $\dot{x}(t)$ from Eq. (12.5), Eq. (12.20) is modified as

$$C_s [A x(t) + B(u(t) + Dd(t))] = -\frac{k_1 s_0}{s_0 + |s(t)|} s(t), \tag{12.21}$$

Further simplification gives

$$C_s Ax(t) + C_s B(u(t) + Dd(t)) = -\frac{k_1 s_0}{s_0 + |s(t)|} s(t), \tag{12.22}$$

Further, solving earlier Eq. (12.22), the control law can be expressed as

$$u(t) = -(C_s B)^{-1} \left[C_s Ax(t) + \frac{k_1 s_0}{s_0 + (t)} s(t) \right] - Dd(t). \tag{12.23}$$

Once the control law is computed, the next step is to derive the stability criteria for closed-loop system in Eq. (12.5) using a sliding surface in Eq. (12.9) and control law in Eqs. (12.8) and (12.17), respectively.

12.4 Stability Analysis

In this section, the stability analysis of the closed system in Eq. (12.5) is proved for the designed controller in Eqs. (12.8) and (12.17) in the presence of matched uncertainty and sliding surface mentioned in Eq. (12.9).

12.4.1 Stability Analysis Using Exponential Reaching Law

Consider the quadratic Lyapunov function as

$$V_s(t) = \frac{1}{2} s^T(t) s(t), \tag{12.24}$$

Taking time derivative of Eq. (12.24), we have

$$\dot{V}_s(t) = s^T(t) \dot{s}(t), \tag{12.25}$$

Referring Eq. (12.9), Eq. (12.25) can be written as

$$\dot{V}_s(t) = s^T(t) \left[C_s \dot{x}(t) \right], \tag{12.26}$$

Substituting the value of $\dot{x}(t)$, we have

$$\dot{V}_s(t) = s^T(t) \left[C_s [Ax(t) + B(u(t) + Dd(t))] \right], \tag{12.27}$$

Substituting the value of $u(t)$ from Eq. (12.16), we have

$$\dot{V}_s(t) = s^T(t)\left[C_s\left[Ax(t) + B\left(-(C_sB)^{-1}\left[C_sAx(t) + \frac{k}{N(s)}sign(s)\right] - Dd(t) + Dd(t)\right)\right]\right],$$

(12.28)

Further simplification leads to

$$\dot{V}_s(t) = -s^T(t)\left[\frac{k}{N(s)}sign(s(t))\right] < 0.$$

(12.29)

The term $\dot{V}_s(t)$ is negative as k, p, α and δ_0 are strictly positive integers. So, the closed-loop system is asymptotically stable in the presence of matched uncertainty.

12.4.2 Stability Analysis Using Nonswitching Reaching Law

Considering the same Lyapunov function in Eq. (12.24) and substituting the value of $u(t)$ from Eqs. (12.23) to (12.27), we have

$$\dot{V}_s(t) = s^T(t)\left[C_s\left[Ax(t) + B\left(-(C_sB)^{-1}\left[C_sAx(t) + \frac{k_1s_0}{s_0 + |s(t)|}s(t)\right] - Dd(t) + Dd(t)\right)\right]\right],$$

(12.30)

Further simplification gives

$$\dot{V}_s(t) = -s^T(t)\left[\frac{k_1s_0}{s_0 + |s(t)|}s(t)\right] < 0.$$

(12.31)

The term $\dot{V}_s(t)$ is negative as k_1 and s_0 are positive constants as well as $\left[\frac{k_1s_0}{s_0 + |s(t)|}s(t)\right] > 0$. So the closed-loop system is asymptotically stable in the presence of matched uncertainty.

Thus, for both the cases, the stability of closed-loop system in Eq. (12.5) is proved for proper values of positive constants and user-defined variables.

12.5 Results and Discussions

In this section, the effectiveness of the proposed control algorithms are validated on the quadruple tank system in the presence of matched uncertainties. The simulation results are discussed in two parts: (i) Figures 12.2–12.5 show the nature of state variables, sliding variables, and control signals for exponential reaching law and (ii) Figures 12.6–12.9 show the nature of state variables, sliding variables, and control signals for nonswitching reaching law. The maximum input voltages required to drive pump-1 and pump-2 are $V_1 = 12.74\,V$ and $V_2 = 14.36\,V$, respectively. The estimated plant parameters for simulation setup are given by

$$D_1 = D_2 = D_3 = D_4 = 9.2 \text{ cm,}$$

$$h_{max} = 50 \text{ cm,} \ h_1{}^0 = h_2{}^0 = h_3{}^0 = h_4{}^0 = \frac{h_{max}}{2} = 25 \text{ cm,}$$

$$A_1 = A_2 = A_3 = A_4 = 66.44 \text{ cm}^2, d_1 = 0.2 \text{ cm,} d_2 = 0.9 \text{ cm,}$$

$a_1 = a_2 = 0.0314 \text{ cm}^2$ crosssection of outer hole of tank-1 and tank-2, $a_3 = a_4 = 0.64 \text{ cm}^2$ crosssection of outer hole of tank-3 and tank-4, $\gamma_1 = 0.33$, and $\gamma_2 = 0.307$ such that the following condition $0 < \gamma_1 + \gamma_2 < 1$ holds true.

Substituting the earlier parameters, the matrices A, B, C, and D are represented as

$$A = \begin{bmatrix} -0.0021 & 0 & 0.042 & 0 \\ 0 & -0.0021 & 0 & 0.042 \\ 0 & 0 & -0.042 & 0 \\ 0 & 0 & 0 & -0.042 \end{bmatrix}, \ B = \begin{bmatrix} 0.0367 & 0 \\ 0 & 0.0320 \\ 0 & 0.0722 \\ 0.0745 & 0 \end{bmatrix},$$

$$C = \begin{bmatrix} 1 & 0 & 0 & 0 \\ 0 & 1 & 0 & 0 \end{bmatrix}, \ D = \begin{bmatrix} 0 & 0 \\ 0 & 0 \\ 0 & -0.111 \\ -0.104 & 0 \end{bmatrix}.$$

The sliding gain C_s is computed using LQR method by proper selection of $Q = diag(1000,0)$ and $R = diag(1,0)$ matrices having $C_s = \begin{bmatrix} 2.1702 & 20.5584 & -8.8711 & 23.7822 \\ 20.5323 & 0.4145 & 24.1863 & -9.8814 \end{bmatrix}$. In both cases, the initial conditions are assumed to be $x(t) = [1-2-3-4]^T$, respectively.

The system states, control signals, and sliding variables in exponential and nonswitching reaching law show satisfactory response for $k_1 = 100, s_0 = 0.05$, $\alpha = 20$, $p = 1$, $k = 0.5$, and $\delta_0 = 0.1$.

12.5.1 Results of Exponential Reaching Law

Figures 12.2–12.5 show the responses of the state variables, sliding variables, and control signals for the specified initial conditions in the presence of

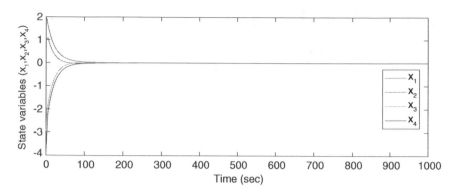

FIGURE 12.2
State variables (x_1, x_2, x_3, x_4).

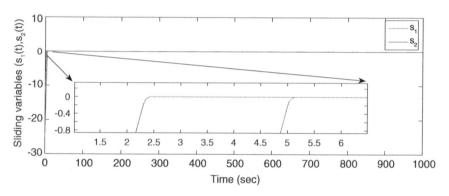

FIGURE 12.3
Sliding variables $(s_1(t), s_2(t))$.

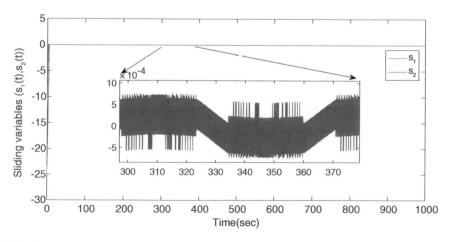

FIGURE 12.4
Magnified sliding variables $(s_1(t), s_2(t))$.

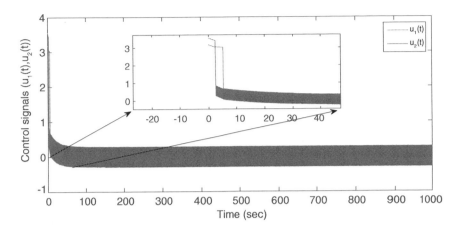

FIGURE 12.5
Control signals $(u_1(t), u_2(t))$.

matched disturbances. It is observed from Figure 12.2 that all the state variables would converge to zero from their given initial condition. The existence of sliding surfaces is shown in Figure 12.3. The convergence of the system states to zero proves the attractiveness of the sliding surface. The sliding surfaces $s_1(t)$ and $s_2(t)$ converge to zero at 2.5 and 5 s, respectively. It is shown in the magnified window of Figure 12.3. Figure 12.4 shows the response of the sliding variables with a magnified band. It can be noticed that both the sliding variables drive towards the origin over a finite interval of time with high level of chattering, having values $\pm 5 \times 10^{-4}$. The same effect of chattering is also observed in control signals $u_1(t)$ and $u_2(t)$ as shown in Figure 12.5. When these control signals are applied to drive the pumps, it may deteriorate the performance and reduce the efficiency of pumps. Moreover, higher the values of k faster the convergence rate, but at the same time, the magnitude of chattering level increases in the control signal, which is not suitable to drive the pumps.

12.5.2 Results of Nonswitching Reaching Law

Figures 12.6–12.9 show the response of the state variables, sliding variables, and control signals for a specified initial condition in the presence of matched disturbances using nonswitching reaching law. It is observed from Figure 12.6 that all state variables would converge to zero from their given initial conditions. Figure 12.7 shows the existence of sliding surfaces $(s_1(t), s_2(t))$. It is observed that both the surfaces converge to zero at a faster rate compared with exponential reaching law. The sliding surfaces $s_1(t)$ and $s_2(t)$ converges to zero at 0.25 and 0.4 s, which is approximately $\left(\dfrac{1}{10^{\text{th}}}\right)$ times faster than the exponential reaching law. It is shown in the magnified window of Figure 12.7.

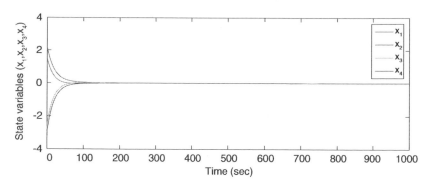

FIGURE 12.6
State variables (x_1, x_2, x_3, x_4).

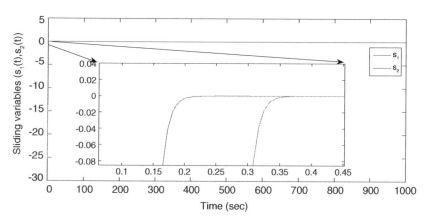

FIGURE 12.7
Sliding variables $(s_1(t), s_2(t))$.

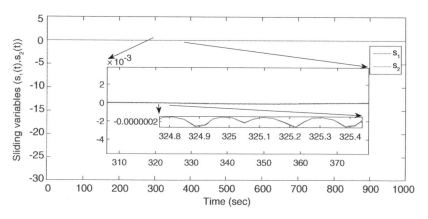

FIGURE 12.8
Magnified sliding band for sliding variables $(s_1(t), s_2(t))$.

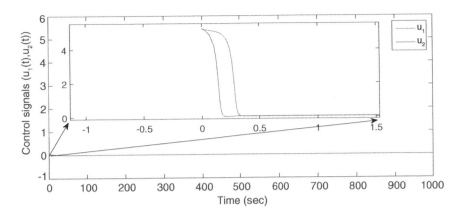

FIGURE 12.9
Control signals $(u_1(t), u_2(t))$.

TABLE 12.1

Comparative Analysis of Sliding Variables and Control Signals

Reaching Laws	$(s_1(t), s_2(t))$		$(u_1(t), u_2(t))$	
	Chattering	Convergence Time (s)	Chattering	Convergence Time
Exponential	5×10^{-4}	2.5 and 5	High	10
Nonswitching	0.2×10^{-7}	0.25 and 0.4	Negligible	0.3

Figure 12.8 shows the magnified response of the sliding band. It can be noticed that both the sliding variables slide towards the origin over a finite interval of time with negligible chattering, having values $\pm 0.2 \times 10^{-7}$. The same effect of chattering is also observed in control signals $u_1(t)$ and $u_2(t)$ as shown in Figure 12.9. When these control signals are applied to drive the pumps, neither does it deteriorates the performance nor it reduces the efficiency of pumps. Thus, from earlier results, it is proved that the control law derived using nonswitching law is more robust than an exponential reaching law for the quadruple tank application, as it provides faster convergence rate and less chattering in the presence of matched disturbances without increasing the amplitude of control signal.

The comparative analysis of sliding variables and control signals are shown in Table 12.1.

12.6 Conclusion

In this chapter, robust controllers are developed for quadruple tank MIMO system in the presence of matched uncertainties. The SMCs are designed using two different approaches, i.e. (i) exponential reaching law and (ii) non-switching reaching law. The stability of closed-loop MIMO system is assured through Lyapunov approach. The robustness of derived control algorithms is proved with respect to fast convergence, chattering level, and disturbance. The efficacy of both control algorithms is compared on quadruple tank MIMO system. The simulation results proved that the SMC designed using nonswitching type reaching law is more robust than the control algorithm derived using exponential reaching law, as it provides faster convergence and negligible chattering on the control signal without increasing its amplitude. This results in improved performance of the system and efficiency of the pumps in the presence of matched disturbances. The simulation results show that the control algorithm derived using nonswitching reaching law provides better performance compared with other methods of SMC proposed in the literatures. In future, the proposed control strategy can be explored considering the effect of process delays and computational delays on the system.

References

1. Albertos, P., Antonio, S. *Multivariable Control Systems: An Engineering Approach.* Science and Business Media, Springer, Singapore, 2006.
2. Ozkan, S., Kara, T., Arici, M. Modelling, simulation and control of quadruple tank process. *IEEE International Conference on Electrical and Electronics Engineering (ELECO)*, Turkey, pages 866–870, 2017.
3. Johansson, K.H. Relay feedback and multivariable control. PhD Thesis, 1997.
4. Johansson, K.H. The quadruple-tank process: A multivariable laboratory process with an adjustable zero. *IEEE Transactions on Control Systems Technology*, 8(3): 456–465, 2000.
5. Azam, S., Jorgensen, S. Modelling and simulation of a modified quadruple tank system. *IEEE International Conference in Control System, Computing and Engineering (ICCSCE)*, Malaysia, pages 365–370, 2015.
6. Grebeck, M. A comparison of controllers for the quadruple tank system. Department of Automatic Control, Lund Institute of Technology, Lund, Sweden, Tech. Rep., 1998.
7. Johansson, K.H., James, B., Bryant, G., Astrom, K. Multivariable controller tuning. *IEEE Proceedings of American Control Conference*, Philadelphia, PA, pages 3514–3518, 1998.
8. Recica, V. Automatic tuning of multivariable controllers. Lund Institute of Technology, MSc Thesis, 1998.
9. Tunyasrirut, S., Suksri, T., Numsomran, A., Gulpanich, S., Tirasesth, K. The auto-tuning PID controller for interacting water level process. *International Journal of Electrical and Information Engineering*, 1(12): 134–138, 2007.
10. Shehu, I., Wahab, N. Applications of MPC and PI controls for liquid level control in coupled-tank systems. *IEEE International Conference on Automatic Control and Intelligent Systems*, Malaysia, pages 119–124, 2016.

11. Gouta, H., Said, S., M'sahli, F. Model-based predictive and back stepping controllers for a state coupled four-tank system with bounded control inputs: A comparative study. *Journal of the Franklin Institute*, 352(11): 4864–4889, 2015.
12. Gouta, H., Said, S., M'sahli, F. Observer-based back-stepping controller for a state-coupled two-tank system. *IETE Journal of Research*, 61(3): 259–268, 2015.
13. Basci, A., Derdiyok, A. Implementation of an adaptive fuzzy compensator for coupled tank liquid level control system. *Journal of Measurement*, 91(1): 12–18, 2016.
14. Benayache, R., Chrifi-Alaoui, L., Bussy, P., Castelain, J. Design and implementation of sliding mode controller with varying boundary layer for a coupled tanks system. *17th Mediterranean Conference in Control and Automation*, Greece, pages 1215–1220, 2009.
15. Almutairi, A., Zribi, M. Sliding mode control of coupled tanks. *IEEE Transactions on Mechatronics*, 16(7): 427–441, 2006.
16. Basci, A., Sekban, H., Can, K. Real-time application of sliding mode controller for coupled tank liquid level system. *International Journal of Applied Mathematics, Electronics and Computer*, 4(1): 301–306, 2016.
17. Biswas, P., Srivastava, R., Ray, S., Samanta,A. Sliding mode control of quadruple tank process. *Mechatronics*, 19(4): 548–561, 2009.
18. Prusty, S., Seshagiri, S., Pati, U., Mahapatra, K. Sliding mode control of coupled tanks using conditional integrators. *IEEE Indian Control Conference (ICC-2016)*, Hyderabad, pages 146–151, 2016.
19. Latosinski, P. Sliding mode control based on the reaching law approach-A brief survey. *International Conference on Methods and Models in Automation and Robotics*, Miedzyzdroje, Poland, pages 519–524, 2017.
20. Bartoszewicz, A. A new reaching law for sliding mode control of continuous time systems with constraints. *Transactions of the Institute of Measurement and Control*, 37(4): 515–521, 2015.
21. Shah, D., Mehta, A. *Discrete-Time Sliding Mode Control for Networked Control System*. Springer Nature, Singapore, 2018.

13

Discrete-Time Sliding Mode Control for Nonlinear System Adulterated by Network Irregularities

Dipesh Shah
Sardar Vallabhbhai Patel Institute of Technology

Axaykumar Mehta
Institute of Infrastructure Technology Research and Management

CONTENTS

13.1 Introduction

In any closed-loop system, communication networks are employed for an exchange of information between control system components such as actuator, sensor, and controller, which are defined as networked control systems (NCSs) [1]. Due to numerous advantages such as cheaper cost, compactness, simplicity in installation, and easier maintenance, NCS has received wide popularity in the field of control and industrial application such as automobile sectors, robotics, aerospace, medical sectors, and many others. The dynamic characteristics of NCS create different challenges such as fixed or variable delays, loss of single or multiple data packets, misalignment of data packet, allocation of resource, and sharing of bandwidth for the researchers [2]. Among all these constraints, time delay issue needs to be handled properly to improve stable performance of closed-loop NCSs.

To compensate the effect of network-induced delays, different control algorithms are developed by researchers in the past few decades. Initially, during developments for random time delay models in NCS, various difficulties were confronted. Therefore, random delays were replaced with constant delays for approximation studies. These delays are also referred as deterministic delays. Initially, various concepts based on buffering techniques were introduced for the compensation of deterministic delays in continuous-time domain [3]. The major drawback of the proposed technique was that the validation was valid for a deterministic type of network delays. So, to overcome this drawback, a time stamp approach was introduced to compensate the effect of time-varying network-induced delays in continuous-time domain [4]. Later, Pade approximation [5], memory feedback controller [6], Smith predictor [7], and multiple step delay compensator [8] techniques were developed for the compensation of deterministic and time-varying network-induced delay in continuous-time domain.

With the quick advancement in digital controllers, numerous techniques such as zero-order-hold [9], Kalman predictor [10], Luenberger output feedback observer [11], Tustin approximation [12], bilinear transformation [13], and time delay estimation [14] are proposed for compensation of deterministic as well as random network-induced delay in discrete-time domain. In all aforementioned techniques, researchers have assumed that network-induced delay possesses integer type of values in a discrete-time domain. Thus, to overcome this issue, the concept of fractional network-induced delay having noninteger values in discrete-time domain was introduced [15].

In NCS, the characteristics of network-induced delay are dependent on network topology, type of communication medium, and the structure of NCS. In direct structure NCS, the communication is carried out using lease line concept in which network-induced delays are deterministic. While in shared or hierarchical structure NCS, generally, the network nodes are shared among different controllers. Thus, the network-induced delays have random characteristics. When these delays are changed into discrete-time domain, they generally possess noninteger values that are defined as fractional network-induced delay.

In all the aforementioned research works [1–13] till date, researchers have worked to nullify the effect of fractional and nonfractional delay in the discrete- and continuous-time domain for linear systems. Very few research papers [16–18] in the continuous domain are available that discuss the compensation of network-induced delay for nonlinear NCSs. As per the author's best knowledge, the issue related to the compensation of fractional delay for nonlinear NCS in discrete-time domain is not yet addressed. This motivates the authors to explore discrete-time sliding mode control for compensating the effect of fractional delay in the presence of system uncertainties and disturbances. The principle benefit to design the discrete-time sliding mode control (SMC) is to observe each sampling instant for the control signal. This feature rejects the disturbances and makes the system more robust. In NCS, as the communication occurs in digital form, SMC is designed in discrete-time domain.

This chapter is organized as follows: Section 13.2 describes direct structure NCS. Section 13.3 describes the problem statement. Section 13.4 describes the design of nondelayed sliding surface followed by discrete-time sliding mode control for nonlinear NCS in Section 13.5. Sections 13.6 and 13.7 focus on the key results and the conclusion.

13.2 Networked Control System

Figure 13.1 shows the direct structure NCS with fractional delay compensator. In this structure, communication network connects a sensor and actuator with the plant. As shown in Figure 13.1, the control signal is communicated from controller to actuator across the same network, which is called a forward channel. The data signal communicated from sensor to controller across the network is defined as a feedback channel. The state information transmitted through feedback channel experiences sensor to controller delay (τ_{sc}) while the control information transmitted through forward channel experiences controller to actuator delay (τ_{ca}). These delays are deterministic in nature as there are no other controller nodes shared through

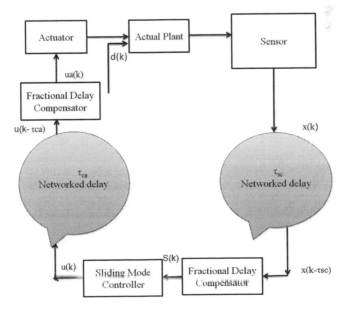

FIGURE 13.1
Direct structure NCS with fractional delay compensator.

the communication network. If the tolerance limit is lesser than the effect of these network-induced delays, it may damage the plant. This leads to inferior plant performance.

13.3 Problem Statement

Consider the discrete-time nonlinear single input single output system as

$$x_i(k+1) = x_{i+1}(k); \quad i = 1, 2, \dots, n-1.$$

$$x_n(k+1) = f(x(k)) + g(x(k))[u(k) + d(k)], \tag{13.1}$$

$$y(k) = x_1(k).$$

where f and g are unknown nonlinear functions, $x(k) \in R^n$ represents the system state vector, $u(k) \in R$ represents the control input, $y(k) \in R$ represents the system output, and $d(k)$ signifies the matched uncertainty applied at the input side of the system. $d(k)$ varies slowly followed by a constant interval $kh \le t \le (k+1) h$ [19].

In NCS, due to the presence of real-time network in both the channels, the data packets will experience the total network-induced delay (τ) during transmission. Thus, consider the network-induced delay system in Eq. (13.1) which is represented as

$$x_i(k+1) = x_{i+1}(k); \quad i = 1, 2, \dots, n-1.$$

$$x_n(k+1) = f(x(k)) + g(x(k))[u(k - \tau') + d(k)], \tag{13.2}$$

$$y(k) = x_1(k).$$

where $\tau' = \dfrac{\tau}{h}$ represents the fractional part of total network-induced delay.

Together, sensor to controller fractional delay (τ'_{sc}) and controller to actuator fractional delay (τ'_{ca}) is known as total fractional network-induced delay. Considering negligible computational delay, mathematically, it is represented in normal form as

$$\tau' = \tau'_{sc} + \tau'_{ca}, \tag{13.3}$$

where $\tau'_{sc} = \dfrac{\tau_{sc}}{h}$ and $\tau'_{ca} = \dfrac{\tau_{ca}}{h}$.

Assumption 1: In this work, it is assumed that the nature of total network-induced delay is deterministic, and it satisfies the following condition

$$\tau < h. \tag{13.4}$$

Remark 1: Apart from network-induced delays, there exist processing and computational delays. In this paper, both delays are neglected as their values are $\left(\dfrac{1}{4}\right)^{\text{th}}$ times negligible, which are subsequently low compared with networked delay.

Assumption 2: The bounds for disturbance $d(k)$ are described as

$$d_l \le d(k) \le d_u, \tag{13.5}$$

where d_l and d_u denotes lower and upper bound of disturbance.

The major contribution in this work is to design robust discrete-time sliding mode controller for the system mentioned in Eq. (13.2) in the presence of deterministic fractional network-induced delay $\left((\tau'_{sc}),(\tau'_{ca})\right)$ and matched uncertainty satisfying conditions in Eqs. (13.4) and (13.5), respectively.

The strategy of sliding mode controller depends on the design of sliding surface and control law, which drives the system states in the direction of the designed surface. In the later section, the design of nondelayed sliding surface that nullifies the cause of network-induced sensor to controller fractional delay in discrete-time domain is proposed.

13.4 Design of Sliding Surface for Nonlinear NCS

In discrete-time domain, bilinear transformation and Tustin approximation are widely used techniques for compensating the network-induced delay having integer values, while in the case of noninteger values, Thiran's approximation is the most appropriate approach to compensate the network-induced delay. In this work, feedback channel delay (τ'_{sc}) is compensated in sliding surface using a novel approach of approximation that is represented in **Lemma 1** as follows:

Lemma 1: A nondelayed sliding surface designed for system in Eq. (13.2) in the presence of network-induced fractional delay between sensor to controller satisfying the conditions in Eqs. (13.4) and (13.5) is given as

$$s(k) = \sum_{i=0}^{n-1} c_i \left[x_i(k) - \beta x_i(k-1) \right] + x_n(k) - \beta x_n(k-1). \tag{13.6}$$

where $\beta = \dfrac{\tau'_{sc}}{1 + \tau'_{sc}}$ and c_i is the roots of the polynomial within the unit circle.

Proof: The sliding surface in its normal form with sensor to controller fractional delay (τ'_{sc}) is described as

$$s(k) = C_s x(k - \tau'_{sc}),\tag{13.7}$$

where $x(k - \tau'_{sc})$ is the delayed state variable, $C_s - [c_1, c_2, \ldots, c_n]$ is the sliding gain, c_n is assumed to be 1, and c_i $(i = 1, \ldots, n)$ are chosen such that the roots of the polynomial $r(z) = c_1 + c_2 z + \cdots + c_n z^{n-1}$ are inside the unit circle.

Thus, the sliding surface for system in Eq. (13.2) is rewritten as

$$s(k) = \sum_{i=0}^{n-1} c_i x_i (k - \tau'_{sc}) + x_n (k - \tau'_{sc}).\tag{13.8}$$

Applying z-transform to Eq. (13.8), we get

$$s(z) = \sum_{i=0}^{n-1} c_i x_i(z) z^{-\tau'_{sc}} + x_n(z) z^{-\tau'_{sc}}.\tag{13.9}$$

The discrete form of sensor to controller network-induced fractional delay $\left(z^{-\tau'_{sc}}\right)$ based on Thiran approximation [20] is denoted as

$$z^{-\tau'_{sc}} = \sum_{k=0}^{n} (-1)^k \binom{n}{k} \prod_{j=0}^{n} \frac{2\tau'_{sc} + j}{2\tau'_{sc} + k + j} z^{-k}.\tag{13.10}$$

where n specifies the order of approximation.

Further, applying first-order approximation with $n = 1$ to earlier Eq. (13.10), we have

$$z^{-\tau'_{sc}} = \sum_{k=0}^{1} (-1)^k \binom{n}{k} \prod_{j=0}^{1} \frac{2\tau'_{sc} + j}{2\tau'_{sc} + k + j} z^{-k}.\tag{13.11}$$

The earlier Eq. (13.11) is further expanded as

$$z^{-\tau'_{sc}} = \left[(-1)^0 \binom{1}{0} \left\{ \frac{2\tau'_{sc}}{2\tau'_{sc}} * \frac{2\tau'_{sc} + 1}{2\tau'_{sc} + 1} \right\} z^0 + (-1)^1 \binom{1}{0} \left\{ \frac{2\tau'_{sc}}{2\tau'_{sc}} * \frac{2\tau'_{sc} + 1}{2\tau'_{sc} + 2} \right\} z^{-1} \right].$$

On simplifying, we have

$$z^{-\tau'_{sc}} = 1 - \beta z^{-1},\tag{13.12}$$

where $\beta = \dfrac{\tau'_{sc}}{1 + \tau'_{sc}}$.

Substituting Eq. (13.12) in Eq. (13.9), we get

$$s(z) = \sum_{i=0}^{n-1} c_i x_i(z) \left[1 - \beta z^{-1} \right] + x_n(z) \left[1 - \beta z^{-1} \right]. \tag{13.13}$$

On further simplification and applying inverse z-transform to Eq. (13.13), the nondelayed sliding surface is written as

$$s(k) = \sum_{i=0}^{n-1} c_i \left[x_i(k) - \beta x_i(k-1) \right] + x_n(k) - \beta x_n(k-1). \tag{13.14}$$

This completes the **Proof**.

Eq. (13.14) states that the effect of fractional delay (τ'_{sc}) in feedback channel at the sliding surface is nullified. This is done by two parameters. First, present state variables available through the network, and second, past state variables that are multiplied by the parameter "β" approximated using Thiran's approximation. After computing non-delayed sliding surface, sliding mode control law is designed that directs the system states towards the designed sliding surface in Eq. (13.14).

13.5 Discrete-Time Sliding Mode Control for Nonlinear NCS

In this section, the robust control algorithm and stability of closed-loop system in Eq. (13.2) are proved using nondelayed sliding surface in Eq. (13.14) in the presence of system uncertainties.

Theorem 13.1

The system states in Eq. (13.2) would reach in a finite time onto the non-delayed sliding surface proposed in Eq. (13.14) in the presence of network-induced delays (τ'_{sc}, τ'_{ca}) satisfying (Eq. 13.4) and matched uncertainty in Eq. (13.5), provided the control law is designed as

$$u(k) = -g(x(k))^{-1} \left[f(x(k)) + \sum_{i=0}^{n-1} c_i \left[x_i(k+1) + \beta x_i(k) \right] \right.$$

$$\left. - \beta x_n(k) - \left\{ 1 - q[s(k)] \right\} s(k) + d_c(k) - d_1 \right] - d(k). \tag{13.15}$$

Proof: Considering the reaching law proposed in Ref. [21] in the presence of network-induced fractional delay as

$$s(k+1) = \{1 - q[s(k)]\} s(k) - d_c(k) + d_1, \tag{13.16}$$

where

$q[s(k)] = \dfrac{\varphi}{\varphi + |s(k)|}, d_c(k)$ represents the compensated disturbance, $d_1 = \dfrac{d_u + d_l}{2}$

indicates the mean value of $d(k)$, and $d_2 = \dfrac{d_u - d_l}{2}$ indicates deviation in the value of $d(k)$. Here, φ is the designer's constant that satisfies the condition $\varphi > d_2$.

Remark 2: The disturbance $d(k)$ in Eq. (13.16) is applied through the communication network via feedback channel. Thus, the compensated value of disturbance based on Thiran's approximation [22] is given as

$$d_c(k) = d(k) - \beta d(k-1). \tag{13.17}$$

The reaching law in Eq. (13.16) indicates that the state vector always approaches towards the sliding band and is shown as

$$|s(k)| \leq \dfrac{\varphi d_2}{\varphi - d_2}. \tag{13.18}$$

Referring to Eq. (13.14), Eq. (13.16) is written as

$$\sum_{i=0}^{n-1} c_i \left[x_i(k+1) - \beta x_i(k) \right] + x_n(k+1) - \beta x_n(k)$$

$$= \{1 - q[s(k)]\} s(k) - d_c(k) + d_1. \tag{13.19}$$

Remark 3: Eq. (13.14) indicates that the fractional delay present in the feedback channel (τ'_{sc}) is nullified at the sliding surface while the fractional delay present in the forward channel (τ'_{ca}) is nullified at the actuator side. Therefore, while computing the control signal, there is no effect of delay at the controller side. Hence, the control signal in Eq. (13.2), is given as

$$u(k - \tau') = u(k). \tag{13.20}$$

Substituting $x_n(k+1)$ in Eq. (13.19), we get

$$\sum_{i=0}^{n-1} c_i \left[x_i(k+1) - \beta x_i(k) \right] + f(x(k)) + g(x(k))[u(k) + d(k)] - \beta x_n(k)$$

$$= \{1 - q[s(k)]\} s(k) - d_c(k) + d_1. \tag{13.21}$$

Further simplification gives

$$g(x(k))[u(k)+d(k)] = f(x(k)) - \sum_{i=0}^{n-1} c_i \left[x_i(k+1) - \beta x_i(k) \right] + \beta x_n(k)$$

$$+ \left\{ 1 - q[s(k)] \right\} s(k) - d_c(k) + d_1. \tag{13.22}$$

Rearranging the terms, the control law is expressed as

$$u(k) = -g(x(k))^{-1} \left[f(x(k)) + \sum_{i=0}^{n-1} c_i \left[x_i(k+1) + \beta x_i(k) \right] \right.$$

$$\left. -\beta x_n(k) - \left\{ 1 - q[s(k)] \right\} s(k) + d_c(k) - d_1 \right] - d(k). \tag{13.23}$$

This concludes the **Proof**.

After computing the control law, the sufficient condition for the stability of closed-loop system in Eq. (13.2) is derived such that the system states should remain within specified band using nondelayed sliding surface (Eq. 13.14) and control law (Eq. 13.23).

13.5.1 Stability Analysis

For given positive scalars (τ'_{sc}) and (τ'_{ca}) with total network-induced delay (τ) satisfying Eq. (13.4) and controller Eq. (13.23), sliding system is approached by closed-loop system's trajectories. Thus, the condition stated in Eq. (13.24) becomes feasible:

$$0 \le \varnothing < s^T(k)s(k). \tag{13.24}$$

Proof: The nondelayed sliding surface is given by

$$s(k) = \sum_{i=0}^{n-1} c_i \left[x_i(k) - \beta x_i(k-1) \right] + x_n(k) - \beta x_n(k-1). \tag{13.25}$$

Selecting the Lyapunov function as

$$V_s(k) = s^T(k)s(k). \tag{13.26}$$

Writing forward difference of the earlier Eq. (13.26),

$$\Delta V_s(k) = s^T(k+1)s(k+1) - s^T(k)s(k). \tag{13.27}$$

Substituting the value of $s(k+1)$ using Eq. (13.25), we get

$$\Delta V_s(k) = \left[\sum_{i=0}^{n-1} c_i \left[x_i(k+1) - \beta x_i(k)\right] + x_n(k+1) - \beta x_n(k)\right]^T$$

$$\left[\sum_{i-0}^{n-1} c_i \left[x_i(k+1) - \beta x_i(k)\right] + x_n(k+1) - \beta x_n(k)\right] - s^T(k)s(k)\right]. \quad (13.28)$$

Substituting the value of $x_n(k+1)$,

$$\Delta V_s(k) = \left[\sum_{i=0}^{n-1} c_i \left[x_i(k+1) - \beta x_i(k)\right] + f(x(k)) + g(x(k))[u(k) + d(k)] - \beta x_n(k)\right]^T$$

$$\left[\sum_{i=0}^{n-1} c_i \left[x_i(k+1) - \beta x_i(k)\right] + f(x(k)) + g(x(k))[u(k) + d(k)] - \beta x_n(k)\right]$$

$$- s^T(k)s(k).$$

Referring Eq. (13.23), $\Delta V_s(k)$ is given as

$$\Delta V_s(k) = \emptyset - s^T(k)s(k). \quad (13.29)$$

where

$$\emptyset = \left[\left(1 - q[s(k)]s(k)\right) - d_s(k) + d_1\right]^T * \left[\left(1 - q[s(k)]s(k)\right) - d_s(k) + d_1\right].$$

The term "$\emptyset$" can be tuned close to zero by appropriately selecting the parameter "φ." If "$\emptyset$" is close to zero, then $s^T(k)s(k)$ will be larger than "$\emptyset$." Thus, for any small parameter "ρ'," we have

$$\emptyset - s^T(k)s(k) < \rho's^T(k)s(k). \quad (13.30)$$

Thus, by tuning the parameter "φ," we have $\Delta V_s(k) < \rho's^T(k)s(k)$, which guarantees the convergence of $\Delta V_s(k)$ and implies that any trajectory of the system in Eq. (13.2) will be driven onto the sliding surface and maintain on it.

Similarly, the effect of fractional delay betweenthe controller and actuator (τ'_{ca}) is nullified at the actuator end. The nondelayed control signal thatis applied to the plant is given by

$$u_a(k) = u(k) - \beta'u(k-1), \quad (13.31)$$

where $\beta' = \dfrac{\tau'_{ca}}{1 + \tau'_{ca}}$ and $u_a(k)$ is the nondelayed control signal calculated at the actuator side.

Eq. (13.31) states thattheeffect of fractional delay from the controller to an actuator is nullified through past and present control information, and the parameter "β'''" is approximated based on Thiran's approximation at each sampling instant.

13.6 Simulation Results

In this section, a numerical example is presented for the validity of control scheme proposed in this chapter.

Example 13.1: Consider the following second-order nonlinear SISO (single input single output) system:

$$\dot{x}_1(k) = x_2(k), \tag{13.32}$$

$$\dot{x}_2(k) = f(x(k)) + g(x(k))(u(k-\tau') + d(k)), \tag{13.33}$$

$$y(k) = x_1(k). \tag{13.34}$$

where $f(x(k)) = 0.1\sin(x_2)$, $g(x(k)) = 0.01\cos(x_1, x_2)$, and $d(k) = 0.02\sin(0.086k)$.

Figures 13.2–13.10 show the nature of state variables, control signal, sliding surface, and compensated control signal for $\tau = 12.8$, 25.6, and 28 ms. It is assumed that the control system components (such as plant, sensor, actuator, and controller) are connected through a same communication network and have same characteristics. The state information (x_1, x_2) will experience a delay of $\tau_{sc} = 6.4$, 12.8, and 14 ms, respectively, from sensor to controller. The control signal $u(k)$ computed at controller side will suffer from a controller to actuator delay of $\tau_{sc} = 6.4$, 12.8, and 14 ms, respectively. With the sampling interval of $h = 30$ ms, the fractional part of sensor to controller delay and controller to actuator delay is computed to be $\tau'_{sc} = \tau'_{ca} = 0.213$, 0.426, and 0.466, respectively. Based on the available information on control access network (CAN) and switched ethernet network medium [23], the minimum total network-induced delay that can occur is 12.8 and 25.6 ms. The 28 ms is chosen to demonstrate the effectiveness of the proposed control strategy under amaximum delay that can occur in the sampling time of 30 ms. To prove the robustness of the proposed control method, input side of the channel is applied with a slow time-varying disturbance. The parameters designed for the stable response of the nonlinear system are $c_1 = -0.8$, $\beta = 0.00635$, 0.0118, and 0.0138, $\varphi = 100$, and $d_1 = 0.05$. The sliding band is computed as $|s(k)| \leq \pm 0.04$.

Figures 13.2 and 13.4 show the nature of state variables (r_1, r_2) for a specified network-induced delay (τ) with initial conditions $x_1 = 1$ and $x_2 = 0.5$, respectively. It is observed that, with the presence of fractional delays, the state variables have convergence to zero in finite time. Figures 13.3 and 13.5

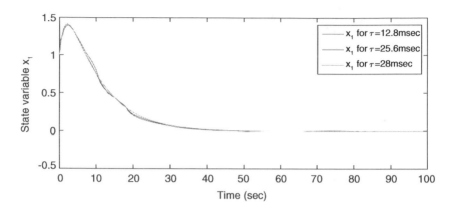

FIGURE 13.2
State variable $x_1 = 1$ for $\tau = 12.8$, 25.6, and 28 ms.

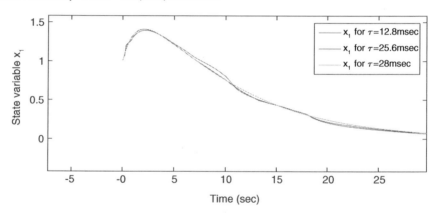

FIGURE 13.3
Magnified results of state variable $x_1 = 1$ for $\tau = 12.8$, 25.6, and 28 ms.

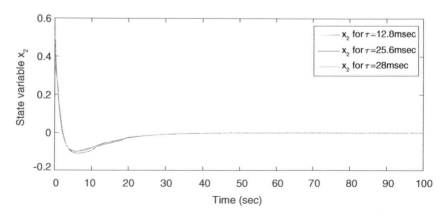

FIGURE 13.4
State variable $x_2 = 0.5$ for $\tau = 12.8$, 25.6, and 28 ms.

show the magnified results of state variables. Also, both the channels experience acompensated effect of network-induced delay. Here, the system state variables are calculated from the first sampling instant. Figure 13.6 describes the nature of sliding surface computed at the controller side. Observing the results, it can be inferred that the system state slide remains within the specified band along the predefined surface in Eq. (13.18). Thus, the sliding surface based on Thiran's approximation functions properly even in the presence of sensor to controller delay (τ_{sc}). Figures 13.7 and 13.9 describe the behavior of control signal computed at the controller side and actuator side, respectively. Noticing the results in Figures 13.8 and 13.10, the effect of fractional delays in both channels is nullified using Thiran's approximation rule.

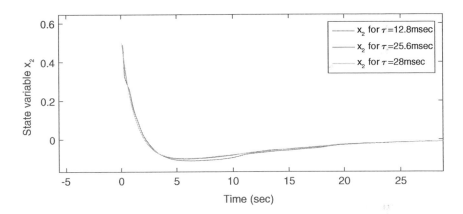

FIGURE 13.5
Magnified results of state variable $x_2 = 0.5$ for $\tau = 12.8$, 25.6, and 28 ms.

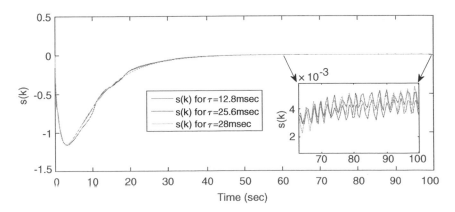

FIGURE 13.6
Nondelayed sliding surface $s(k)$ for $\tau = 12.8$, 25.6, and 28 ms.

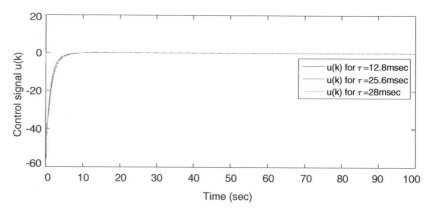

FIGURE 13.7
Control signal $u(k)$ computed at controller end for $\tau = 12.8, 25.6$, and $28\,\mathrm{ms}$.

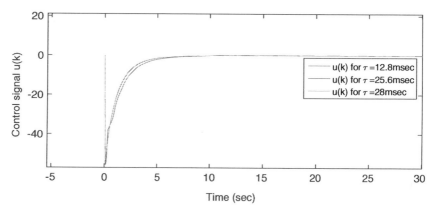

FIGURE 13.8
Magnified results of control signal $u(k)$ computed at controller end for $\tau = 12.8, 25.6$, and $28\,\mathrm{ms}$.

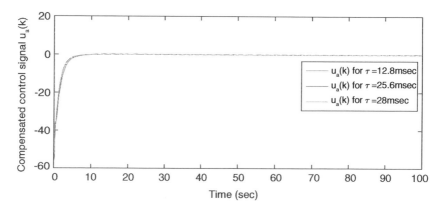

FIGURE 13.9
Nondelayed control signal $u_a(k)$ computed at actuator end for $\tau = 12.8, 25.6$, and $28\,\mathrm{ms}$.

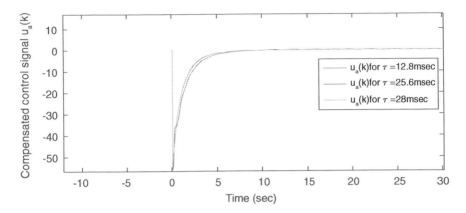

FIGURE 13.10
Magnified results of nondelayed control signal $u_a(k)$ computed at actuator end for $\tau = 12.8$, 25.6, and $28\,\text{ms}$.

Based on the obtained results, the designed controller is proved to be more robust for a given nonlinear system as it is insensible to network nonidealities such as matched uncertainty and network-induced delays.

13.7 Conclusion

In this chapter, discrete-time sliding mode controller is designed for nonlinear NCS that nullifies the effect of network-induced discrete-time domain delays. The network-induced delays in the forward and feedback channel are compensated through Thiran's approximation rule. The effect of feedback channel delay is nullified at the sliding surface such that the system states slides along the predefined surface and remains within the specified band. The results proved that discrete-time sliding mode controller designed using the proposed sliding surface proves to be more robust as it is insensitive to matched uncertainty and network-induced delays. The stability of closed-loop nonlinear NCS is secured by Lyapunov function. Lastly, the efficacy of proposed control strategy considering real-time communication delays (such as CAN and switched ethernet) is validated using a real-time application. The proposed control strategy is applicable to unnamed aerial vehicles (UAV) and flight control dynamics where delays and systems behavior have deterministic and nonlinear nature, respectively. The proposed method can be extended for random fractional delays considering the single data packet loss as well as multiple data packet loss condition in future.

References

1. Hespanha, J., Naghshtabrizi, P., Xu, Y. A survey of recent results in networked control systems. *Proceedings of the IEEE*, 95(1): 138–162, 2007.
2. Gupta, R., Chow, M. Networked control system: Overview and research trends. *IEEE transactions on Industrial Electronics*, 57(7): 2527–2535, 2010.
3. Luck, R., Ray, A. Delay compensation in integrated communication and control systems: conceptual development and analysis. *IEEE Proceedings of American Control Conference (ACC90)*, San Diego, CA, pages 2045–2050, 1990.
4. Nilsson, J., Bernhardsson, B., Wittenmark, B. Stochastic analysis and control of real time systems with random time delays. *Automatica, Elsevier*, 34(1): 57–64, 1998.
5. Cac, N., Hung, N., Khang, N. CAN-based networked control systems: A compensation for communication time delays. *American Journal of Embedded Systems and Applications*, 2(3): 13–20, 2014.
6. Chan, H. Closed loop control of a system over a communication network with queues. *International Journal of Control*, 62(3): 593–601, 1995.
7. Hikichi, Y., Sasaki, K., Tanaka, R., Shibasaki, H., Kawaguchi, K., Ishida, Y. A. Discrete PID control system using predictors and an observer for the influence of a time delay. *International Journal of Modelling and Optimization*, 3(1): 1–4, 2013.
8. Yu, Z., Chen, H., Wang, Y. Control of network system with random communication delay and noise disturbance. *IEEE Proceedings of Control and Decision Conference*, China, pages 518–526, 2000.
9. Khanesar, M., Kaynak, O., Yin, S., Gao, H. Adaptive indirect fuzzy sliding mode controller for networked control systems subject to time varying network induced time delay. *IEEE Transactions on Fuzzy Systems*, 23(1): 205–214, 2015.
10. Montestruque, L., Antsaklis, P. On the model-based control of networked systems. *Automatica*, 39(10): 1837–1843, 2003.
11. Yue, D., Han, Q.L., Lam, J. Network-based robust H∞ control of systems with uncertainty. *Automatica*, 41(1): 999–1007, 2005.
12. Yang, W., Fan, L., Luo, J. Design of discrete time sliding mode observer in networked control system. *IEEE Proceedings of Chinese Control and Decision Conference*, Xuzhou, China, pages 1884–1887, 2010.
13. Comanescu, M., Korlinchak, C. Discrete time integration of observers with continuous feedback based on Tustin's method with variable prewarping. *6th IET International Conference on Power Electronics, Machines and Drives (PEMD 2012)*, Bristol, UK, pages 1–6, 2012.
14. Jacovitti, G., Scarano, G. Discrete time techniques for time delay estimation. *IEEE Transactions on Signal Processing*, 41(2): 525–533, 1993.
15. Shah, D., Mehta, A.J. Fractional delay compensated discrete-time SMC for networked control system. *Digital Communication and Networks, Elsevier*, 3(2): 112–117, 2016.
16. Wu, C., Liu, J., Jing, X., Li, H., Wu, L. Adaptive fuzzy control for nonlinear networked control systems. *IEEE Transactions on Systems, Man, Cybernetics: Systems*, 47(8): 2420–2430, 2017.
17. Bsili, I., Ghabi, J., Messaoud, H. Discrete time sliding mode control of nonlinear uncertain systems based on estimation of uncertainties and external disturbances. *International Journal of Applied Engineering Research*, 11(9): 6782–6786, 2016.

18. Li, H., Yang, H., Sun, F., Xia, Y. Sliding-mode predictive control of networked control systems under a multiple-packet transmission policy. *IEEE Transactions on Industrial Electronics*, 61(11): 201–221, 2014.
19. Mehta, A., Bandyopadhyay, B. Multirateoutput feedback based stochastic sliding mode control. *Journal of Dynamic Systems, Measurement, and Control*, 138(12): 124503(1–6), 2016.
20. Shah, D., Mehta, A.J. Discrete-time sliding mode control using Thiran's delay approximation for networked control system. *43rd Annual Conference of the IEEE Industrial Electronics Society, (IECON-2017)*, Beijing, China, pages 3025–3031, 2017.
21. Bartoszewicz, A., Lesniewski, P. Reaching law approach to the sliding mode control of periodic review inventory systems. *IEEE Transactions on Automatic Science and Engineering*, 11(3): 810–817, 2014.
22. Shah, D., Mehta, A.J. Discrete-time sliding mode controller subject to real-time fractional delays and packet losses for networked control system. *International Journal of Control, Automation and Systems, Springer*, 15(6): 2690–2703, 2017.
23. Shah, D., Mehta, A.J. *Discrete-Time Sliding Mode Control for Networked Control System*. Springer Nature, Singapore, 2018.

Index